Mechanical Fault Diagnosis

and condition monitoring

R. A. Collacott

B.Sc. (Eng), Ph.D., C.Eng., F.I.Mech.E., F.I.Mar.E.,

*Director, U. K. Mechanical Health Monitoring Group,
Leicester Polytechnic*

Chapman and Hall
London

A Halsted Press Book
John Wiley & Sons , New York

First published in 1977
by Chapman and Hall, Ltd.,
11 New Fetter Lane,
London EC4P 4EE

© 1977 R.A. Collacott

ISBN 0 412 12930 2

Typeset by Alden Press Oxford, London & Northampton and
printed at University Printing House, Cambridge

Distributed in the U.S.A.
by Halsted Press, a Division
of John Wiley and Sons, Inc., New York

Library of Congress Cataloging in Publication Data

Collacott, Ralph Albert.
      Mechanical fault diagnosis.

      1. Machinery—Testing.   2. Machinery—Reliability.
I. Title
TJ148.C57   621.8'028   77-397
ISBN 0-470-99095-3

Contents

Preface

Although the most sophisticated fault diagnosis and condition monitoring systems have their origin in the aerospace and nuclear energy industries, their use is by no means restricted to such areas of 'high technology'. Modern machinery in most industrial plants is now so complex and expensive that mechanics find it increasingly difficult to detect failure by, for instance, recognising changes in sound 'signatures', and few plants can afford the luxury of regular 'stripping down'. Increasingly, therefore, early-warning devices are being employed in an effort to prevent catastrophic breakdown.

This book provides the first co-ordinated compilation of fault diagnosis and condition monitoring devices. It proceeds in three logical steps. The early chapters deal with those conditions which contribute to deterioration and the consequent likely development of faults. The middle part of the book considers the various techniques of monitoring and discusses the criteria for their selection in different situations. The final chapters provide a guide to the interpretation of the information signals deriving from monitoring, relating to reliability science and the mathematics of probability, and thus providing decision data on which management can act.

Through establishing the UK Mechanical Health Monitoring Group, I have appreciated the enormously wide industrial interest in this field. It is my hope that this book will serve the needs of both plant engineers and senior executives in providing a quick source of reference to those devices most suited to their requirements. It should also prove valuable to mechanical materials and control engineers whose particular interests include manufacture and maintenance of complex machinery.

1976   R. A. Collacott

I Failure types, investigation and occurrences

1.1 Introduction

Failure is defined in the BS Maintenance Glossary as 'The termination of the ability of an item to perform its required function.' This can involve such failure categories as follows:

(1) Catastrophic failures which result in an immediate inability of a system to achieve its function;

(2) Performance failures associated with a reducing performance of the equipment.

(3) When the operator deliberately takes the equipment out of service, even though it is producing, at that time, its specified output.

1.2 System failure and component failure

The extent to which any particular system failure or component failure has an effect on the 'normal function' or 'mission' depends on the function. There is a great difference between the failure of a pump on a machine tool and on an aeroplane or space-craft.

When considering the failure of a ship's machinery, Bridges [1] stated that certain systems and equipment may only be required during part of the mission. If, during the passage of a ship to the main mission area, systems or equipments which are required in the main mission fail, and they cannot be repaired before passage completion, then the mission is deemed to have failed.

Typically, consider a mission by a fishing vessel. It might have two shafts both of which are needed during fishing but only one shaft is needed for the passage to or from the fishing grounds. Therefore if one shaft set of propulsion equipment failed during the passage to the fishing grounds, and it could be repaired before fishing commenced, the mission would still continue. If, however, the failure could not be repaired, then the whole mission for the purpose of the analysis would be assumed to be aborted. Similarly if a propulsion system failure occurred during the fishing phase the mission would then be deemed to have failed.

1.3 Failure decisions

The events which lead up to a plant or system failure do not fall into a neat sequence and the planning network is very complex. A typical planning network based on a nuclear plant operation involves the following steps:

(1) Knowledge of the plant characteristics;

(2) Analysis of methods of operation (steady-rate and transient);

(3) Establishment of safety limits for all plant characteristics;

(4) Consideration of the available signals;

(5) Setting control trips (allowing for instrumentation inaccuracies);

(6) Design of operation interlocks;

(7) Selection of automatic control and operation overrides;

(8) Survey protection systems, involving (a) automatic shutdown, (b) restriction of plant operation, (c) use of auxiliary protection systems;

(9) Formulate recovery methods based on the consequences of the protection systems;

(10) Establish the cause of the operation of the protective systems.

1.4 Failure classifications

The terms 'failure' or 'fault' may be viewed from different angles according to the effect which the lack of performance has on the overall functional capability. Such aspects as economic viability, safety, engineering complexity, speed, causal influences all provide classifications leading to a description of failure.

1.4.1 Engineering failure classifications

There are two distinct classes of failure:

(1) Intermittent failure: failures which result in a lack of some function of the component only for a very short period of time, the component reverting to its full operational standard immediately after failure;

(2) Permanent failure: failures which result in a lack of some function which will continue until some part of the component is replaced.

1.4.2 Degree of failure classification

Permanent failures may be further subdivided into the following two types:

(1) Complete failure: failure which causes the complete lack of a required function. (It should be noted that in certain cases the limit when a lack of function is said to be complete is open to interpretation, which depends upon the application);

(2) Partial failure: failure which leads to a lack of some function but not such as to cause a complete lack of the required function.

1.4.3 Speed of failure classification

Both complete and partial permanent failure may be further classified according to the suddenness with which the failure occurs:

(1) Sudden failure: failure which could not be forecast by prior testing or examination;

(2) Gradual failure: failure which could have been forecast by testing or examination.

1.4.4 Degree and speed of failure classification

Both failure forms can be combined to give the following further classifications:
    (1) Catastrophic failures: failures which are both sudden and complete;
    (2) Degradation failures: failures which are both partial and gradual.

1.4.5 Cause-of-failure classification

According to the manner by which failure develops, so it be further classified:
    (1) Wear-out failures: failure attributable to the normal processes of wear as expected when the device was designed;
    (2) Misuse failure: failure attributable to the application of stresses beyond the item's stated capabilities;
    (3) Inherent weakness failure: failure attributable to a lack of suitability in the design or construction of the system or component itself when subjected to stresses within its stated capabilities.

1.4.6 Hazard classification

Possible faults (major or minor failures) may be divided into two broad hazard groups, namely dangerous-failures or safe-failures.
    (1) Dangerous faults: (a) protection system — failure to protect when needed, (b) machine tool — failure causing damage to work and/or operator, (c) traction system — failure to brake;
    (2) Safe faults: (a) protection system — failure to operate when not needed, (b) machine tool — failure to start, (c) traction system — failure of brakes to apply when not needed.

1.5 Types of failure

The three types of failure recognized in system and component reliability studies, namely infant mortality, random and time-dependent were considered by Davies [2] in relation to fault diagnosis.

1.5.1 Infant mortality (early failures)

Figure 1.1 shows the realistic situation in aerospace engineering where the margins between stress and strength are prescribed. The quality of components in the total population should be normally distributed about the design point, but occasionally there will arise a sub-population of 'weak' components, i.e. having a strength below the operational stress value.
    The term 'strength' is used here loosely to describe quality. This problem arises when a new type of defect escapes through the quality control filter, or when assembly methods lead to unsatisfactory construction. For electronic components it

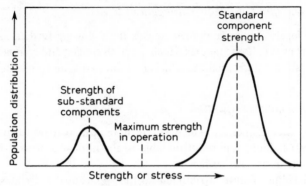

Figure 1.1. *Infant mortality failure probability*

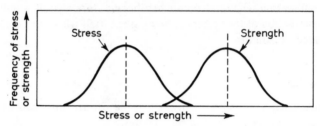

Figure 1.2 *Random failure probability*

is usual to burn-in the equipment on rigs and, of course, engine test after overhaul relates to this problem.

1.5.2 Random-failures

These are basic engineering problems, as shown in Figure 1.2 the margin between stress and strength is closely prescribed (a typical aviation situation with the conflicting requirements of performance and weight). There will be a Gaussian distribution such that a component on the lower limit of quality acceptance might occasionally experience a high stress limit in service and therefore fail. There are two ways of improving the situation. First, by increasing the separation of the two means, i.e. increasing the strength and lowering the stress; and second, by reducing the standard deviation of the respective patterns, i.e. imposing tighter limits on quality control and restricting the operating range of the equipment.

1.5.3 Time-dependent failures

Two examples are shown in Figure 1.3. Curve A is the classical case where there is a mean time to failure O-M, about which there is a normal distribution. The standard deviations (σ) embrace definite percentages of the whole so that a life can be declared which can control the number of failures allowed to occur. For example, using 3σ should give about 98% success.

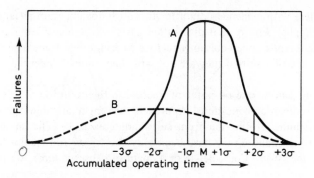

Figure 1.3 *Time-dependent failures*

In practice, the situation shown by curve B is often found. This can be said to be truly time-dependent, but has a very broad distribution to the extent that time control becomes ineffective and very extravagant. Actuarial failure analysis therefore should always be associated with distribution before making decisions on life limitations.

1.6 Failure investigations

The technical investigation of crashed aircraft has for long been regarded by the aircraft industry as an essential procedure from which information can be gained into causes of failure. *The Manual of Aircraft Accident Investigation* [3] not only deals with the procedures which it recommends investigators to follow but also includes consideration of causal factors and the basis for further investigations as well as a comprehensive interpretation of various types of structural failure.

Despite the cost of failures, surprisingly little effort is allocated by non-industrial organizations to the study of engineering failures. Certification societies such as Lloyd's Register of Shipping and The British Engine Boiler and Electrical Insurance Company have a wealth of accumulated experience, but once again little of this is generally available.

In the U.S.A. the Product Liability Law has made it necessary to analyse failures and monitor the reliability of products. In the U.K. the Jost Report [4], led to the establishment of Centres of Tribology, much of the activity of which is concerned with the diagnosis of failures and ways of preventing their recurrence.

1.6.1 Failure investigation principles

The following main principles of procedure were submitted by Professor Meyer of City University [5].

(1) The sooner an investigation is started, the greater are the chances of determining the real cause or causes;

(2) Do not destroy evidence. Do not disturb or tidy up the scene of failure or accident, especially do not touch fracture surfaces or their immediate surroundings;

(3) Interfere with evidence only after thorough documentation (report, description, photographs). Ensure that dismantled parts can be identified individually, reassembled, and repositioned correctly relative to each other. Handle and pack pieces of evidence carefully so that they are not scratched, rubbed, indented, or deformed accidentally;

(4) Do not concentrate on point of fracture to the exclusion of its surroundings and of environmental conditions. Approach the origin of failure gradually after drawing maximum information from surrounding evidence. The immediate origin and cause of local failure may have been only the trigger for a major failure determined by other causes. Coincidence of a chain of causes is more the rule than the exception;

(5) Do not guess or draw easy conclusions. Collect all the facts and then eliminate inessentials. Rely on site photographs, notes and sketches rather than memory. A cause has been established not when it becomes obvious but only when all other possibilities have been eliminated.

(6) Try to obtain the true history from objective evidence and from interviews. Do not take any statements or opinions on trust, especially not your own. Human perceptions, judgements, and decisions are all fallible and subject to subconscious prejudices.

1.7 Failure case studies (complete systems)

The perversity of nature may prevent a full and complete listing of even typical failures but some guidance can always be obtained from a review of failures and faults which have been identified. Many (but not all) serious failures originate from quite trivial sources, the failure to tighten a bolt, a washer omitted, a component fitted the wrong way round, a careless operation of a valve.

1.7.1 U.S. Army helicopter failures

The worldwide accident causes for failure of the FY69 helicopter during the fiscal year 1969 were given [6] as due to the following:

Crew error 49.5% (failure of crew members to cope successfully with an emergency situation)
Material failure 23.6%
Part or component malfunction 1.7%
Other 16.8% (error by non-crew personnel-training deficiency; inadequate landing areas, design of aircraft controls, orientation, etc.)

Technical failures arising from accidents involving materials or components were as follows:

(1) Power plant failures: predominant causes were breakage of the 4th stage compressor of an axial-flow turbine;

(2) Power train failures: flexible couplings in the shaft between the power plant

and the main and/or tail rotor failed to provide rotational stability and became excessively displaced;

(3) Driveshaft coupling/support failures: a poor design involving many parts and this low reliability, coupled with inadequate lubrication, led to failure of the couplings and supports. Parts were fitted incorrectly so that oil flow was impeded;

(4) Rotor system failures: catastrophic failures at the interface between main and tail rotor blades due to high stress concentrations;

(5) Tail rotor centre hub yoke failures: in 29 accidents 32 fatalities occurred as a result of a tensile failure of the yoke's threaded end. Failure originated in an area of high stress concentration at the thread root and 'safety pin' hole intersection;

(6) Main rotor spar fatigue failure: the juncture of the blade spar with the main rotor hub failed due to high stress concentration and the inclusion of metal (drilling) chips at the bolt and cuff interface;

(7) Hydraulic systems and flight control failures: groups of such failures arose from the following causes: (a) hydraulic tubing failure, (b) push-pull flight control tube failures, (c) primary and secondary hydraulics drive failure;

(8) Fasteners: the most common loss involves the loosening of a castellated nut from a bolt due to the omission of the cotter pin, in some instances re-use fibre nuts have failed as the fibres lost their grip.

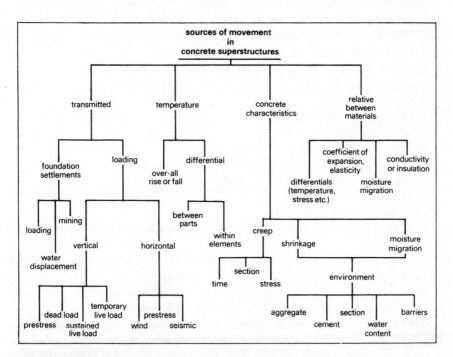

Figure 1.4 *Causes of movement in concrete structures. Major causes are ground tremors and subsidence, but the unstable properties of the concrete are a contributory factor*

1.7.2 Buildings

The failure of concrete buildings is unpredictable and can result from many causes. A survey made by the Cement and Concrete Association, Wrexham Springs, Slough, Bucks., included the algorithmic fault tree, (Figure 1.4) showing that while major causes of failure are ground tremors and subsidence, the unstable properties of concrete are a contributory factor.

1.7.3 Marine steam turbines

Typical sources leading to the unreliability of marine steam turbine systems were attributed by Donald [7] to the following:

Boilers 49%
Condensers and circulating pumps 12%
Boiler feed pumps 11%
Main turbines 7%
All others 21%

The contributory factors were listed as follows:
   (1) External steam leaks;
   (2) External lubricating oil leaks;
   (3) Inefficient turbine case lagging causing distortion;
   (4) Inefficient steam pipe lagging causing distortion;
   (5) Pipe fixings too rigid, restricting expansion when heated;
   (6) Erosion of blading;
   (7) Fouling of blades with deposits;
   (8) Rubbing of labyrinth glands;
   (9) Rubbing of moving blade tip seals;
   (10) Blockage of bled steam drains or water-extraction drains;
   (11) Failure of blade lacing wire brazing;
   (12) Failure of stator blading or nozzles;
   (13) Warp of horizontal joints;
   (14) Hull distortion causing misalignment particularly of tubine/gear/gear systems;
   (15) Blade and wheel vibrations — a compound turbine may contain up to 3000 blades any of which can produce catastrophic failure;
   (16) Temperature effects on blades reducing Young's Modulus (E);
   (17) Temperature effects on blades reducing the fatigue limit;
   (18) Fretting fatigue at blade roots;
   (19) File marks or scratches on blades producing stress notches;
   (20) Oil 'whirl' or 'whip' causing a circular or elliptical motion of the main shaft;
   (21) Unreliability of marine steam turbine systems due to the complex bleed and/or reheat cycles and higher operating pressures and temperatures.

1.7.4 Aircraft structural failures

An analysis of failures over a period of 20 years which resulted in fatal or serious accidents, reported by Williams [8] Chief Structures Engineer, Air Registration Board, Redhill gave the following:

(1) Static failure: unlikely to occur unless the aircraft is subjected to loads (particularly from thunderstorms) exceeding the maxima for which they were designed;

(2) Fatigue failure: of 19 aircraft examined fatigue cracks were found as follows:

Front spar fittings, lower fourth step 5 aircraft with cracks
Both wings 3 aircraft with cracks
One wing 2 aircraft with cracks
Scarf joints, fretting corrosion several aircraft

1.7.5 Aircraft component defects

Excluding defects of engines, tyres, wheels and brakes the following data was given of premature component removals in a fleet of 12 standard VC 10 aircraft by Redgate [9] Engineering Superintendent, British Overseas Airways Corporation, Hounslow:

Total no. of components on aircraft (excluding engines, tyres, wheels and brakes) 1573
No. of different components 547
No. of premature removals 537
No. of different premature removals 191

A summary of the workshop findings on these aircraft for a typical 3-month period in terms of components were:

Electrical failures 84
Worn or scored 48
Seal failure 44
Mechanical breakage 29
Bearing/lubrication failure 22
Striction and erosion 21
Contamination 14
Calibration errors 10
Corrosion 9
Erosion 7
Diaphragm and capsule failures 4
Other causes 43

Technical examination indicated the following common failure sources:
(1) Electrical failures: Microswitches were the main problem and caused by:

(a) incorrect adjustment causing undertravel or overtravel, (b) moisture in glass causing internal condensation, (c) seizure of plungers due to swelling following moisture absorption;

Solenoids and solenoid valves failed as a consequence of: (a) kinked wires in coils, (b) inadequate support of lead-outs, (c) build-up of dust in the armature tunnel, (d) hammering of valve-seats;

(2) Wearing or scoring: (a) unsuitable materials, (b) unsuitable surface treatment, (c) inadequate bearing areas, (d) poor lubrication, (e) vibration;

(3) Seal failure: (a) incorrect seal configurations and materials fitted, (b) inadequate bearing support for sliding shafts causing the seals to be over-loaded, (c) contamination of seals by debris from normal bearing wear, (d) overheating leading to disintegration;

(4) Bearing failure: (a) life-limitation of sealed grease-packed bearings owing to the drying-out of the grease, (b) bearing closure or breakdown from high temperature, (c) inadequate and ineffective in-service lubrication facilities, (d) bearing seizure when using a dry lubricant such as molybdenum disulphide owing to its lack of corrosion resistance, (e) blockage of oil holes;

(5) Striction and erosion: caused by either inadequate fluid filtration or by cavitation erosion;

(6) Contamination: (a) contaminants in servo air systems are moisture and dirt, (b) moisture is the most common, it arises either from physical soaking over a period of time or from condensation, (c) moisture also affects fire-detector elements and fuel tank contents probes which operate on capacitance principles, (d) contaminants in hydraulic systems are dirt;

(7) Corrosion: occurs in very component, ranging from skin and fasteners in primary structures to the internal corrosion of electric relays;

(8) Diaphragm and capsule failures: these are particularly susceptible to batch defects caused by a drop-off in manufacturing technique, poor quality control or bad materials.

1.7.6 Road vehicle defects

As a result of examinations carried out on behalf of the former Ministry of Transport (now part of the Department of Trade and Industry), Perring [10] listed some defects which had been raised by the Ministry resulting in modifications by the manufacturers (Table 1.1).

1.7.7 Power station commissioning faults

From the records of one particular power station completed during the 1960's and equipped with 120 MW generating units, Morton [11] listed the following incidents:

(1) Economizer-tube weld failures occurred in all boilers owing to an error in manufacturing procedure;

(2) Gross overheating and failure of radiant-superheater tubes in two boilers, in

Table 1.1

Type	Defect	Result
Bus chassis	Steering box failure, due to transfer tube failure allowing recirculating balls to escape.	Transfer tube modified.
Articulated tractor	Brake failure, due to brake valve failure.	Campaign change to modify all valves.
Light van	Fire caused by petrol pipe coming off (push-on type).	Campaign change to introduce pipes with nuts and olives on all vehicles produced since January 1965.
Diesel truck	Dribble pipes from injectors made of plastic caught fire owing to proximity to exhaust manifold.	Campaign change to Bundy tubing being carried out by makers at M.O.T. suggestion.
Plastic brake hose	Instances of fracturing of coiled brake hose 'Susies'.	Campaign change from Nylon No. 6 to Nylon No. 11 carried out by manufacturer. Other manufacturers followed suit.
Sports car	Brake pedal fracture at weld of boss.	Tighter inspection procedure instituted by makers.
Semi-trailer	Shearing of upper turn-table security bolts on tanker semi-trailer	Redesign of upper turn-table security undertaken by manufacturer at M.O.T. suggestion.

one case a heater failed caused by bad steam distribution between tubes in parallel, together with thermal distortion;

(3) At 105 MW load the reheat-steam temperature was 90°F below design, unless the gas-recirculating fan was used. Use of this fan raised the temperature of the superheater-tube metal still further;

(4) On-load corrosion in the walls of boiler-water tubes, apparently associated with weld intrusion into the tube bore;

(5) Slip joints in the satety-valve escape pipes had to be replaced with flexible pipes;

(6) Drum water-level gauges rapidly became obscured in service owing to deterioration of mica;

(7) Remote water-level indication was incorrect because a tapping was badly sited;

(8) Pulverized-fuel mills did not achieve the specified fineness of grinding;

(9) All mill foundations vibrated badly at the gear-tooth contact frequency. Failures of holding-down bolts occurred and nuts became loose;

(10) Scuffing of mill girth gears;

(11) Mill seals were unsatisfactory and alternative packing ineffective. Remachining of the housings and replacement of the packings with synthetic rubber hose eventually produced the desired result;

(12) Hold-ups of raw coal occurred in the mill supply chutes. Fitting larger vibrators and air stimulators proved to be only palliatives — the chutes had finally to be redesigned;

(13) Pulverized-fuel classifiers rejected an excessive percentage of fines and the air-float system for conveying these back to the mill inlet was unreliable;

(14) Pulverized-fuel burners had to be redesigned to improve combustion stability;

(15) Oil-burner design change necessary to eliminate carbon build-up at the taps;

(16) Thrust bearings of primary air fans failed when thrust collars became slack;

(17) Main feed pump and boiler drum pressure drop was considerably above design;

(18) Main feed-pump running clearances deteriorated rapidly owing to scoring caused by foreign matter;

(19) Feed-pump internal joint faces severely eroded necessitating various modifications including the use of harder material;

(20) Labyrinth gland material of the main feed pump had to be changed from white metal to bronze owing to the small margin between the final feed temperature and the melting point of white metal;

(21) Standby feed pump water loss necessitated the installation of a condensate-recovery system;

(22) Under some load conditions the discharge pressure of the freely cavitating main extraction pump was insufficient to overcome the leak-off pressure from the feed-pump labyrinth glands, resulting in reverse flow of hot water from the glands. An additional valve with an automatic-control system had to be fitted in the low-pressure feed line to maintain extraction-pump discharge pressure;

(23) The original arrangement of the emergency and make-up feed connections resulted at times in flooding of the deaerator and at other times in an inadequate supply of make-up. Re-routing of large lines, and revised control arrangements, were necessary to overcome this;

(24) The rotary element of the mechanical filter in the booster-pump suction line seized soon after commissioning. Bearings were eased, but a few months later the element broke up and debris entered the pump, which in turn failed. After further incidents it was eventually necessary to fit automatic control of the filter rotating and back-flushing system;

(25) Numerous tube and weld leaks occurred in the high-pressure feed heaters, with corrosion pitting of tube plates. The heaters were redesigned to alter the steam path and reduce thermal stressing;

(26) The air suction temperature of the condenser was very high, because inadequate internal baffling allowed steam to bypass the air-cooler section;

(27) Cooling-tower performance was poor owing to faulty water distribution;

(28) Corrugated asbestos packing collapsed in one of the cooling towers, necessitating the repacking of all three towers with a different design of packing.

1.7.8 Boiler failures

Particulars of some of the failures experienced in marine water-tube boilers were given in a paper by Hodgkin [12] Chief Marine Project Engineer, Babcock and Wilcox (Operations) Ltd., as follows:

(1) Leakages of gases from the combustion through the boiler casing – obviated by admitting combustion air around the casing;

(2) Heavy bonded slag deposits which ultimately bridged across between tubes in high gas temperature zones when residual fuel was used which produced high-vanadium ash;

(3) Clean-tube areas in slag coated boilers failed due to the increased gas speeds involving high heat transfer rates and higher tube temperatures;

(4) Tube wastage due to attack of the hot surfaces by the vanadium-bearing slag;

(5) Tube scouring from silica by silica picked up from the refractory linings by the combustion gases;

(6) Air heater corrosion due to attacks from sulphur-bearing fuels;

(7) Surface fouling of air heaters requiring increases in fan power;

(8) Hydraulic servo-systems used to operate automatic controls contaminated and blocked (very high clinical cleanliness with filtration below $5 \mu m$ needed in dirty boiler spaces);

(9) Carbon and iron-in-steam fires associated with inefficient combination;

(10) Total destruction of an economiser through the inability of the oil-burning equipment and combustion control equipment to function satisfactorily at low load;

(11) Overheating due to extended flames at full power in the presence of soot and slag build-up;

(12) Flame impingement on boiler walls melting the refractory;

(13) Flame instability;

(14) Water shortage through feed-pump failure;

(15) Solids build-up in boilers due to the presence of dissolved solids.

1.7.9 Diesel engine defects

An analysis of stoppages prepared from reports published by the Diesel Engineers and Users Association [13] based on the year 1970 with respect to industrial diesel engines and following 410 recorded events attributed these stoppages to a variety of causes (Table 1.2).

1.8 Human factors in failure incidents

There is always a reason for a mechanical failure. In a review of marine machinery failures, Batten [14], Principal Surveyor, Lloyds Register of Shipping commented that while there may be 'Acts of God' this phrase was never intended to cover man's ineptitude or materials insufficiency.

That the 'ineptitude of man' – the human factor – is a highly relevant factor in accidents is well-known to organizations concerned with insurance, especially

Table 1.2

Class of defect	% Occurrence
Fuel-injection equipment and fuel supply	27.0
Water leakages	17.3
Valves and seatings	11.9
Bearings	7.0
Piston assemblies	6.6
Oil leakages and lubrication systems	5.2
Turbocharges (excluding damage by intruding foreign bodies)	4.4
Gearing and drives	3.9
Governor gear	3.9
Fuel leakages	3.5
Gas leakages	3.2
Breakages and fractures, other than under other specific headings	2.5
Miscellaneous	2.5
Foundations	0.9
Crankshafts	0.2
	100.0

motor-vehicle insurance. The identification of possible influences of human error led the Canadian authorities to list the consequences of human factors needing attention during aircraft crash investigations (Table 1.3) [15].

Table 1.3

Consequence	Cause (inherent or temporary)
Errors of judgement	Lack of experience
Poor technique	Poor reaction
Disobedience of orders	Physical state
Carelessness	Physical defect
Negligence	Psychological state

In the author's experience [16, 17, 18] of marine operations involving lengthy voyages with limited crews under conditions of isolation it would appear that increasing attention should also be given to the effects of boredom, group interaction, lack of instruction, incorrect motivation, ill-defined areas of responsibility, poor control, group-interaction hostilities, poor communications and the many similar causes of irrational human behaviour.

References

1   Bridges, D.C. (1974), 'The application of reliability to the design of ship's machinery', *Trans. I. Mar. E.*, **86**, Pt 6.
2   Davies, A.E. (1972), 'Principles and practice of aircraft powerplant maintenance', *Trans. I. Mar. E.*, **84**, 441–447.

3 —— *Manual of aircraft accident investigation,* International Civil Aviation Authority (I.C.A.A.), Montreal , Canada.
4 Jost, H.P. (1966), *A Report on Lubrication (Tribology),* Department of Education and Science, H.M.S.O., London.
5 Meyer, M.L. (1969) 'Methodic Investigation of Failure', *Safety and Failure of Components Conference,* I. Mech. E.
6 Darragh, Jr. J.T. & Haley, Jr. J.L. (1970), 'U.S. Army rotary wing mishap failures', *Proc. 9th. Reliability and Maintainability Conference,* 9, A.S.M.E.
7 Donald, K.M.B. (1973), 'Marine steam turbines – some points of design and operation', *Trans. I. Mar. E.,* 85, Pt. 6.
8 Williams, J.K. (1969), 'Major causes of failure – civil aircraft structures', *Safety and Failure of Components Conference,* I. Mech. E.
9 Redgate, C.B. (1969), 'Component defects – an airline's experience', *Safety and Failure of Components Conference,* I. Mech. E.
10 Perring, H. (1969–70), 'Vehicle defects as a contributory cause of road accidents', *Proc. I. Mech. E.,* 184, Pt. 38.
11 Morton, A.J. (1971), 'Pre-service component testing of steam power and process plant', *Proc. I. Mech. E.,* 185, (19).
12 Hodgkin, A.F. (1973), 'Marine boiler development over the past ten years', *Trans. I. Mar. E.,* 85, Pt. 6.
13 —— (1971), 'Analysis of Stoppages', *Mar. Eng. Naval Architect,* April.
14 Batten, B.K. (1972), 'Marine machinery faults' *Trans. I. Mar. E.,* 84, Pt. 9.
15 Darragh, Jr. J.T. & Haley, Jr. J.L. (1959), 'Manual of aircraft accident investigation', Document 6920-An855/3, International Civil Aviation Authority (I.C.A.A.), Montreal, Canada; H.M.S.O.
16 Collacott, R.A. (1974), 'Suicide on the high seas', *Leicester Chronicle,* February 15.
17 Collacott, R.A. (1972), 'A masterate degree for ship's officers?' *The Motor Ship,* August.
18 Collacott, R.A. (1970), 'Engineers are human', *British Engineer,* January.

2 Causes of failure

2.1 Introduction

Of the many causes of mechanical failure those due to delayed time-dependent failure lend themselves to diagnosis and condition monitoring.

Defects may be broadly classified into those which result in a fracture and those for which fractures do not occur. These may in turn be sub-classified according to the causes due to chemical, thermal and mechanical influences. Further sub-classification results in the identification of creep, corrosion, fatigue and mechanical fracture with a considerable further sub-classification of such causes of failure.

The classification of these faults in metals is shown in Figure 2.1 originally prepared by George [1]. Extending this classification to the many non-metals which are incorporated in modern components would involve various types of instability (apart from buckling): elastic plastic, dynamic strength together with more detailed consideration of welding failures and their causes; faulty joining by gluing, trapped air, instability of glue, softening by humidity, embrittlement, overheating.

2.2 Service failures

The failures experienced most frequently in machinery are fracture, excessive deformation and surface failure, particularly corrosion deterioration. A typical review of failures and their causes are given in the following table.

Table 2.1 *Service failures and their causes*

Cause	Components
Corrosion	Plain bearings [2]; gas turbine blades [3]; Hydraulic systems [4]; sparking plugs [5].
Contamination	Controlled pitch propellors [6]; hydraulic systems [7].
Fatigue	Plain bearings [2]; helicopter rolling element transmission bearings, gears [8]; crankshafts, crankpins, steam turbine blades [6]; alternator pole dovetailed root fixings [9].
Overheating	Plain bearings [6]; sparking plugs [5].
Overstressing	Gear teeth, holding-bolts [6].
Seizure	Plain bearings [2]; Steam drum ligament [6].
Wear	Plain bearings [2]; splines, clutches, seals spacers, bearing liners [8]; main shafts, stern gear [6]; hydraulic systems [7]; diesel exhaust valve seats [5].

The causes of failure have been expressed in very general terms. For example, corrosion includes sulphidation and the effects of deposits, while fatigue includes spalling or surface fatigue and wear includes fretting. There are also cross-related causes such as corrosion fatigue.

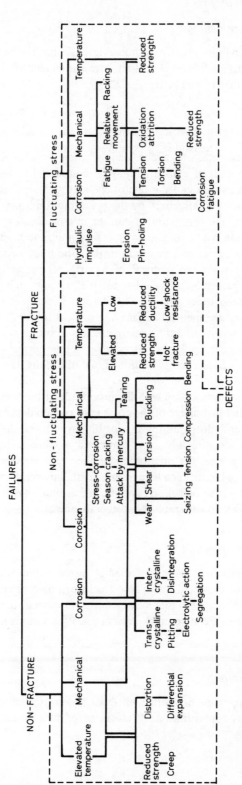

Figure 2.1 *Sub-classifications of failure causes in metals*

2.2.1 Failure surface recognition

Fractographic examination provides a valuable basis for the post-mortem determination of the causes of failure. As a first stage, visual identification can be aided by an 'atlas' of fracture surfaces [10].

2.3 Fatigue

This occurs under the action of cyclic loading when a crack initiates and grows. Although the fatigue limit of a material may be considered as a material property, the condition of this surface and the rate of crack growth are factors of great practical significance, which are influenced by the operating conditions.

Fatigue starts with the formation of surface microcracks (whether by surface roughening, grain boundary cracking or by cracking around hard inclusions) with subsequent extension across and penetration into the body of the metal. The increased life resulting from the removal of a surface layer at frequent intervals throughout a test, irrespective of whether the life is many millions or only a few thousand cycles, demonstrates that crack initiation is confined to surface grains.

The direction of development of surface microcracks is initially that of the operative slip planes. It remains so until a micro-crack is of such a size that the amount it opens and closes, under resolved cyclic stresses acting normal to the crack faces, is sufficient to affect a large enough volume of material along its edge for it to grow as if in a continuum.

The growth process is associated with the magnitude of the tensile strain range in the volume of material just ahead of the crack edge (which depends on the amount the crack is opened and closed during the loading cycle), and the growth direction becomes normal to the direction of the nominal maximum cyclic tensile stress (i.e. for a given crack size, the orientation of the crack becomes that giving the maximum opening), being independent of any crystalline slip directions. The crack may then be termed a macrocrack; its growth rate is much faster than that of the initial microcrack and it spreads rapidly through the metal.

2.3.1 Fatigue crack propagation

Small cracks in a component can be detected by non-destructive techniques (NDT) at an early stage in its expected life. These cracks can be regularly inspected such that a knowledge of the growth rate may enable a component to have a long useful life before replacement. Relative crack growth rates for various metals are given in Table 2.2 [11].

2.3.2 Stress concentrations

Failures from fatigue frequently originate from geometrical shape changes which cause intensive stress concentrations. The extent of these stress intensities for a

Table 2.2 *Relative fatigue crack growth rates*

Material	Relative rate of fatigue crack growth
Mild steel	1
Low-alloy steel	1
Copper	4
Titanium	10
Aluminium	15
5% Mg–Al alloy	50
$5\frac{1}{2}$% Zn–Al alloy	150

particular shape can be determined either by tests on plastic models under the polarized light illumination (when the varying strains associated with stress variations produce different colour) or by mathematical techniques such as the finite element technique.

A stress analysis at the root of a bolt thread by the finite element method, compared with experimental values using a copper-electroplating method, was reported by Maruyama [12]. The load distribution at the bolt thread which is screwed in a nut thread was computed by the finite element method combined with the point matching method, stress concentration under this load distribution was computed. A typical element division is shown in Figure 2.2.

2.3.3 Stress raisers

Changes in shape which produce high stress concentrations may be deliberate as a consequence of design or may be unintentional due to some manufacturing defect. These are known as 'stress raisers' in that they influence sensitivity at the onset of fatigue. Among the more common stress raisers are:

(1) Notches: any change of section which alters the local stress distribution is termed a 'notch' and therefore includes keyways, circumferential grooves, holes, contour changes, scratches, threads etc., hard and high tensile metals are particularly notch-sensitive;

(2) Decarburization: carbon lost from the surface of a ferrous alloy following heating in a reactive medium considerably reduces the fatigue resistance locally on the surface;

(3) Corrosion: pits on the surface act as notches with the same deletereous effects; corrosive activity accelerates failure as during the tensile parts of the fatigue cycle corrosion cracks are opened and exposed to corrosive attack;

(4) Fretting corrosion: when two parts are clamped, press-fitted or shrunk together and subjected to alternating flexure an oxide is produced at the inter-surface causing failure similar to a corrosion product;

(5) Inclusions: metals are not homogeneous and may contain 'foreign particles' or even macroscopic discontinuities which act as inferior stress raisers;

(6) Internal stresses: residual stresses combined with those arising from the applied fluctuating stresses result in high local stresses which weaken the material.

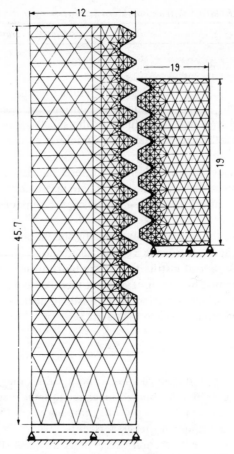

Figure 2.2. *Element division of a bolt-nut joint for stress analysis*

Such residual stresses may arise from heat treatment (including quenching), drawing, rolling, cold working, cold straightening, un-normalized weld assemblies, heat treated welds etc.

(7) Clamping and press fits: typically, this occurs to a shaft with a collar clamped to it, or, a press-fit assembly without effective distribution of local stress. These reduce the endurance limit such that the life of a smooth shaft may be halved as the result of clamping a collar. In many instances the initial failure crack is formed, undetected, in the press fit.

2.3.4 Fatigue fracture identification

The appearances which characterize a fatigue failure surface during a post-mortem examination are:

(1) Little permanent deformation;

(2) Break marks showing the progression of the crack can be seen under a microscope;

(3) The break marks are smooth as a result of rubbing;

(4) The fractures propagate in a direction normal to the principal tensile axis (where combine stresses are involved).

The brittle appearance of a fatigue fracture results from the macrocrack growth stage, the cyclic plastic strain occurring at any instant (at stress levels less than those causing general yielding) is confined to the small volume of material just ahead of the crack front. Characteristic fatigue fractures appear distinctively when examined microscopically. There is little 'necking' typical of a ductile static failure.

The fracture surface comprises two distinct regions, one smooth and velvety (the fatigue zone), the other coarse and crystalline (the instantaneous zone). The smooth velvety appearance of the fatigue zone is caused by rubbing the mating surfaces as the crack opens and closes under repeated loading.

It is not uncommon for several fatigue cracks to propagate from a fatigue zone. In each fatigue zone the origin of the fatigue crack can be found by locating the centre of radiation of fatigue waves. These fatigue waves are variously known as 'clamshells', 'oyster shells' or 'stop marks' and are particularly noticeable in aluminium alloy panels. (Laboratory fatigue tests rarely show these markings since they arise from the stopping of crack development due to different stress levels and repetition rates experienced in actual service).

2.3.5 Fretting fatigue

Fretting corrosion combined with crack-initiated fatigue can reduce the strength (endurance limit) of metals very considerably. An example of fretting fatigue failure in aluminium alloy wing spars on an aircraft [13] in which lug and pin joints joined the main spars of the outer wings to the inner wings and the wings to the central spar. If the pins were cadmium-plated the joints gave no trouble (apparently the soft cadium acted as a lubricant and also prevented galvanic corrosion). Many cases of fretting and corrosion, sometimes accompanied by fatigue in the aluminium alloy spars, were observed on aircraft with bare pins.

Another case of fretting which led to fatigue occurred in the flexible coupling of a low pressure turbine when the pinion sleeve unwound and shattered its cast-iron casing. Post-mortem examination showed that severe fretting had occurred on the side of all the coupling's teeth and on the ends of some of the teeth. The loss of metal by fretting was such that few teeth would be in contact and the load carried on only two or three teeth. This caused high local stresses and out-of-balance forces. Metallographic examination showed that fatigue cracks were present, indicating that corrosion probably initiated the fatigue failure. The rust was mainly β-$Fe_2O_3 \cdot H_2O$ which forms only in mildly acid conditions [14] (generally induced by chloride ions) leading to the conclusion that the prime cause of failure was severe fretting of the coupling teeth causing out-of-balance forces such that seawater corrosion lowered its fatigue strength and caused fatigue cracks.

2.3.6 Subsurface fatigue

Although most fatigue failures initiate from surface defects a number may originate from below the surface. The common sources of origin of such failures are: (a) inclusions due to a lack of homogeniety in the material; (b) hardened surfaces; (c) rolling loads.

(1) Inclusions: the presence of discontuities in the material introduces possible sites for stress-raising. Investigations of the effects of biaxial or triaxial stress concentrations suggest that high subsurface shear stresses play an important part. The effects of inclusions are particularly critical if they are made of hard and nonductile materials and located in areas of high stress;

(2) Hardened surfaces: the transition area between a hardened case and softer inner inner core have been observed frequently to be the sites from which fatigue may develop. A change of microscopic structure together with the presence of residual stresses is believed to be responsible for this failure;

(3) Rolling loads: normal working conditions for a ball bearing, the track of an overhead crane and a rail under a locomotive produce the following typical shear stresses which will fluctuate with the rolling action which takes place; (Table 2.3).

Table 2.3 *Typical rolling stresses*

	Diameter *in*	*Width* *in*	*Shear stress* *lbf in*$^{-2}$ *in*	*Depth* *1/1000* *in*
Ball bearing	0.25	—	135 000	5
Crane track	8.00	1.00	10 000	40
Rail	30.00	1.00	24 800	243

Cracks form below the surface in each of these instances, with ball bearings they result in pitting, spalling and flaking; with rails 'shelling' of the track has been a frequent form of failure.

2.3.7 Metallurgical influences on fatigue strength

Metallurgical changes have a limited effect on fatigue life, the principal factors being grain size, microstructure and orientation.

2.3.8 Service loads causing fatigue

Vibratory stresses are the principal cause of fatigue failure. Typical examples of some operational loads are as follows:

(1) Aircraft loadings: a summary of the sources of fatigue loading in aircraft presented by Ward and Parish [15] Stress Engineers, British Aircraft Corporation Ltd., Preston, Lancs, included;

Manoeuvre	Ground testing
Atmospheric turbulence	Taxiing
Cabin pressurization	Take-off roll and ground-to-air transition
Fuel tank pressurization	Landing impact
Flying controls	Braking
Buffet	Reverse thrust
Noise (engine and boundary layer)	Thermal stresses

(2) Marine loadings: a survey by Smedley [16], Senior Principal Surveyor, Lloyds Register of Shipping, Crawley, of some causes of fatigue failure in service included; (a) wave action and drilling vibration (*Sea Gem* drilling barge, 1965), (b) rudder vibrations due to a ship's wake, (c) propellor vibrations of a critical 10th-order from a 5-bladed propellor, (d) cyclic bending of engine connecting rods, (e) bending fatigue of crankshats due to misalignment following (i) wear (ii) maloperation, (f) cyclic bending of slack bolts due to movement of the loose castings, (g) cyclic torsion of propellor shafts exposed to sea-water corrosion attack.

2.4 Excessive deformation

Loads which impose stresses in excess of the elastic limit may result in functional failure. This may not necessarily involve a fracture although fracture failure is most usual.

The loads applied to a component may be a combination of three possible modes (i) static, (ii) repeated and (iii) dynamic.

Static loads may be applied gradually so that at any instant in time all parts are essentially in equilibrium. Such short-time static loading typically arises when the load slowly and progressively increases to its maximum service value which is held for a limited time and then not re-applied often enough to introduce fatigue considerations. Static loads may, alternatively, be applied and held at the maximum service value for such a lengthy period that the creep or flow characteristics of the material are influential in determining ultimate life.

Repeated loads are generally associated with fatigue as the stress is applied and wholly or partially removed or increased many times in rapid succession.

Dynamic loads involve a state of movement so that except for the terminal positions there is no static equilibrium for long periods of time. Classes of dynamic load are (i) sudden loads (ii) impact loads.

Sudden loading occurs when a mass or 'dead load' not in motion is suddenly applied to a body. This can be proved to create a stress approximately twice as great as if the mass were applied gently.

Impact loads are associated with motion as one body strikes another such that exceptionally high stresses can be generated as the impact kinetic energy is transferred to strain energy. Materials which ordinarily fail in a ductile manner under static loads often fail in a brittle manner if the rate of loading is high.

2.4.1 Load stresses (Hertz)

Although Hertzian theories on the elastic contact between curved surfaces do not directly relate to service failures which involve plastic deformation causing mutual indentation of the contacting surfaces, Hertzian stress analysis continues to be widely used in engineering analysis as it is effective up to the final failure condition.

Hertz [17] derived equations for the magnitude and type of stresses imposed on and below the surface during compression contact. By assuming normal pressure and static conditions the contact surface between two elastic cylinders with parallel axes is a narrow rectangle. The pressure distribution across the width of this rectangle is semi-elliptic. Timoshenko [18] showed that the compressive stresses decreased with depth but that while the shear stress was zero at the surface layer it increased to a maximum value at a distance below the surface.

When cylinders roll, a torque is applied in addition to the load forces which was shown by Smith and Liu [19] to distort the semi-elliptic pressure distribution. For a coefficient of friction of 0.33 the maximum compressive stress is increased 39% and the maximum shear stress 42%.

Surface breakdown limits the life and load-bearing capability of contacting surfaces as indicated by the results shown in Table 2.4 [20].

Table 2.4 *Grey iron casting, ASTM A-48-48 class 30, heat-treated*

% sliding	Cycles to failure
50 (driving roll)	88.711×10^6
101.2 (driven roll)	51.009×10^6
303.3 (driven roll)	29.204×10^6

2.4.2 Indentation

Dents, pits, grooves and distortion commonly result from the application of excessive stress. Typical examples of mutual indentation are shown in Figure 2.3 from Atkins and Felbeck [21] where the specimens are of identical shape and similar

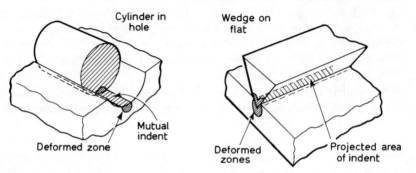

Figure 2.3 *Types of mutual indentation*

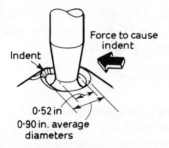

Figure 2.4 *Indentation damage – automobile tie-rod*

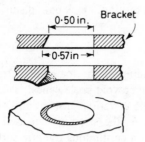

Figure 2.5 *'Belling' of a bolt-hole*

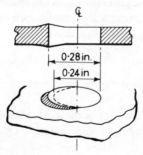

Figure 2.6 *'Ovalling' of a bolt-hole*

mechanical properties so that the resulting impressions are congruent (not only are they mating pairs but their response to deformation should be exactly the same). With specimens of different mechanical properties, but of the same geometry, the indentations would differ with the softer material undergoing greater deformation.

Plastic flow theory is frequently applied to calculate stresses causing indentation [22, 23]. Plastic deformation begins when the mean pressure is 1.1 Y (where $Y =$ yield pressure of the material under uniaxial compression) and full plasticity at 2.8 Y.

Typical examples in which failure has resulted from indentation include an automobile tie-rod, Figure 2.4 involving the outer surface in a pressure of 12 000 lbf in^{-2} (84.7 kg mm^{-2}, 832 MN m^{-2}) the 'belling' of a bolt hole subject to lateral forces,

Figure 2.5 involving stresses of 86 000 lbf in^{-2} (60.7 kg mm^{-2}, 595 MN m^{-2}) and ovalling of a bolt hole, Figure 2.6 due to the shear force on a bolt and stresses of 60 900 lbf in^{-2} (42.9 kg mm^{-2}, 420 MN m^{-2}).

2.4.3 Cleavage fracture

This is a failure which takes place along well-defined crystallographic planes within the grains. Basically, a cleavage fracture surface contains large areas which are relatively smooth and featureless separated by numerous other characteristic features — such as cleavage steps, cleavage feathers, river markings and cleavage tongues — which are a direct result of crack path disturbance.

2.4.4 Ductile failure

Basically caused by overloading of the material. A transgranular break occurs by means of a process called 'microvoid coalescence'. This means that the metal pulls apart under the applied stress at various discontinuities such as inclusions, precipitates and grain boundaries. As the stress is increased these microvoids grow and eventually coalesce to form a continuous fracture surface consisting of numerous cup-like depressions generally referred to as dimples. In a ductile failure due to pure tensile stresses the dimples are generally equiaxed and more or less circular, whereas under shear conditions they assume a parabolic form. In pure shear the dimples on the mating faces point in opposite directions, while tensile tearing produces parabolic dimple pairs having the same sense.

2.4.5 Static load fractures

Such fractures result from the application of a single load causing stressing in excess of the ultimate strength of the material. Characteristic fractures involve permanent deformation of the failed part resulting from a stress application in excess of the yield point.

2.4.6 Tension

This produces a local deformation or 'necking' and the fracture surface is made up of planes of separation inclined at about 45° to the direction of the load. Two parts of a rod broken by axial tension may resemble a 'cup and core' with 45° bevelled edges. Pure tension fractures part cleanly without any rubbing between the two halves.

2.4.7 Compression

Failures due to compression occur in two general forms, (a) block compression and (b) buckling. Block compression occurs in short heavy sections which separate on oblique planes as in tension except that there is rubbing of the two halves of the fracture during separation. Buckling occurs in long sections and involves typical bending change of shape.

2.4.8 Bending

Bending moments applied to a material are resisted by complementary tensile and compressive stresses set up in the material. As a consequence, bending failure of metals produces a tension-type fracture on the outside of the bend and the compression type fracture on the inside.

2.4.9 Shear

Two types of failure under shear can be identified, (a) block shear and (b) buckling. With block shear the two halves of the fracture slide one across the other and the surface will appear rubbed, polished or scored — the direction of scoring indicates the direction of the applied shear corce. Buckling occurs in metal sheets, usually in a diagonal fashion such that the direction of the wave troughs or crests is the tension diagonal of the sheared panel. When rivets, screws or bolts fail in shear the hole elongates and an open moon-shaped space appears behind the fixing.

2.4.10 Torsion

This is a form of shear. The two halves of the fractured metal specimen retain some permanent twist — the fracture surfaces often exhibit tensile fracture surfaces oblique to the angle of twist.

2.4.11 Impacts

Collisions and explosions create stresses under impulsive action due to the effects of stress waves. A typical example of the progressive effects of a centrally applied shock force is seen in Figure 2.7 based on failure examples by Johnson [24] showing that the counteraction of the reflected stress waves tears off a spall at the lower edge of the material while deforming the body of the material with a central cavity.

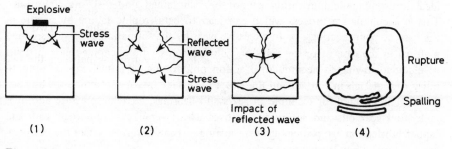

Figure 2.7 *Generation of a stress wave fracture*

2.5 Wear

Most life-deterioration of a machine is the consequency of wear between two surfaces between which there is a relative motion. Wear is a process of surface damage producing wear debris which may have secondary effects. When two surfaces are in rubbing contact the complex processes which occur in the surface layers may involve one or more of the following activities:

(1) Microcutting: in which abrasive particles of wear debris gouge out tiny particles of metal or cause surface deformations under plastic flow;

(2) Plastic and elastoplastic deformation: which occurs to specific areas of the surface as a result of high (Hertzian) local pressures, produced as a result of contact between microasperities when the surfaces come into contact;

(3) Surface fatigue: following repeated elastic deformations to the surface as a result of fluctuating forces;

(4) Local heating: the result of insufficient heat transfer due to friction when a combination of high pressure and speed is involved. Local temperature rises can be so high as to cause phase changes and melting of the surfaces (weld junctions);

(5) Oxidation: causing the formation of films of solid solutions and oxides which are removed. Other chemical interactions, produce a similar erosion/wear phenomenon;

(6) Molecular interaction: when surfaces become bound together under intense pressures and low speeds, this results in the cold welding of individual sections and the transfer of metal particles from one surface to the other;

(7) Rehbinder effect: a loosening effect arising from the lubricant filling up microcracks and causing an increase in pressure which leads to damage of the surface layers;

Each of these activities may occur simultaneously so that the actual form of wear may be classified as follows:

(1) Abrasive wear: caused by the ploughing or gauging of hard particles against a relatively soft working surface. This is probably the most serious single cause of wear in engineering practice [25]. Lubricant filtration and efficient sealing of bearings are important matters requiring attention in order to minimize this source of deterioration.

(2) Scuffing: an activity in which the mating surfaces come into contact when ideal hydrodynamic lubrication cannot be maintained under sliding conditions. This is a cumulative process which can lead to catastrophic failure by seizure in the case of plain bearings, or to excessive wear by the formation of work-hardened material in the case of piston rings or cylinder bores. The phenomenon is difficult to elucidate owing to accumulative action obliterating its initial stages. Compatibility of materials in contact and the effect of surface treatments and lubricants can be most important for example in gears and bearings. Extreme pressure additives in hypoid gear oils and anti-wear additives used in crankcase oils to aid cam and tappet lubrication can protect against scuffing but can adversely affect the performance of some plain bearing materials;

(3) Fatigue (pitting) wear: which usually occurs in rolling friction and is caused by fatigue of the surface layers. If there is also relative sliding of the surfaces, wear resulting from the fatigue of the microasperities is also possible. With this form of wear there is fragmentation of the surface layers, but this is characterized by low wear rates;

(4) Molecular wear (galling or adhesive wear): is characterized by the development of local metal joints and the removal from or adherence of particles to the rubbing surface. This type of wear occurs with high pressures and, as a rule, develops at a rapid rate. If there is considerable heating in the sliding zone, this type of wear is known as thermal wear. Thermal wear occurs if the surfaces have relative high sliding velocities;

(5) Mechanical-corrosion (oxidation) wear: occurs if oxidation processes are significant. The plastically deformed and oxygen-saturated surface layer of the component fractures as a result of repeated loading, and fresh sublayers of metal become exposed;

(6) Cavitation wear (cavitation erosion): occurs when the surface is acted on by a fluid flow with local hydraulic impacts in the cavitation zone. If the component operates in a flow of hot gas, the surface is softened and oxidizes, and other processes may occur, as a result of which, particles are removed from the surface and carried away by the fluid flow (gas erosion). Consequences of wear are mainly the loss of engagement between mating surfaces such as splines, screw threads etc. (lack of initial tightness may be the primary cause of the wear itself). An indirect result of wear may be excessive clearance permitting the ingress of corrosive or abrasive material which in turn may accelerate the wear and give rise to surface fatigue or stress corrosion failure.

2.5.1 Load-bearing surfaces

The topography, chemical nature and physical structure of the moving surfaces and the environment in which they operate influences their friction and wear [26].

Any true engineering surface will be covered by a thin oxide layer, angstroms in thickness, [27] consisting essentially of chemi-sorbed oxygen but preceded, in most cases, by rapid adsorption from the atmosphere. This is not, by any means, a uniform layer but a heterogeneous zone of oxide in metal, related to the method of surface preparation. The effect of the contaminant layer is to protect, initially, the underlying metal but under certain conditions it is not totally regenerative, being dispersed by the wearing action to promote true metallic contact.

During the initial stages of lubricated sliding an equilibrium condition is set up whereby the surface layers are plastically deformed in order to spread the full operating load [28]. This surface deformation, mostly continued to the softer bearing material, can occur through occasional intermetallic contact but also through loads transmitted via the oil film. This then increases the hardness of the surface layer and also changes metal micro-structure. Where metallic contact does occur, this will produce a potentially highly reactive virgin surface available to interact with both lubricating oil and additives [29].

Environmental effects on the lubrication of load-bearing surfaces are influenced by the surface energy state relating the catalytic properties of surfaces and surface reactions. Experiments suggest that the resistance of some solids to strain and deformation is reduced by the presence of a surrounding medium containing surface active substances, the Rehbinder effect. The effect is associated directly with reversible adsorption lowering surface energy. The surface deformations occurring during metallic contact lead to the formation of fresh surface. Such surfaces are very sensitive to the adsorption of surface active materials which greatly facilitate the deformation process, often accelerating material destruction more effectively than a chemical reaction.

2.5.2 Wear-rate variations with time

There are three distinct wear-rate phases of a component during its working life [30].

During phase I the components are run in, i.e. the micro and macro-geometries of the surfaces change. To reduce the normal level of wear, represented by phase II the rate of wear developed during phase I should be kept to a minimum and therefore transition from phase I to phase II should occur as quickly as possible.

Phase II wear is the normal erosion which may result from any of the several processes of 'wear'. In most cases it is possible to assume a linear relationship between wear rate and time. The main factors determining this wear are the specific pressure and relative rubbing velocity. For abrasive wear conditions linear wear is directly proportional to the specific pressure on the rubbing surface and the sliding path. If the lubricant contains abrasive particles, the wear rate is often a power function of the sliding velocity.

Phase III results from a change in the surface structure coinciding with stress variations and the metallurgical strength of the material such that fatigue develops at a catastrophic rate.

2.5.3 Wear-rate equations

Several equations and a number of different mathematical models have been evolved in order to provide a theoretical evaluation of wear rate.

(1) Abrasive wear: for surfaces which have been properly run-in it is possible to obtain good experimental and theoretical correlation using equations derived by Barwell and Strang [31] and by Archard [32].

(2) Microcutting wear: the shape of the asperities cut from the metal has a controlling influence on the resistance to the wear forces and accordingly, to the rate of wear;

(3) Plastic contact wear: experimental studies based on the concept that plastic failure of the aspèrities is related to material fatigue led Kragelskii [33] to derive a wear-rate equation;

(4) Elastic contact wear: from an assessment of the size of single contact spots under the influence of elastic contact coupled with fatigue data, Kragelskii quoted a wear rate equation for this condition.

2.5.4 Wear surface temperature monitoring

Indirect methods which have been used to measure the surface of solid bodies with a high degree of accuracy during sliding may be classified as:

(1) Surface colour observation;

(2) Using selective low-temperature-melting powders on the surface;

(3) 'Thermal' wax pencils;

(4) Surface phase transformation observation using indicator metals of known phase characteristics [34].

Direct methods quoted by Kragelskii [33] include:

(1) Intrinsic thermocouple techniques;

(2) Resistance thermometers;

(3) Sliding thermocouples;

(4) Radiation [35];

(5) Thermography [36] which uses the luminescence of phosphors under an ultra-violet light source.

2.5.5 Scuffing

Gross damage characterized by the formation of local welds between sliding surface normally defines the term 'scuffing'. It originates primarily from an increase in the coefficient of friction between rubbing or sliding surfaces.

During contact Hertzian pressures may apply which rupture any lubricating film separating two surfaces thus enabling them to weld together. Hirst [37] quoted the work of Hamilton and Goodman [38] and referred to the influence of friction in increasing the Hertzian stresses and the introduction of surface flow under plastic deformation, in particular that a 30% increase in the coefficient friction may suffice to produce shear deformation over a whole contact area.

The danger of scuffing increases where both rolling and sliding occur between surfaces: the effect of sliding is greatest. Under such conditions high flash temperatures may develop which increase the plasticity and tendency of the surface metal to melt as the heat generated during the contact cannot be dissipated. Scratches terminating in surface melting are symptomatic of this form of failure.

Severe surface damage may ensue from the high flash temperatures generated during friction when individual points may be in contact for periods of the order of 10^{-7} s. Studies by Bowden and Ridler [39] indicate that flash temperatures rises of 700°C were produced during a Constantan-steel contact of under 10^{-4} s. Theoretical evaluations have been made by Archard [40] of the flash temperature rises under conditions of plastic deformation and at low speed.

Gear teeth and the pistons in internal combustion engines are particularly prone to fail as a result of surface scuffing. Most studies have been concentrated on gear failure (because the sequence of events on the pistons of internal combustion engines can range from cylinder scoring to seizure and bearing failure).

With gear scuffing three sets of conditions appear to exist, namely:

(1) Highly rated reduction gearing as used in aircraft power units in which no scuffing can be tolerated;

(2) Marine transmission units in which a limited amount of scuffing can be permitted;

(3) Low-speed light-duty gearing as in machine tools in which scuffing has no significant effect.

2.5.6 Frictional welding

The frictional welding process related to scuffing becomes apparent under conditions of sliding at high speeds and pressures such that when the load is removed parts of one surface remain welded to the other.

Three stages have been observed in this process:

(1) External friction due to dry or boundary lubrication conditions;

(2) Heating of a surface layer with the destruction of surface impurities and the creation of sites at which seizure can occur;

(3) Extremely high heat generation accompanied by intermolecular welding of the two surfaces.

The welded products form a fine-grained structure free of holes and foreign inclusions. Similar and dissimilar materials may weld together in this manner.

2.6 Corrosion

Corrosive deterioration arises [41, 42] as a result of electrochemical or chemical-erosion attack due to environmental conditions and the presence of anodic bodies either in major macroscopic or microscopic quantities. In addition to the loss of load-bearing material as exemplified by the rusting away of oxidized material, much primary corrosion in the form of rust occurs during erection when the working parts of machines may be exposed for long periods in a moist or wet atmosphere leading to the formation of a considerable amount of iron rust (FeO). Most manufacturers use appropriate rust protectives but if the surface is not correctly prepared these may only cover up a continuing process the products of which will eventually enter the system. Sources of corrosive attack are varied, ranging from the more obvious attack due to the biological deterioration of marine growths experienced in machine engineering and power stations [43] to the less distinctive effects of metallic inhomogeneities in weak electrolytes (such as dew formed in industrial atmospheres).

With plant after erection, corrosive activity related to operating conditions is usually identified as (a) stress corrosion, (b) corrosion fatigue or (c) cavitation.

2.6.1 Stress corrosion

This causes cracking under the action of a steady stress in a corrosive environment. The cracks tend to lie perpendicular to the principal tensile stress without plastic deformation. The cracks follow an irregular route and result in a coarsely textured fracture face which may be transgranular or intergranular depending on the materials and corrosive agents involved [44].

2.6.2 Corrosion fatigue

While greatly influenced by a corrosive atmosphere this arises under a fluctuating stress with a tensile component. Fissures of an 'estuary' form penetrate inwards from the surface and some may extend into the body of the metal as deep cracks. The cracks lie perpendicular to the principal tensile stress with paths which are mainly transgranular, irregular and branched. These cracks are usually filled with corrosion products.

2.6.3 Cavitation

The collapse of minute vapour bubbles involving impingement and associated chemical or electro-chemical surface activity is the cause of cavitation failure. The surface becomes extensively pitted sometimes with a honeycomb pattern. Fatigue cracks are likely to develop from these surface pits. Cavitation is likely to occur in hydraulic machinery under circumstances in which changes of velocity cause associated changes in the pressure of the working fluid [45]. Cavitation damages a wide range of engineering components — ship's propellors and rudders, the waterside of diesel engine cylinder liners, plain bearings, fuel-injection systems, pumps and hydraulic systems.

2.6.4 Sulphur corrosion

Sulphidation is a corrosive activity associated with the combustion of most fuels.

Most fuels contain impurities such as sulphur which produce gas-water vapour mixtures resulting from the burning of hydrogen to form H_2O and sulphur to sulphur trioxide SO_3. When the temperature of such gases reaches approximately $300°F$, water vapour and sulphur trioxide combine to form sulphuric acid at the dew point temperature and a fog of sulphuric acid is formed. This acid condenses on cold surfaces and can also be absorbed on solid particles, such as fly ash, and any unburned hydrocarbons. The result is corrosion of cold section metal particularly in boilers and the formation of deposits with finally, the discharge of air-polluting acid smut.

An alternative problem may arise in marine diesel engines which may sometimes use supplies of fuel with a low sulphur content. These engines occasionally experience high liner and piston ring wear rates usually leading to unscheduled overhauls

[46]. An investigation by Cotti and Simonetti [47] concluded that this high wear was a result of unreacted alkaline material precipitated from highly alkaline lubricants in the presence of fuels containing insufficient sulphur for reaction to occur.

Another problem experienced with marine gas turbines and reported by O'Hare and Holburn [3] concerned the corrosion or sulphidation which destroyed some 90% of the nozzle guide vanes of a Pratt and Whitney gas turbine. It was found that during combustion, particles of molten sodium sulphate are formed from sulphur in the fuel and sodium present in airborne sea salt, or possibly as a fuel contaminant. The molten Na_2SO_4 particles collect on the nozzles being very adhesive in the molten condition. Corrosion starts at the interface. The salt reacts with the protective oxide film on the surface and prevents it from healing. Sulphur crosses the interface and combines preferentially with chromium, destroying its protection. Oxygen in the hot gas stream then oxidizes the unprotected metal in these areas. Once started, this process self propagates and can be stopped only by the complete removal of all oxidized material and all metal sulphides from the metal matrix.

2.6.5 Vanadium deposits

Fuels often contain vanadium which is present as the metallic compound vanadium porphyrin and remains in the carbonaceous ash particles called 'cenospheres' which coke around the fuel injectors of diesel engines, particularly the large engines used for marine propulsion. The deposits may be soft carbon (soot) or hard vitreous carbon, typical low-melting vanadium complex compounds found in these ashes are given in Table 2.5.

Table 2.5 *Fusion temperatures of vanadium products*

Compound	Fusion temperature $°F$
Sodium metavanadate ($Na_2O \cdot V_2O_5$)	1166
Sodium pyrovanadate ($2Na_2O \cdot V_2O_5$)	1184
Sodium orthovanadate ($3Na_2O \cdot V_2O_5$)	1562
Sodium vanadic ($Na_2O \cdot V_2O_4 \cdot 5V_2O_5$)	1157
Sodium vanadate ($5Na_2O \cdot V_2O_4 \cdot V_2O_5$)	995
Ferric metavanadate ($Fe_2O_3 \cdot V_2O_5$)	1571
Sodium sulphate	1616

These compounds can adhere to and fuse onto metal surfaces resulting in the formation of slag and corrosion of the metal. In addition to the destruction of metal surfaces such vanadium deposits decrease the heat transfer efficiency, particularly in boilers and can lead to the spalling of refractory linings [48].

2.6.6 Lead deposits

Exhaust valve and spark plug problems in spark-ignition engines are usually associated with excessive temperatures and corrosive deposits. Lead deposits on exhaust valves can, at temperatures above about 810°C, react with the protective chromium-rich oxide layers, and destroy the oxidation resistance of valves. Coloured compounds seen on exhaust valves indicate that the corrosion reaction is proceeding [49].

Analysis of these deposits [5] indicates the presence of various lead oxysulphates ($PbO \cdot PbSO_4$, $2PbO \cdot PbSO_4$ and $4PbO \cdot PbSO_4$). Specimens of exhaust valve steel exposed at high temperature to lead oxysulphates and the coloured products when subjected to complete chemical and X-ray diffraction analysis showed that lead oxychromates were formed [50]. These lead deposits react with the protective oxide layers on the valve steel and destroy them.

2.7 Blockage sludges

Industrial plant and machinery has to handle difficult materials of various kinds, ranging from powders which arch [51] to low-grade fuels which congeal like tar or even seperate out wax.

Heavy fuels with high velocity are often used in marine engines. Such fuels are the low quality residual products of distillation from which it is not reasonable to extract further petroleum products. Depending upon their source of manufacture such fuels may contain water, dirt, asphaltic precipitates and wax. Failure to remove water and dirt (frequently rust) leads to corrosion of the injector nozzles and their abrasion by the sediment. Asphaltic precipitates prevent the 'handling' of fuel by impeding the pumping in the fuel system; it is for this reason the oil may be treated to encourage higher solvency and detergency. Mixtures of fuels from different sources, for example, one of a light type rich in paraffinics, the others heavy with asphaltenes, do not blend easily resulting in imperfect combustion due to variable injection. Waxes produce plugs of tarry fuel which clog the lines.

2.8 Blockage in cooling systems

Normal conditions existing within heat transfer systems are far from ideal [52], thus the coolant side of internal combustion engines is usually left as-cast with adhering core-sand and rust.

A direct investigation by French [53] showed that in addition to the presence of foreign matter in the coolant spaces the coolant itself could be contaminated with oil from gas leakage at the cylinder head gasket, impurities in the cooling water ranging from algae to lime products as well as soluble oils and ethylene glycol. This dirtiness of the heat transfer surfaces accounts for many of the anomalies found during engine tests.

This blockage of cooling systems and impaired heat transfer leads to thermal stressing, one of the most common sources of engine failure under highly rated conditions. Under extreme conditions as with burning of exhaust valves following exhaust gas blow-by, or burning of piston crowns or cylinder heads, or yet again as a result of 'detonation', metal may actually be removed in appreciable quantities.

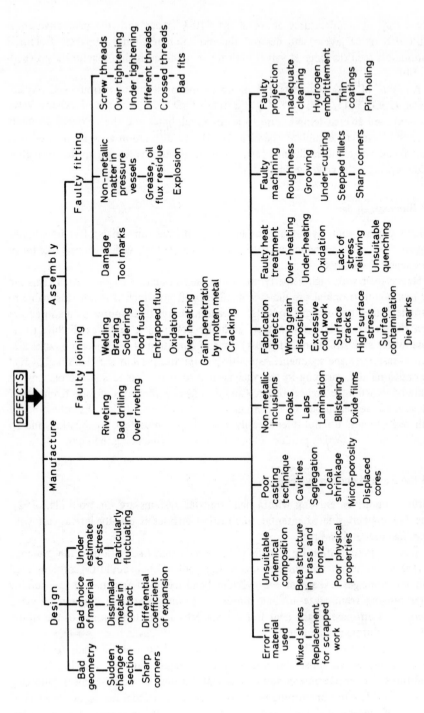

Figure 2.8 *Causes of failure due to deficiencies of design, manufacture and assembly*

More commonly, however, a fatigue crack is formed and this propagates in time through to the coolant space, allowing gas and oil to blow into the coolant under running conditions and/or water to leak into the cylinder when the engine is shut down. The crack normally develops following yield of the metal as described by Fitzgeorge and Pope [52] due to an excessive compressive thermal stress, and under load cycling conditions of operation a tensile fatigue failure occurs.

2.9 Design, manufacturing and assembly causes of failure

Apart from the operational causes leading to delayed time-dependent deterioration a wide range of premature failures may be attributed to deficiencies of design, manufacture and assembly. A chart of such defects is shown in Figure 2.8 [1].

It was held by Marsh [53] that the most common causes of premature failure arose from the existance of defects in minor items of equipment such as drain cocks, screwed couplings, small switches, pipe joints, gland packing. With complex systems the use of multiple redundancy may not increase reliability owing to the influence of common faults although redundance itself may be either active or passive, active redundancy being such that all parts of the system are in operation at full load while with passive redundancy the standby component is idle, the idle passive component may be brought into operation either by witnessing a fault signal or by automatic switch triggered by a fault signal.

Manufacturing faults causing failure can be reduced by the use of quality control techniques (static and non-destructive testing). An analysis of failures would indicate the 'worst six' based on a Pareto curve of occurence.

Incorrect assembly is a critical cause of failure. Incorrect assembly may occur during manufacture or after servicing and is one of the principal arguments in favour of condition maintenance and 'leaving well alone'. Typical instances in which failure has resulted from incorrect assembly are:

(1) In the erection of a steel framework assembly was such that near-failure loads were applied to some components during assembly. An increase in wind loading resulted in failure conditions and collapse;

(2) In the mating of adjacent surfaces of sub-assemblies of a machine, two sub-assemblies were thrown slightly out-of-line when bolted together such that shafts and gears projecting from the interface carried the cantilevered weight of the sub-assembly, a shaft was bent and premature failure followed;

(3) Bearing locknuts and other retention hardware sometimes back off because they have not been tightened sufficiently or old damaged components used with catastrophic results. The occurrence is random and detectable by debris indicators. A high proportion of failures involve maintenance errors and exhibit infant mortality-type characteristics;

(4) Turbine nozzle guide vane failure after only 1890 h operation which caused damage to both the compressor and free turbine, initial examination suggested that the nozzle securing ring had been incorrectly assembled and blocked off cooling air to the turbine 1st stage nozzle guide vanes. This caused overheating and the failure of 17 vanes;

(5) Compressive fracture of the exhaust elbow was detected by the leakage of exhaust gas into the engine room. The defect arose from misalignment between the thin-walled stainless-steel elbow and the adjacent thick-wall exhaust trunking which restricted their mutual free expansion;

(6) Pop rivet shanks which broke off after fitting door seals had not been removed from air intake ducting. These were sucked into the engine causing severe damage, namely,

low pressure compressor	9 vanes, 28 blades
high pressure compressor	76 vanes, 162 blades;

The foreign objects were able to 'snake' through the h.p. compressor as it rotated at 6000 revs/min but could not pass through the h.p. compressor which rotated at 8000 revs/min.

References

1 George, C.W. (1947). 'A review of the causes of failure in metals', J. Birmingham Metall. Soc. 310.

2. Wilcock, D.F. & Booser, E. R. (1967), *Bearing Design and Application,* McGraw Hill, New York.

3 O'Hare, T.L.R. & Holburn, J.G. (1973), 'Operating experience with gas turbine container ships', *Trans. I.Mar.E.,* 85.

4 Olsen, J.H. (1969), 'Flow induced damage to servo valves using phosphate ester hydraulic fluids', Vickers, *Aerospace Fluid Power Conference,* 1969 (re. Pall Corporation, Glen Cove, NY11542, USA).

5 Wilson, R.W. (1972), 'The diagnosis of engineering failures', S.A. Mech. Eng., 22, No. 11.

6 Batten, B.K. (1972), 'Marine machinery failures', *Trans. I. Mar. E.,* 84, Pt. 9.

7 Farris J.A. (1970), 'The control of silt and abrasive wear in hydraulic systems', Field Service Report No. 47, September (Pall Corporation, Glen Cove, New York 11542, USA).

8 Harding, D.G. & MACK, J.C. (1970), 'Design of helicopter transmission for on-condition maintenance', *Proc. 9th Reliability and Maintainability Conference,* A.S.M.E.

9 Morton, P.G., Goodman, P.J. & Kaweck, Z.M. (1969—70), 'Fretting fatigue in keyed dovetailed root fittings', *Proc. I. Mech. E.,* 184, Pt. 3B.

10 Pohl, E.J. (1964), *The Face of Metallic Fractures,* Vols. I & II, Munchener Ruckversicherungs Gesellschaft, Munich.

11 Frost, N.E. & Marsh, K.J. (1969—70), 'Designing to prevent fatigue failures in service', *Proc. I. Mech. E.* 184, Pt. 3B.

12 Maruyama, K. (1974), 'Stress analysis of a bolt-run joint', *Bull. Japan Soc. Mech. Eng.* 17, No. 106, pp. 442—450.

13 Williams, J.K. (1969—70), 'Major causes of failure — civil aircraft structures', *Proc. I. Mech. E.* 184, Pt. 3B., p. 222.

14 Graham, R. (1955), 'Staining of engineering components', *Corrosion Technology*, 2, No. 9.

15 Ward, A.P. & Parish, H.E. (1969), 'The choice of fail safe and safe life fatigue philosophies in aircraft design', Conf. Safety & Failure of Components, I. Mech. E.

16 Smedley, G.P. (1969). 'Some causes of fatigue failures in service' Conf. Safety and failure of Components, I. Mech. E.

17 Hertz, H. *Gesammelte Werke* (Collected Works), Vol. 1., Leipzig (1895).

18 Timoshenko, S. (1934), *Theory of Elasticity*, McGraw Hill, New York.

19 Smith, J.O. & Liu, C.K., 'Stresses due to tangential and normal loads on an elastic solid', A.S.M.E. Paper 52-A-13.

20 Lipson, C. & Colwell, L.V. (1961), *Handbook of Mechanical Wear*, University of Michigan Press.

21 Atkins, A.G. & Felbeck, D.K. (1974), 'Mutual indentation hardness in service failure analysis', *Chartered Mech. Eng.*, June, 78–83.

22 Tabor, D. (1951), *The Hardness of Metals* Clarendon Press, Oxford.

23 Timoshenko, A. & Goodier, W. (1970), *Theory of Elasticity* McGraw Hill, New York.

24 Johnson, W. (1972), *Impact Strength of Materials*, Edward Arnold, London.

25 Collacott, R.A. (1942), 'Piston friction', *Oil Engine.*

26 Lansdown, A.R. & Hurricks, P.L. (1973), 'Interaction of lubricants and materials', *Trans. I. Mar. E*, 85, Series A, Pt. 7.

27 Tomashov, N.D. (1966), *Theory of Corrosion and Protection of metals*, McMillan, London.

28 Hurricks, P.L. (1971), 'Overcoming industrial wear', *Ind. Lub. & Trib.*, p. 345.

29 Morecroft, D.W. (1971), 'Reactions of octadecane and decoic acid with clean iron surfaces', Conf. Ch. Effects at Bearing Surfaces, Swansea Tribology Centre.

30 Pronikov, A.S. (1973), 'Dependability and durability of engineering products', Butterworth. London.

31 Barwell, J.T. & Strang, C.D. (1952). 'Metallic wear', *Proc. Roy. Soc. A* 212, 470.

32 Archard, J.F. (1953), 'Contact and rubbing of flat surfaces', *J. App. Physics*, 24, 981.

33 Kragelskii, I.V. (1965), 'Friction and wear', Butterworth, London.

34 Kragelskii, I.V. & Schvetsova, E.M. (1956), 'Simulation of the processes occurring on rubbing surfaces', Symposiom, Increasing Wear Resistance and Life of Machines (Russian) 12, *Izdrvo Akad. Nauk.* USSR p. 16, Mashgig.

35 Martin, E.J. & Wilson, J.E. (1954), 'Unusual temperatures in brake drums', *Automot. Ind.*, 110, No. 13, 34.

36 Urbach, F., Nail, N.R. & Pearlman, D. (1949), 'Observations of Temperature distributions and of thermal radiations by means of non-linear phosphorus' *J. Opt. Soc. America*, 39, 1011.

37   Hirst, W. (1974), 'Scuffing and its prevention', *Chartered Mech. Eng.*, April 83–92.

38   Hamilton, G.M. & Goodman, L.E. (1966), Influence of surface friction on the Hertzian stress distribution', *J. Appl. Mech.*, 33, 371.

39   Bowden, F.P. & Ridler, K.E.W. (1936), 'Phenomenon of flash temperatures between highly loaded moving surfaces', *Proc. Roy. Soc. A* **154**, 640.

40   Archard, J.F. (1959), 'The temperature of rubbing surfaces' *Wear*, **2**, 438.

41   Evans, U.R. (1960), *The Corrosion and Oxidation of Metals: Scientific Principles and Practical Applications* Edward Arnold, London.

42   Uhlig, H.H. (1971), *Corrosion and Corrosion Control: An Introduction to Corrosion Science and Engineering*, Wiley, New York and London.

43   Rogers, T.H. (1968), *Marine Corrosion*, Newnes-Butterworth, London.

44   Collacott, R.A. (1945), 'Corrosion in pipe fittings', *Marine Engineer*, June.

45   Wilson, R.W. (1965), 'The control of cavitation in diesel engines'. Symposium, Cavitation and Corrosion in Diesel Engines, *British Rail Board Publication*.

46   Baker, A.J.S. & Kimber, J.D. (1974), 'Research engines for low and medium speed application', *Trans. I. Mar. E.*, 87, January, 125–145.

47   Cotti, E. & Simonetti, G. (1969), 'Combating wear in large- and medium-sized diesel engines operating on residual fuels'. *Proc. I.M.A.S.*

48   Collacott, R.A. (1942). 'Refractory linings of furnace walls', *Marine Engineer*, February.

49   Godfrey, D. & Courtney, R.L. (1971), 'Investigations of the mechanism of exhaust valve seat wear in engines run on unleaded gasoline', SAE Paper 710357, January.

50   Godfrey, D. & Courtney, R.L. (1954), 'X-ray diffraction patterns of lead components'.

51   Collacott, R.A. (1971), 'Bulk materials handling – a survey of research and development', *British Engineer*, September.

52   Collacott, R.A. (1941), 'Heat transfer research in engineering', *Power and Works Engineer*, June.

53   French, C.C.J. (1969), 'Problems arising from the water cooling of engine components', *Proc. I. Mech. E.*, **184**, Pt. 1. No. 29, 507–542.

54   Fitzgeorge, D. & Pope, J.A. (1955), 'An investigation of the factors contributing to the failure of diesel engine pistons and cylinder covers'. *Trans. N.E. Coast Inst. Engineers & Shipbuilders*, 71, 163.

55   Marsh, W. (1969–70), 'The philosophy of failure in engineering systems', *Proc. I. Mech. E.*, **184**, Part 3B.

3 Fault detection sensors

3.1. Introduction

Sensors used in the monitoring of machinery with the object of identifying faults are usually related to the identification of physical and/or chemical changes. As a consequence, sensors are frequently similar to those employed for control purposes and, indeed, there is a considerable similarity between the two techniques.

3.2 Contaminant monitoring

A detailed review of this technique is presented in Chapters 8 and 9, it is based on the principle that deterioration products are carried in a fluid such as lubricating oil. Such 'sensing' as is used, consists of sample or debris collection followed by its analysis in terms of constituents, size, shape or quantity. The methods in active use include magnetic (chip) detectors, spectrometric programmes, 'Ferrograph' and various X-ray techniques.

3.2.1 X-ray fluorescence

By the use of X-ray fluorescence techniques the N940 analyser [1] provides data on the elemental constituents of solid, liqued or power samples. A spectrum is obtained by exciting the sample with an isotope X-ray source it is then analysed by an energy responsive proportional counter, usually referred to as a detector. Amplified detector pulses are fed to a pulse height analyser (PHA) for effective element separation then to a counter and a four digit light-emitting diode (LED) display output. When the atomic numbers of the wanted elements are very close together, filters can be inserted to give further discrimination.

3.3 Corrosion monitoring

Since corrosion proceeds as an electro-chemical process the techniques of monitoring are directed to the identification of the electrical currents set up specifically by the corrosive activity.

3.3.1 Corrosometer

This is the proprietary name for an instrument in which the electrical resistance monitoring technique uses a balanced bridge technique to measure the change in resistance of a probe as it thins away under corrosion. Details of this method were given by Bovankovich [2].

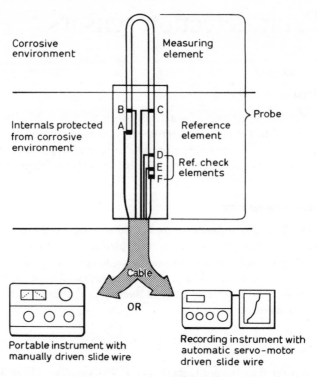

Figure 3.1 *Electrical resistance corrosion measuring system*

The probe, shown in Figure 3.1, comprises an exposed U-shaped upper part which is subjected to corrosive attack and a protected internal part providing the counter balancing resistance elements. The balance readings accordingly record the effective thickness of the exposed probe and thus the implied corrosion rate in MPY (mils per year) where 1 mil = 0.001 in.

3.3.2 Corrator

By detecting the polarization resistance the instantaneous corrosion rate and pitting tendence in a conductive liquid can be monitored and incorporated into an inhibitor injection control system as explained by Britton [3].

When a test electrode is polarized from its free corrosion potential by a small applied voltage, V, either the anodic or cathodic characteristic of the corrosion reaction will be enhanced. These anodic and cathodic characteristics differ from each other; the test currents obtained (I_a for $+ V$, I_c for $- V$) must be combined in the following way to produce useful correlation with true corrosion rate:

$$\text{Corrosion rate} = \frac{K \, I_a \times I_c}{V \, I_a + I_c} \tag{3.1}$$

where: K = a constant relating the units of corrosion rate to those of V and I. Corrosion rate is measured in MPY (mils per year).

3.4 Force monitoring

Force may be measured directly by a variety of methods as also, may specific force (pressure). Only a brief account is given of some of the many techniques.

3.4.1 Load cells

The compressive force applied to a component may be measured by load cells. The load is carried by a cylindrical column of a special steel to which strain gauges are bonded. The load is measured by the interpretation of the strain/stress variations by the use of a Wheatstone Bridge circuit.

3.4.2 Proof ring transducer

Based on the traditional proof ring design the body of this load transducer [4] is a homogeneous elastic member machined all over from a grain-oriented billet of high tensile steel. Deformation of the ring is sensed by a miniature inductive displacement transducer with zero friction, infinite resolution and high electrical output. With a linear relationship between load and displacement it is possible to read the applied load directly. This transducer can be used on both compressive and tensile modes under static or dynamic conditions.

3.4.3 Tourmaline underwater pressure transducers

Used to measure the pressure waves in a liquid medium without disturbing the wave shape, they are built up from discs of piezo-electric tourmaline crystals [5]. These transducers are made up in single, double or quadruple element construction, each gauge pile being coated to seal the pile and maintain high insulation without detracting from the response of the gauge to pressure and frequency.

3.4.4 Strain gauges

Any strain or linear deformation in the structure or component to which a strain gauge is attached will produce a proportional change in the resistance of the wire used to make the strain gauge or stress transducer.

The degree of sensitivity of a strain gauge can be evaluated from the basic equation relating to the change of resistance to strain:

$$\Delta R = R.(GF).\Delta L/L \tag{3.2}$$

where: R = gauge resistance in ohms, L = conductor length, GF = gauge factor based on the material and geometry of the gauge.

The strain, being related to stress by Young's Modulus can be expressed:

$$\Delta L/L = f/E \tag{3.3}$$

where: E = Young's Modulus, f = applied stress.

To measure changes in strain (stress) by the use of strain gauges involving changes of resistance, a Wheatstone Bridge circuit with various refinements is used.

3.4.5 Standard strain gauges

Many types of strain gauge have been developed because of the ease with which they can be readily adapted to remote displays in analogue or digital form and provide an output signal which can be amplified, summated or integrated as required. They may be classified as follows: (i) wire gauges, (ii) foil gauges, (iii) semiconductor gauges.

Wire strain gauges may be either (a) flat wound with a meandering or zig-zag pattern, or (b) wrapped around. Constantan wire approximately 0.001 in diameter wound in a flat form or sandwiched between two pieces of carrier material normally make up a wire strain gauge.

Foil strain gauges may be either (a) etched, or (b) die-cut. Etched grid types are usually manufactured by applying a carrier to the thin metal foil and then etching the grid pattern complete with outlet connecting tabs and alignment references. Die-cut grid gauges are precision cut and pressed into the carrier complete with connecting tabs.

Semiconductor strain gauge elements consist of a thin silicon strip which may be produced by sawing, grinding, lapping and etching silicon monocrystal rods. Silicon responds actively to strain and has a high K factor (DR/R) which may reach a value of 200 as compared with values of 2.0 to 6.0 for metal.

P-type semiconductors increase their resistance under tension (P = positive going). N-type semiconductors reduce their resistance under tension (N = negative going).

Valuable information regarding the choice of bonding cements, protective coatings, surface preparation and advanced strain gauge designs is given by Bailey [6].

3.4.6 Strain gauge (Wheatstone) bridges

Small changes of resistance are normally detected by means of a Wheatstone bridge circuit with four resistors arranged as shown in Figure 3.2(a). When there is no current passing across the centre wire (the galvanometer branch), the relationship of the resistors are

$$\frac{R_1}{R_4} = \frac{R_2}{R_3} \tag{3.4}$$

or

$$R_1 = \frac{R_2}{R_3} \cdot R_4 \tag{3.5}$$

An example of the required accuracy is given by the following: to measure a strain

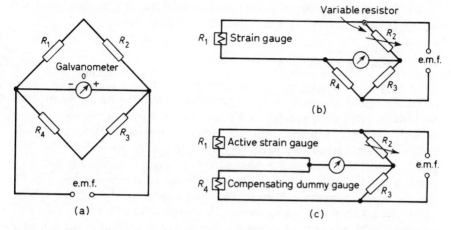

Figure 3.2 *Strain gauge (Wheatstone) bridge circuits (a) Basic Wheatstone bridge.*
(b) Balanced bridge (c) Temperature compensation

of 0.05% (500 microstrain) in a 100 ohm gauge with a gauge factor of 2.0, this
would represent a resistance change of 0.1 ohm. For an accuracy of 1% the measurement must be made to 0.001 ohms in 100 ohms (the switch contact resistance in a
normal Wheatstone Bridge is often greater than this).

In applying a balanced bridge to strain-gauge measurements one of the resistors
(R_1 in Figure 3.2(b)) is the strain gauge. The bridge is then balanced (zero galvanometer reading) by adjusting any of the other resistors, say R_2. The main disadvan-
tage of balanced bridge is that temperature changes and strains cause an imbalance
in the bridge because of unequal resistance effects upon the measuring gauge and its
balanced resistor. A compensating 'dummy' gauge (basically a resistor, R_4 in Figure
3.2(c)) can be bonded to an unstrained block of material of the same coefficient of
expansion as the test structure and exposed to the same environment.

Instead of rebalancing the bridge circuit after loading a structure, the galvanometer reading itself can be taken as an indication of the strain in the area of the
strain gauge. Almost all dynamic recorders employ the unbalanced bridge method.
The term 'dynamic strain' refers to strain which varies appreciably in magnitude
over a short time interval. Dynamic recorders use either a.c. or d.c. power for the
bridge, depending upon the application.

3.5 Gas leakage monitoring

Two basic methods can be applied to test for leaks:

(1) Sniffing: in which the leak tester responds to the gas. With sealed, unpressur-
ized units a tracer gas filling is used. With small pressurized components, they are
placed in a 'bomb' with a pressurized tracer gas and after a 'soaking' removed and
sniffed for tracer leakage.

(2) Hood method: Components are put in a hood and air is pumped round. Gas leakage in the purged gas indicates a leak in the test component.

3.5.1 Auer detector tubes/gas testers

These testers are used by drawing air through the detector tubes which contain chemical reagents sensitive to selective gases. Gas leaks which can be readily detected include: ethyl alcohol; butyl alcohol; methyl alcohol; propyl alcohol; isopropyl alcohol; ammonia; benzene; chlorine; bromine; propane; *n*-butane; acetylene; carbon dioxide; carbon monoxide; ethylene. Other tubes are available for other gases – nitric oxide, nitrogen dioxide, perchloro ethylene, trichlorethylene, phosphine, sulphur dioxide, carbon disulphide, toluene, ethyl bromide, etc. [7]. The selected vapour produces a characteristic discolouration of the detector tube reagent. The length of stain is a measure of the concentration of the substances in the air sample. With most tubes the concentration is read directly off a scale printed on the tubes.

3.5.2 Gas thermal conductivity testers

Changes in the thermal conductivity when gases leak into a small volume provide the basis for leak sensing which responds to such gases as hydrogen, helium, methane hydrocarbons, refrigerant gases, carbon dioxide, argon etc. In fact, this method is suitable for any gas with a different thermal conductivity from air.

3.5.3 Refrigerant leak tester

A heater platinum element is used since this responds to halogenated gases and is sensitive to refrigerant leaks of $1 \times 10^{-6} \, \text{cm}^3 \, \text{s}^{-1}$. The instrument is plugged into an electric supply and allowed to warm up so that the filament has a medium temperature. In normal use the audio output is set to a steady 'tick' and when a leak is found this increases to a high frequency and loud sound.

3.5.4 G.P.O. cable leak tester

To establish the approximate location of a buried cable a low concentration of a tracer gas, such as 5% SF_6 in nitrogen is injected with the normal pressurizing nitrogen into a G.P.O. cable. After the tracer gas has diffused into the surrounding soil its presence, and the location of the cable, is detected by means of a thermal conductivity type leakseeker.

3.5.5 Toxic gas monitor

Chlorine (Cl_2), hydrogen cyanide (HCN) and hydrogen sulphide (H_2S) can be monitored and the concentrations recorded or embodied as the sensor of an alarm device [8].

This device comprises two electrodes immersed in an electrolyte which flows into a porous glass cell. The central electrode provides a reference datum, the outer electrode is active in measuring the gas concentration. As gas diffuses into the glass cell it passes into the electrolyte and generates an electrode current proportional to the concentration of gas present.

3.5.6 Ultrasonic leak testing

Leaks in automobile pneumatic braking systems can be accurately located by means of an ultrasonic detector which gives both an audible and visual indication of the leak. It is a simple hand-held device no bigger than a large torch designed to provide on-site location of air or gas leaks in both pressure and vacuum pipe systems and vessels. Air leaks generate a broad spectrum of ultrasonic energy — sound energy of a frequency too high to be heard by the human ear — in the 40 kHz region. This energy can be detected by a suitable microphone, amplifier and indicator, even in the presence of considerable audible noise from other sources.

The detector incorporates a highly directional ceramic ultrasonic microphone. This is coupled to a transistorized multi-stage amplifier, tuned to 40 kHz ± 4 kHz, whose output is indicated visually on the instrument's moving-coil meter or is audible through detachable head-phones.

3.5.8 Dispersive infra-red gas analysis

Both leaks and pollution can be detected by using a variable-path gas cell [9] which is also claimed to be capable of detecting small concentrations of organic vapours such as vinyl chloride monomers. The single beam analyser uses a Nichrome wire source, the energy from which is focussed through a chopper and circular narrow-band, past an interference filter onto a slit. The pass band can be varied from 2.5 to 14.5 μm by rotation of the filter wheel.

The chopped monchromatic beam passes into the long-path cell where the path length, i.e. distance travelled by the infra-red energy in the sample cell, may be adjusted from 0.75 to 20.25 m in 1.5 m increments, by a control knob. The exit beam then falls onto a pyroelectric detector which generates an electrical signal proportional to the infra-red energy.

3.6 Air pollution monitoring

Faults in a process or its control lead to the contamination of the environment so that monitoring not only provides statutory safeguards but also acts as a sensor of operational defectiveness.

3.6.1 Non-dispersive infra-red monitoring

Two similar Nichrome filaments are used as the source of infrared radiation as shown in Figure 3.3. Beams from these filaments travel through parallel cells; one beam

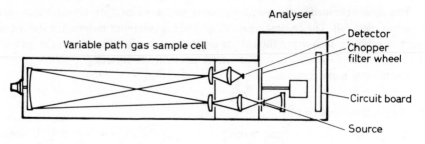

Figure 3.3 *Miran II non-dispersive infrared pollution monitor*

traverses the sample cell, the other beam, the comparison cell. The emergent radiation is directed into a single detector cell. An interrupter, or 'chopper', located between the radiation source and the cells, alternately blocks radiation to the sample cell and the comparison cell. When the infrared beams are equal an equal amount of radiation enters the detector cell from each beam. When the gas to be analysed is introduced into the sample cell, it absorbs (and reduces) the radiation reaching the detector via the sample beam. Consequently the beams become unequal, the radiation entering the detector flickers as the beams are alternated, and the detector gas expands or contracts in accordance with the flicker. This movement of the gas causes the membrane to respond and the condenser microphone capacity which, in turn, generates an electrical signal proportional to the difference between the two radiation beams. The amplified signal is fed to the indicating meter, and/or a potentiometer recorder.

3.6.2 Catalytic combustion monitoring

Electrical resistance change following catalytic combustion is measured by a balanced four-element bridge circuit to provide the basis of this monitor. Each leg of the bridge consists of a small ceramic bead supported on a coiled filament called a Pellement detector. Two of the Pellement detectors are inactive, and two are catalysed. When a test sample passes over the catalysed Pellement units, the combustibles burn catalytically on their surface, raising the Pellement unit temperature and, consequently, increasing their electrical resistance. This resulting imbalance in the bridge creates an output voltage proportional to the concentration of total combustibles in the test sample.

3.6.3 Hydrocarbon monitor

A continuous low-level monitor used to detect and measure trace contaminants can be used to detect hydrocarbons in: automobile exhausts; gas mains; sewers; fuel stores; aerosol stores; paint shops; refrigerant leaks.

The analyser is made up of four elements: (a) sample flow system, (b) combustion gases system, (c) burner assembly, (d) electrometer and power supplies.

Operation of the instrument is based on the ionization of carbon atoms in a hydrogen flame. Normally, a flame of pure hydrogen contains an almost negligible number of ions. Adding organic compounds — even traces — results in a large number of ions in the flame.

The sample to be analysed is mixed with a hydrogen fuel and passed through a small jet; air supplied to the anular space around the jet supports combusion. Any hydrocarbon carried into the flame results in the formation of carbon ions. An electrical potential across the flame jet and an 'ion collector' electrode suspended above the flame produces an ion current proportional to the hydrocarbon count. This is measured by an electrometer circuit whose output then provides an analysis signal for the direct reading meter, or for an optional potentiometric recorder.

3.6.4 Missile fuel and oxidiser monitor

An analyser which was originally developed for the continuous monitoring of toxic hazards from missile fuels and oxidisers can detect minute concentrations of gases and vapours in air or process gas streams [10].

An ionisation chamber detects an aerosol which can be formed from the contaminant vapour. These effect the conductivity of the ion chamber and measure the concentration of the gas or vapour. Continuous ionization of the sample is provided by a mild alpha source the radioactivity of which is approximately that of the luminous dial of a watch.

Formation of particles for a wide range of materials is accomplished by either (a) chemical reactions; (b) direct pyrolysis; (c) a combination of both (a) and (b); (d) without sensitization if the gas (such as oxygen) is electronegative in a non-electronegative process stream, such as argon. Trace concentrations of water vapour, carbon monoxide, and carbon dioxide will be detected under these conditions.

A change in the circuit occurs only when some material in the sample reacts to form an aerosol. This changes the conductivity and resistance which can be detected by a Wheatstone Bridge circuit. This resulting unbalance of the bridge measures the contaminant.

3.6.5 Partial pressure oxygen analyser

Net oxygen concentration in flue gases from combustion processes may be measured by the Hagen-probe without the use of a sampling system [11]. The principle involves measurement of the oxygen partial pressure by an electrochemical cell in which the output increases logarithmically when the oxygen content decreases. Because there is no sampling the response is fast, measured in milliseconds.

The cell is fabricated from stabilized Zirconia and functions as an electrochemical equilibrium cell with the property of conducting electrical current exclusively by oxygen ions. At the cell operating temperature of 155°F, the e.m.f output is a function of the difference in oxygen partial pressure on either side of the cell. A reference gas is introduced to the probe internally. The resulting cell output is inversely proportional to the logarithm of the sample oxygen partial pressure as effectively described by the following Nernst Equation:

$$\text{e.m.f.} = \frac{RT}{4F} \ln (P_1/P_2) + C \qquad (3.6)$$

where: e.m.f. = cell output, volts,
     R = gas constant,
     F = Faraday constant,
     P_1 = reference gas (air) partial pressure,
     P_2 = sample gas partial pressure,
     C = cell constant,
     T = absolute temperature.

3.6.6 Chemiluminescent pollution monitoring

Chemical reactions produce chemiluminescence of the type experienced by the white 'glow' of phosphorus when exposed to air. The use of this phenomenon has been exploited by van Heusden [12] by the use of image-intensifier techniques to monitor gas traces. The spectrum range and peak wavelengths for a number of gases is shown in Figure 3.4 together with the active range of a XP1002 photo-multiplier tube.

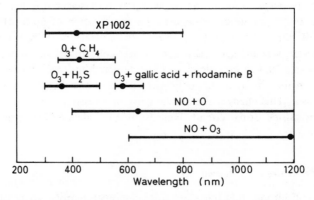

Figure 3.4 *Situation and spectrum range of a number of chemi-luminescent reactions together with the spectral region of the multiplier tube*

3.6.7 Smoke monitor

Particulate matter is satisfactorily sensed by optical response techniques.

A typical monitor involves a projector, receiver and control unit. The projector throws a beam of light across the chimney or exhaust so that the amount of light which penetrates the smoke provides a measure of the smoke density. A cadmium sulphide photoconductive cell is extremely sensitive to changes in light and acts as the 'receiver' in that the light which falls onto the cell is converted into electrical energy — the current varies with the amount of light transmitted through the smoke. The sensitivity is such that a range from 0 to 100% obscuration smoke density can be sensed and a signal may be calibrated to express the intensity of smoke present in terms of the Bacharach Index [13].

3.6.8 Generator core overheating monitor

Local overheating resulting in the failure of the insulation has caused the close-down of a number of generators. Such overheating can be detected by sensing the thermal decomposition of the organic materials.

Such a sensor [14] is known as a 'core monitor' and is extremely sensitive; detecting overheating before any effects on the operation of the machine are evident. It consists of an ion chamber and can detect condensation nuclei down to 0.001 μm.

3.7 Liquid contamination monitoring

Continuous surveillance of the quality of process water such as that used in boiler feed and cooling systems may be performed by the multi-channel analysis of critical physical and chemical properties of water such as dissolved oxygen, pH conductivity, temperature, turbidity.

Monitoring is based on specially designed sensing probes which provide an input to corresponding signal conditioning modules. Such modules amplify and linearize the signal from the sensor to provide the desired output. A combination of a sensing probe and a signal conditioning module is termed a 'parametric' system'. The following may be involved in a multiple-channel system [15].

(1) Dissolved oxygen parametric system: Using an 'International Biophysics' dissolved oxygen probe, this sensor operates by detecting the amount of oxygen which diffuses across a thin Teflon membrane and is measured by a polarographic technique where a current is passed between a gold cathode and a silver anode;

(2) pH parametric system: Employs a pH probe assembly which consists of three elements — a glass sensing electrode sensitive to hydrogen ion concentration, a solid state reference electrode with extremely good antifouling characteristics and a temperature sensor which automatically compensates for changes in sample temperature; a preamplifier unit is also located in the probe assembly;

(3) Conductivity parametric system: Employing a 4-electrode probe-cell assembly in which 2 electrodes impose a known electrical current through the sample while the other 2 electrodes are connected to a high impedance sensor network and by determining the volt drop through a given amount of solution they determine the conductivity of the water;

(4) Temperature parametric system: Incorporating a platinum resistance thermometer which utilises the 4 wires connected to the conductivity sensor, the first pair impressing a small current through the sensor itself, the second pair measuring the voltage drop at the terminals of the platinum electrode. Since the measuring wires

carry no current and are connected to a high impedance differential amplifier line interferences and voltage drops are virtually eliminated.

(5) Selective ion parametric system: Utilizes selective ion probes with a membrane which permits only ions of a particular species to pass into the sensing compartment of the probe. Once past the membrane ions create a change of potential at an electrochemical sensor. An antifouling reference electrode and a temperature compensating electrode are incorparated in the assembly. Specific ion measurements are offered for: ammonia, bromide, cadmium, calcium, fluoride, cupric, cyanide, iodide, lead, nitrate, perchlorate, potassium, silver, sodium sulphide, sulphur dioxide.

(6) Oxidation reduction potential (ORP) parametric system: which is similar to the pH sensor, the significant difference being that the glass sensing electrode is replaced by either a gold or platinum ORP electrode and a solid state antifouling reference electrode. The oxidation reduction potential is a measure of the concentration and strength of the various oxidizing and reducing ions present in the speciemn – ions such as hypochlorite, sulphide, permanganate, ferric greatly affect the oxidation reduction potential.

(7) Turbidity parametric system: Either a Hack surface scatter four turbidimeter, or a Monitek forward scatter turbidimeter may be used.

The Hack turbidimeter measures the amount of light reflected from the surface of a water sample due to the effect of the turbidity – creating suspended particles in the bulk solution. The sample is pumped into the analyser and flows up an inclined large-bore open tube to overflow across the top into a drain. A narrow beam of light is directed onto the smooth flowing surface of the liquid, part of the beam enters the liquid and is reflected by the suspended particles. A photo-cell above the surface detects the amount of reflected light – as the numbers of particles in the water increases, so does the 'turbidity' and the amount of reflected light.

The Monitek turbidimeter passes a beam of light through liquid flowing through a sensing cell. The intensity of the direct beam and the light energy scattered in the forward direction by suspended solids are both measured and a ratio between the two computed to provide a continuous measurement of total suspended solids irrespective of changes of colour, refractive index or light transmittance. The output signal is a linear function of turbidity and the signal is auto matically compensated for changes in source light intensity and dirt accumulation on the viewing windows.

(8) Dissolved oxygen parametric system: To prevent the fouling of the probe a thallium metal electrode is used, operating on the principle that the corrosion potential of thallium metal is a function of the oxygen concentration in the liquid.

3.7.1 Liquid density monitoring

Electromagnetic suspension of a totally immersed plummet of determined buoyancy provides the basis of a liquid density monitor as described by Sage [16].

A spherical plummet is positioned inside the probe unit and varies with the liquid density, its location can be sensed by a pair of search coils fed with a high frequency supply. Since the plummet is slightly more dense than the liquid it tends to sink. The electro-magnetic force from a solenoid situated above the plummet prevents it from sinking and the search coils allow just sufficient electric current to flow through the solenoid coil to maintain the plummet in a central position. Measurement of this current provides a direct analog to the liquid density and can be used either quantitatively or as a part of a monitoring system.

3.7.2 Sludge density monitoring

An ultrasonic monitor, marketed by Ronald Trist Controls Ltd., under the trade name Mobrey-Sensall, features an ultrasonic sensor and an electric control unit. The sensor is hermetically sealed in its housing and contains a piezoelectric crystal transmitter and receiver separated by a gap. The control unit generates an electrical signal which is converted to an ultrasonic signal by the transmitter crystal. When the gap is filled with liquid the ultrasonic signal reaches the receiver crystal and is fed into the control unit where it is amplified. If sludge is present the signal does not pass and the control remains inoperative. The build-up of slurries including china clay, metallic oxides, yeast, and ion resins as well as build-down from the surface of oil, ice, floating liquids or chips can be detected by this method.

3.8 Non-destructive testing techniques

Visual examination is an important method for appraising the condition of critical surfaces for which optical aids have been evolved. Other methods of assessment are covered in more detail in Chapter 2.

3.8.1 Beta scatter thickness gauge

The amount of wear or deterioration of thin coatings and films may be measured by this method.

In a typical instrument [17] beta particles enter a material after radiation from an isotope. The mean path of these particles is such that they pass through the material and are either (a) transmitted (b) lose energy and become absorbed or (c) reverse direction and emerge (backscatter) from the surface into which they were bombarded. The proportion either (a) transmitted, (b) absorbed, or (c) backscattered depends upon the thickness of the surface (also, source emission density, mass density, bombardment area). The thickness gauge involves (i) a radioactive isotope source, (ii) a table, (iii) a detector.

A Geiger—Müller tube is used as the detector and converts the beta particles scattered back from the test surface into electrical pulses which can be processed into a direct voltage which is proportional to the pulse count and can be related to the thickness.

3.8.2 Acoustic pulse crack detectors

Pulses of sound are generated in material from which their velocity is measured [18]. This velocity is related to the physical condition of the material and provides a means for detecting the existence of voids, cracks or inclusions.

Two piezoelectric probes are used, one generates the sound, the other acts as a receiver and detects the sound after it has passed through. The probes may be used several metres apart which means that very large structures may be tested.

A pulse generator causes sound to travel through the material, at the same instant it energizes a timing circuit. When the sound reaches the receiver probe the timing circuit is de-energized. Between these two events the timing circuit counts pulses from another source at the rate of 10^6 pulses per second. Thus, for a period of 1 ms between transmission and reception the timing circuit will have counted 1000 pulses. This information is passed to the display circuit.

Defects in the material or structure absorb the sound and thus change the pattern of returning pulses thus providing a means of detecting their presence by re-positioning the receiver probe.

3.8.3 Ultrasonic signature analysis

Moving systems do not necessarily produce audible sounds under all conditions but they all produce ultrasonic signals. For this reason, ultrasonics present an extremely wide field of actual and potential applications. Probably the best-known use of ultrasonics is in the monitoring of the condition of bearings. A number of portable testers have been developed, with both contact and proximity probes. Other applications of ultrasonics include:

(1) Piston blow-by which produces a characteristic ultrasonic hiss;

(2) Compressors in which cracked and warped plates produce ultrasonic pulsing or whistling;

(3) Surface and subsurface cracks;

(4) Fatigue of structures;

(5) Fluid and gas leaks in a seeded system;

(6) High temperature, high pressure leaks in exotic metallic gas systems used in the production of electronic components;

(7) Arcing in high-voltage electrical systems;

(8) Radio and T.V. interference;

(9) Circuit failures;

(10) Capacitor and transformer failures.

3.8.4 X-ray laminography

The laminographical examination technique, pioneered by Dr Moler of Illinois Institute of Technology, radiology is used to detect broken conductors or poorly etched areas on multilayer printed circuit board assemblies. This operates on the principle of averaging the examination of an unwanted image over a large area while keeping the required image in focus.

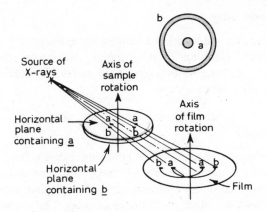

Figure 3.5 *X-ray laminography — objects in focus are sharp (a) while those out of focus are blurred (b)*

This can be achieved by rotating the specimen under observation. As shown in Figure 3.5 a disc containing the letters 'a' and 'b', and a flat piece of X-ray film. Both the plate and specimen are rotated at synchronous speed around parallel axis that are co-planar with the source, as shown in the diagram. As the plates rotate, image 'a' will be sharp but image 'b', because it is on a different plane, will produce a smeared image. By adjusting the relative positions of the same specimen film, and source, it is possible to get a sharp image of 'b' or any other plane.

It is possible to achieve sharply detailed laminographs of layers as closely spaced at 0.003 in, in which details as small as 0.0005 in are easily observed. If the X-ray energy is increased from 20 to 50 keV, to 300 keV a variety of materials and samples of much greater thickness can be tested. Instead of using individual X-ray films for each exposure, laminographic scanning could be employed.

A television camera fitted with a zoom lens can generate an image of the sample on a fluorescent screen. The zoom lens then magnifies small areas of the sample, to overcome the line-resolution problem inherent in the normal television camera.

3.9 Optical examination

Viewing ports are provided in some machines to enable the unaided eye to examine the condition of components or to provide an access for viewing devices. A typical application is the examination of turbine blades and combustion chamber liners in gas turbines. The device may be a simple angled mirror, or a flexible light piple and a source of illumination. Small television cameras are sometimes used.

Refinements to visual examination sometimes include X-radiography, which can be very effective in detecting cracks in combusion chambers or liners. The main shaft of the RB211 engine is drilled to allow insertion of a radioisotope for radiography purposes.

3.10 Temperature sensing

Heat transference rates and associated temperature gradients are closely related to the performance and condition of prime-movers and process plant. Various techniques for monitoring temperatures have been evolved in addition to the refinements in thermometry which abound.

3.10.1 Thermography

All substances radiate a characteristic temperature pattern. Once the 'thermal signature' of a healthy system has been established, it can serve as a standard for comparing the performance of any similar system. Heat pictures obtained by thermography may be used for plant condition monitoring. The routine surveillance of blast furnace linings, stoves, hot blast and dirty gas mains, steel mixing vessels and BOS vessels by the use of thermography has been used in the steel industry and reported by Rogers [19].

Thermal imaging is carried out by the use of an infrared camera which scans the target system and converts its infrared radiation into a proportional electrical signal. This signal is amplified and displayed on a CRT or permanently on photographic film. The temperature can be determined and compared to the known emissivity of the target and indicated internal conditions of the operating systems.

3.10.2 Computer-stored temperature readings

A system to monitor the thermal discharge from a nuclear power plant and the temperature of the discharge into the river uses temperature sensors together with a salinometer and dewpoint sensor in conjunction with a digital data acquisition system and a digital stepping magnetic tape recorder [20].

Each sensor is sampled and digitized by the data acquisition system and then recorded on magnetic tape every 1.2 seconds thus providing 50 data points/minute or 3000 data points/sensor/hour. This data is recorded in IBM compatible format and is interfaced directly with an IBM 360 computer.

3.10.3 Piston temperature radio telemetry

A system using thermocouples embedded in the crown of a diesel piston and the temperature transmitted through a radio frequency link was developed at the University of Manchester Institute of Science and Technology [21]. A block diagram of the whole system is shown in Figure 3.6. The transmission principle depends on the multiplexing of all transducer channels onto a common frequency modulated transmitter. The radio frequency signal was received at a stationary point and the output derived from a stationary receiver. Power was continuously transferred from a stationary point to a pick-up on the connecting rod.

The nature of temperature signal made conditioning (data processing) necessary.

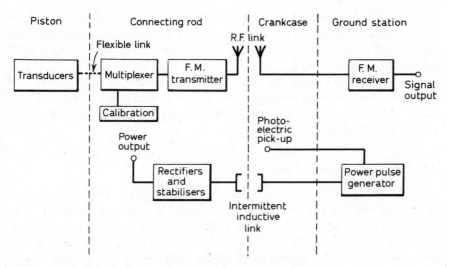

Figure 3.6 *Piston radio-telegraphy temperature monitoring circuit*

A high mean temperature existed with a small amplitude ripple component super-imposed at a frequency of half engine speed with a peak when the fuel ignited.

3.10.4 Turbine blade pyrometer

Infrared measurement of the temperature of turbine blades during operation is provided by the Kollsman system [22, 23] which allows for individual blade surface measurement giving (i) average mean temperature, (ii) average peak temperature, (iii) maximum peak temperature over a range from 600 to 1250°C. Applications for this pyrometer include:

(1) Turbine operation at optimum performance temperature;

(2) Protection against hot spots;

(3) Diagnostic inputs.

Infrared energy emitted by the turbine blade surface is collected by the pyrometer's optical head sapphire lens. This radiation is, in turn, passed through a fibre optic light guide and onto a silicon cell detector located within the head amplifier.

Current generated from the silicon cell is amplified and then conditioned by signal data processing. During this processsing the signal current is analysed and recorded as a digital readout of the maximum blade temperature. The outputs are available coaxially or as a BCD function to tape recorders, oscilloscopes, fuel control systems, etc. The raw data output, (in linear or nonlinear form) as supplied from the buffer amplifier to the detector, is also available as a coaxial output for blade profile analysis.

3.10.5 Flame failure sensor

The absence (or alternatively, the presence) of a flame can be detected by sensing the temperature of the air surrounding that flame. A temperature-sensitive cell uses thermal radiation effects to sense the air temperature, typically a Hird-Brown sensing head has a maximum sensitivity of 250°C with a 100% emissivity target, 350°C with a 50% emissivity target.

The viewing head senses the air around the flame and initiates positive 'ON' switching when temperatures exceed 250°C, this is usually associated with 'ON' 'OFF' connections in the amplifier-control unit normally mounted at some position remote from the flame.

3.10.6 Templugs — temperature sensors

A special technique for assessing temperature levels in inaccessible locations relies on the fact that certain alloys change their hardness in a reproducible manner when heated [24]. One alloy (a plain, high-carbon steel) can be calibrated to measure temperatures between 70°C and 720°C for times between 1 second and 1000 hours; another (an 8% chromium, 3% silicon steel) records temperatures between about 550°C and 950°C. Under ideal conditions, that is when the time of exposure is known accurately, these temperature-sensitive alloys can attain an accuracy of ± 5°C.

The alloys are usually employed as small flat-ended grub screws with internal hexagonal socket heads, the most popular size is 6 BA. Such grub screws called 'Templugs' can be fitted to any item of equipment, left in position during operation and subsequently removed for hardness testing.

A number of Templugs fitted into a diesel engine exhaust valve, cylinder head and piston crown (there were almost 300 in one six-cylinder engine) were used by Wilson [25] to provide data on the heat balance in the engine, with particular reference to cracking problems on the cylinder head, where thermal gradients of the order of 400°C per cm were recorded.

3.10.7 Temperature-indicating labels and paints

A range of temperature-sensitive labels are available which are hermetically sealed and coated so as to react precisely when exposed to rated temperatures by irreversibly changing colour from silver to jet black [26].

3.11 Particle sensing

Solid particles present in liquids or gases are probably deterioration debris which, when monitored, provide a basis for the assessment of plant condition and, when examined, provide the data for failure prognosis.

3.11.1 Dust classification

Air-borne particles vary considerably in physical shape, range and distribution of particle size, density, regularity and adhesibility. One form of classification is related to their ease of filtration ranging from high permeability dusts which are easy to filter down to low permeability dusts which present the greatest control problems. Four classifications are:

(1) High permeability (easily filtered): Particles which are fibrous, irregular in shape and having a high length/breadth ratio. Air contaminated by such dusts involve low filter resistance and low pressure drops. Typical substances are: sanded wood dust, cork dust, grain, dry brick dust.

(2) Medium permeability: Possibly granular and of regular shape with free airspace between individual particles. Typical substances are: iron grindings, grain dusts, mixed animal feeds.

(3) Fair permeability: Possibly granular and of regular shape with free airspace between individual particles. Typical substances are: rubber dust, coal dust, plastics dust.

(4) Low permeability (difficult to filter): Commonly, dusts of fine particles with a flake-like appearance. Typical substances are: metal oxides, carbon black, mica, hygroscopic materials.

3.11.2 Particle mass monitor

Real-time particulate mass monitoring can be carried out to yield an output representing mass concentration. A simple quartz crystal microbalance (QCM) principle is employed [27]. An internal pump passes the gas sample through a pipe and particles in the sample strike the adhesive surface of a quartz crystal which acts as one side of the microbalance. The added mass of the particles on the quartz crystal shifts the frequency of an oscillator circuit in which the crystal is the frequency-determining element. A reference circuit produces a frequency from an isolated reference crystal. The output difference between the particle-impacted crystal and the reference crystal produces a beat frequency which is a function of the mass concentration.

3.12 Proximity monitors

By the use of proximity techniques the distances from a probe to a surface can be finitely measured or incorporated into rate-function circuitry to measure velocities. Thus reference to the use of a thin film sensor to monitor the wear of stern tube bearings, diesel engine cylinder liners etc. was made by Knudsen [28]. As shown in Figure 3.7, the resistor is particularly thin so that wear produces a significant and measurable change of the resistance.

3.12.1 Displacement monitoring

Axial and rotary movements of a rotor assembly can be monitored by non-contract displacement measuring transducers mounted to suitable surfaces on the rotor. The

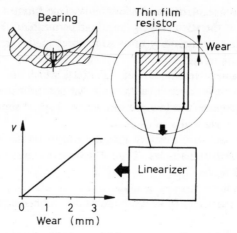

Figure 3.7 *Resistance change wear monitor*

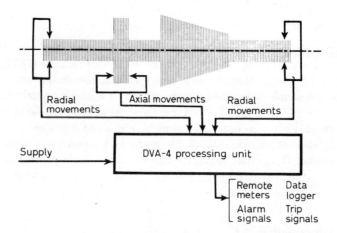

Figure 3.8 *DVA—4 displacement monitoring system applied to a rotor*

signals from the transducers are converted by the electronic modules into readily interpreted data relating to either radial vibration or axial movements of the rotor relative to the machine stator assembly. Monitoring provides a means for the automatic indication of changes in vibration level or displacement due to such factors as bearing deterioration or increased out of balance.

The schematic diagram, Figure 3.8 shows this typical application. Radial monitoring in one plane only is shown for reasons of clarity but measurements can be made on mutually perpendicular axes by adding the appropriate electronic units to the modular racking systems [29].

The modules needed for the DVA—4 displacement system include:

(1) Power supply module: Which provides the appropriate stabilized outputs to drive the oscillator, amplifiers and switching modules;

(2) Oscillator module: Which provides the carrier frequency power to energise the transducers;

(3) Transducer amplifier module: In which the signal from the transducers is amplified and detected and a corresponding d.c. output provided to drive recording equipment or oscilloscopes.

An extension to the displacement monitor is a facility for vibration monitoring by which the peak to peak displacement signal from the transducer amplifier is shifted to a reference level above which the amplitude is sensed.

3.13 Sound monitoring

Human operators are normally highly sensitive to the detection of defects as a result of sudden changes of sound. Studies by the author [30] show that the loosening of components either as a result of wear or the slackening of fastenings is particularly susceptible to such forms of monitoring.

3.13.1 Microphones

The three most widely available microphones for sound are piezoelectric (including ceramic), the moving coil and the condenser. The poor linear frequency response and variable phase shift (introducing waveform distortion), of the moving coil microphone, make it generally unsuitable for diagnostic purposes and the choice would normally lie between a ceramic or a condenser microphone, the latter usually being preferred.

Due to its high acoustical impedance a condenser microphone may be fitted with probes and used to measure the sound pressure at positions which might otherwise be inaccessible or environmentally unsuitable.

Where the frequency range of interest extends up to 10 kHz, it is advisable to restrict the diameter of any condenser microphone to $\frac{1}{2}$ in (12.5 mm) to avoid corrections and uncertainties due to diffraction effects.

3.13.2 Microphone sensitivity

Microphone transducer sensitivity is rated in terms of the reference sensitivity of 1 V output for a pressure of 1 N/m^{-2} or in terms of the older reference, 1 V/ubar. The apparent sensitivity levels for these two ratings differ by 20 dB; a pressure of 1 N m^{-2} corresponds to 94 dB re 20 N m^{-2}, whereas 1 ubar corresponds to 74 dB re 20 u N m^{-2}. A typical microphone open-circuit (unloaded) sensitivity level is -40 dB re 1 V per N m^{-2}.

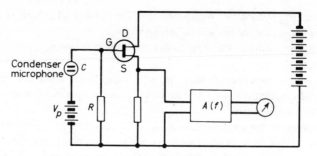

Figure 3.9 *Simplified sketch of a pre-amplifier circuit*

3.13.3 Pre-amplification

Thermal noise inherent in a microphone is generated by Brownian movement of the membrane. In order that very weak noises (sound pressures) can be detected a pre-amplified circuit is used to amplify the electrical detection.

An idealized pre-amplified circuit is shown in Figure 3.9 which incorporates a junction field effect transistor. Over a frequency range of 20 to 20 000 Hz sound pressure as low as 11 dB can be detected, with a smaller bandwidth even smaller sounds can be detected.

3.14 Vibration transducers

Transducers for the measurement of vibration employ electromagnetic, electro-dynamic, capacative, piezoelectric or strain gauge principles of operation. Of these, the most widely used in recent years for vibration work is the piezoelectric accelero-meter, largely by virtue of the fact that it is self-generating, is small in size and weight, can be designed to be free of resonances over a wide frequency range, has good stability, low sensitivity to strain, temperature variations, airborne sound and magnetic fields, a large dynamic range and is not easily damaged.

3.14.1 Accelerometers

Compression type sensors are normally used for condition monitoring and basically consist of a number of piezoelectric discs on which is placed a relatively heavy mass. The assembly is mounted onto a base preloaded by means of a stiff spring and sealed in a metal case as shown in Figure 3.10.

On being subjected to vibration, the mass exerts a force onto the piezoelectric discs which generate an electrical signal directly proportional to the force applied and therefore to the acceleration of the mass. The signal can also be displayed in velocity or displacement units by the use of integrating circuits which are built into some analysers.

The usable frequency range of many accelerometers is in the region of 10—40 kHz although a 10 kHz range should be suitable for most machinery dignostic work.

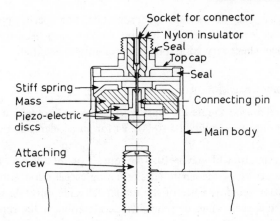

Figure 3.10 *Piezoelectric accelerometer*

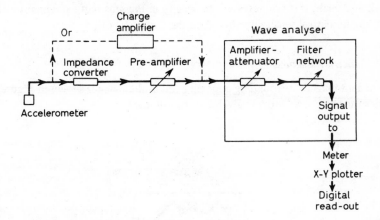

Figure 3.11 *Block diagram of vibration measuring rig*

As with the velocity type pick up, the range is considerably influenced by the method of attachment: the most efficient being attachment via a steel stud direct onto a rigid part of the machine. Attachment by other methods such as special cement compounds, plasticine, magnet or hand-held probe lowers the upper frequency limit to between 2–5 kHz.

In order to amplify the signals it is possible to use techniques of either charge amplification or combined impedance converter and preamplifier as shown in Figure 3.11.

3.14.2 Piezoelectric crystals

Early makes of piezoelectric accelerometers employed Rochelle Salt or barium titanate as the active material. These materials have been largely replaced by lead zirconate titanate due to its superior properties. For special applications new materials such as lead niobate and lithium niobate have been developed whilst quartz is still used where very good long-term stability is essential.

3.14.3 Accelerometer sensitivity

The terms generally used to rate the sensitivity of an accelerometer are the open circuit (unloaded) output in millivolts (0.001 V) for an acceleration that corresponds to that of gravity (g).

Since the acceleration of gravity (0.90665) m s^{-2} or 386.09 in s^{-2} is always directed towards the centre of the earth, an alternating acceleration rated in terms of g units is to be interpreted as a use of the numerical value only. As such, it is sometimes used for the peak value of sinusoidal acceleration or the rms value (0.707) times the peak for a sinusoidal vibration. As long as it is recognized that the electrical output and the acceleration are to be measured in the same way, rms and rms, or peak and peak, it is not necessary to specify in the sensitivity statement which is meant; the numerical sensitivity should be the same for both.

3.14.4 Velocity transducers

A permanent magnet surrounding a moving coil in the manner shown in Figure 3.12 provides the basis of an inductive velocity transducer.

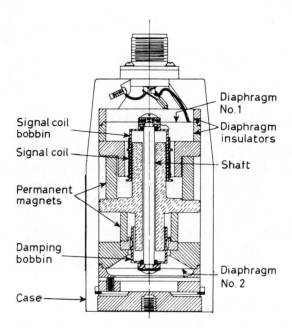

Figure 3.12 *Inductive vibration velocity transducer*

Machine vibration causes the magnet to move within the coil, thus converting the motions of the machine into an electrical signal. The signal is proportional to vibrational velocity. Most instruments also indicate displacement by passing the signal through an integrating amplifier. The usable frequency range is in the region of 10—1000 Hz being considerably influenced by the method adopted to attach the transducer to the machine.

3.14.5 Stroboscope

A broad-spectrum frequency assessment such as the location of sources of maladjustment, misalignment and wear can be made by stroboscopic examination.

The stroboscope permits rotating or reciprocating objects to be viewed intermittently and produces the optical effect of slowing down or stopping motion. For instance, an electric fan revolving at 1800 revs/min will apparently stand still, if viewed under a light that flashes uniformly 1800 times per minute. At 1799 flashes per minute the fan will appear to rotate forward at 1 rev/min, and at 1801 flashes per minute it will appear to rotate backwards at 1 rev/min.

The amplitude of vibration can be assessed if a fine reference line is scribed on the vibrating part. This technique has been used to confirm the calibration of vibration calibrators, and automotive engineers have used it to measure crankshaft whip and vibration.

3.14.6 Magnetic tape recording

Tape recordings offer the facility of being able to store signals for later reproduction and analysis and of expanding or compressing time scales. Multichannel recording can also preserve the time and phase relationships of a number of signals.

Two types of magnetic recordings are in widespread use, direct recordings (with high frequency bias) and frequency modulated (F.M.) recording.

When the subsequent signal processing is to be a simple spectrum analysis of each signal independently and the frequency range of interest falls between 30 Hz and 10 kHz then a direct recording technique is adequate and economical. If the signals to be processed are of a lower frequency or the signal processing is such that the phase and time relationships of a number of signals are relevant, the F.M. recording is far superior and usually essential.

3.14.7 Laser-holographic vibration surveillance

Holography is a method by which the phase and amplitude of the wavefront scattered from an object illuminated by coherent light such as that from a gas laser (e.g. He—Ne at 632.8 nm and argon ion at 514.5 nm) are recorded in the form of an interference pattern on a photographic plate. The original wavefront can be reconstructed by illuminating the photographic plate with coherent light to form a complete three-dimensional image of the object. Considerable expertise in the use of this technique for condition monitoring has been developed by Professor J. Butters, University of Technology, Loughborough.

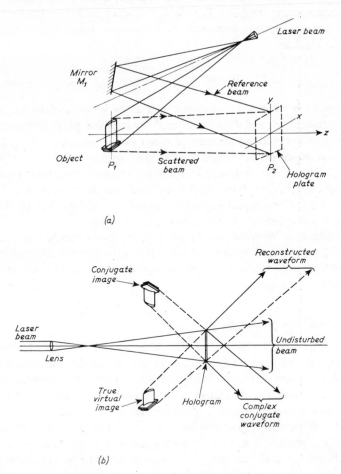

(a)

(b)

Figure 3.13 *Schematic diagram of the holographic recording and reconstruction systems employed, (a) Hologram recording system. (b) Hologram reconstruction system*

A schematic diagram of a hologram recording system is shown in Figure 3.13(a). The object is illuminated by part of an expanded laser beam and some of this light interferes with the light directly reflected on to the plate by the mirror M and the resultant interference pattern is recorded by the photographic plate. The plate is developed and placed in the arrangement shown in Figure 3.13(b) where it will produce a true three-dimensional image, in the position shown, when illuminated by coherent light. A full analysis of this process can be obtained [31].

The effects of a vibrating object on a holographic system were first investigated by Powell and Stetson [32]. Measurement of the vibration characteristics involves the study of the first fundamental modes plus some higher modes by a quick scan of a number of modes using information which can be obtained holographically using time-averaged fringe techniques.

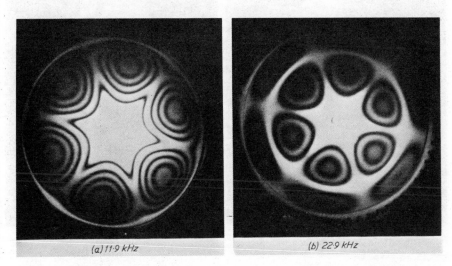

(a) 11·9 kHz   (b) 22·9 kHz

Figure 3.14 *Hologram reconstructions of a specific turbine disc in vibration*

Examples of vibration holographs obtained by Hockley [33] show the vibrational nodal lines as bright fringes and the relative amplitude contoured by the higher order fringes (Figures 3.14, 3.15). The vibrational amplitudes are normally chosen to give a maximum of approximately 15 fringes. In some cases, however, up to 70 fringes have been observed by over-exposing the film for the low order fringes.

3.15 Telemetry

Weak signals from a monitoring source must be communicated to signal processing equipment, there may also be a need to transmit electrical power to the rotor to feed transducers and process their signals. The techniques which have been effectively established are:

(1) Slip-rings: Brushes running on slip-rings offer the most straightfroward way of transmitting electrical signals. Interface phenomena are complex, they represent a resistance in series with a voltage source, variations can be reduced by choice of circuit configurations but there are limits to what can be achieved in this way.

The rubbing-speed involved is important, (i) for a given brush load and coefficient of friction, heat generation is proportional to speed; (ii) lack of circularity in the ring produces radial accelerations of the brush, requiring a contact force proportional to speed. In one application [34], signals from large diameter slip-rings at

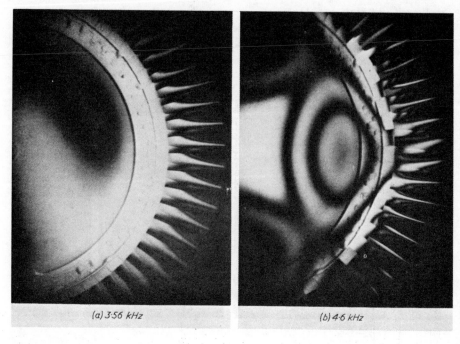

(a) 3·56 kHz   (b) 4·6 kHz

Figure 3.15 *Time-averaged fringe hologram reconstructions of a 10 in diameter specimen turbine disc with a full set of blades*

3000 revs/min were transmitted by the solid-liquid contact of a part moving in mercury.

(2) Transformer: This uses stationary primary winding and rotating secondary winding [35]. Compared with rotating contacts, these have longer life and much greater independence of atmosphere but the circuits are more complex than for slip-rings and require precision bearings.

(3) Capacitor: This provides an alternative scheme with one electrode stationary and the other rotating. Low level signals at high frequency can easily be transmitted in this way. A system which uses pulse modulation is commercially available and operates effectively across the very small capacitance between a band on the rotor and a plate 12 inches or more away.

(4) Radio links: A radio telemetry system for the rotors of large operational steam turbines was developed by the Central Engineering Research Laboratories, C.E.G.B., Leatherhead [36]. Power was fed in at 50 kHz, using a transformer with a few millimetres gap in its ferrite core, thus allowing relative movement between primary and secondary windings. The outgoing signal as a frequency modulation of several carriers in the region of 20 MHz, the relative movement required being accomodated between a loop and an adjacent 'transmission line' to which it was coupled. A block schematic of a single channel — shown for a strain gauge transducer — is given in Figure 3.16.

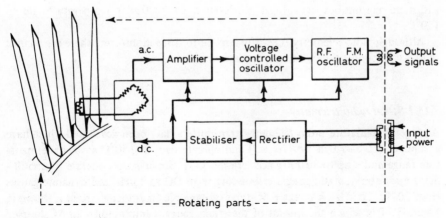

Figure 3.16 *Turbine blade strain gauging by signal telemetry (CEGB)*

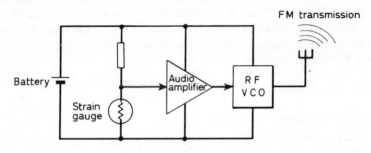

Figure 3.17 *Simple dynamic strain transmitter*

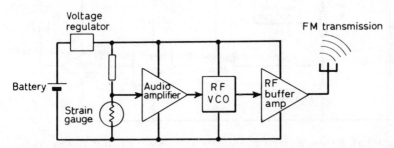

Figure 3.18 *Improved dynamic strain transmitter with voltage regulator and RF buffer amplifier*

This system was used experimentally at Brighton A Generating Station. Systems based on it have subsequently been developed for Blyth, Cottam, Didcot, Drax, Fawley, Meaford, Northfleet, Rugeley and West Thurrock power stations – in several cases by the turbine manufacturers concerned. The circuitry was able to withstand continuous centrifugal acceleration of 5000 g at temperatures up to 120°C.

Although radio telemetry is usually applied to force monitoring there are records of its use in connection with temperature monitoring.

3.15.1 Strain radio telemetry

A range of miniature radio telemetry transmitters have been developed which have a volume of 0.5–1.0 in³ (8–16 cm³) and operate over 0–150°C ambient temperature range under up to 30 000 g acceleration [37]. Strain gauges operate from 500–5000 microstrain, static gauges responding from DC to 1 kHz and dynamic gauges from 20 Hz–20 kHz. A direct F.M. type of dynamic transmitter circuit is shown in Figure 3.17 in which the output of the strain gauge is amplified by an AC coupled audio amplifier (response 30 Hz–20 kHz) which frequency modulates a radio frequency oscillator (100 MHz), this is the method used for standard FM broadcasting.

An improved circuit, Figure 3.18 overcomes drawbacks to the direct F.M. circuit due to battery voltage changes by the introduction of a voltage regulator. Proximity interferences effects which set up spurious signals are eliminated by the introduction of a R.F. buffer amplifier.

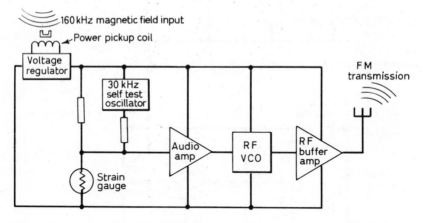

Figure 3.19 *Dynamic strain transmitter with induced power system for jet engine installation*

A further refinement used in the monitoring of jet turbine blades is shown in Figure 3.19, where a non-rotating magnetic field radiates a 160 kHz magnetic field which is picked up by a tuned coil within the transmitter, rectified and regulated to provide a stable D.C. output; a 30 kHz self-regulating test oscillator continuously verifies the transmitter performance.

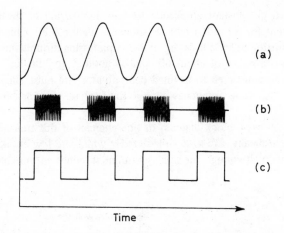

Figure 3.20 *Waveforms in the torsionmeter. (a) Wire position. (b) Radiated signal (c) Received signal after demodulation and shaping*

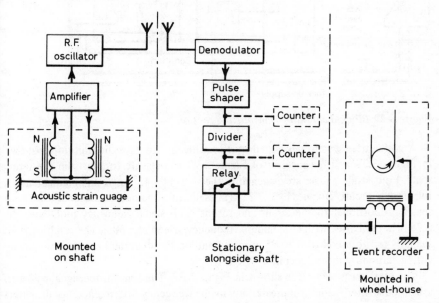

Figure 3.21 *Diagram of one channel of the torsionmeter as used for measured-mile trials*

3.15.2 *Torque radio telemetry*

To monitor the propellor shaft (typically 18–24 feet diameter) of a large ship Shell Research Ltd. [38] used an installation comprising an r.f. keyed by the signal so that it oscillated at constant amplitude for one half-period of the wire vibration and was quiescent for the other half, as shown in Figure 3.20. This r.f. signal was fed to a transmitting aerial consisting of a complete ring around the shaft. The receiving aerial was a sector of about 60° placed about 1 to 2 inches from the transmitting aerial and connected to a tuned demodulator and pulse shaper. The signal from this had the same frequency and a fixed phase relationship to the wire vibration.

Figure 3.21 shows a block diagram of one channel of this transmitter. The r.f. frequencies are typically 275 kHz and 480 kHz which are low enough to raise no circuit problems, high enough for small component values and insensitive to interference.

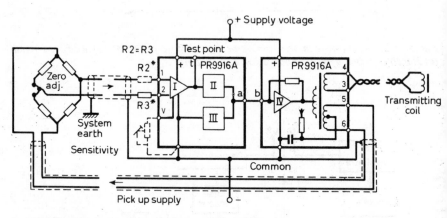

Figure 3.22 *Block diagram of transmitting section*

A contactless system developed by Pye Unicam Ltd. uses the transmitter shown in Figure 3.22 and the receiver of Figure 3.23 to measure torque from a rotating shaft. Four strain gauges are attached to the shaft at a 45° angle and connected in a full bridge circuit, energies by a 6.75 kHz oscillator. The output diagonal provides a signal proportional to the strain and torque. This signal frequency modulates the 6.75 kHz oscillator, with a maximum frequency sweep of ±30%. The resulting F.M. signal is fed to a coil of 5 to 25 turns wound on the shaft, and couples inductively with a stationary receiving coil.

The stationary receiver is shown in Figure 3.23. The transmitted signal passes to a bandpass filter, which suppresses any mains frequency interference signals, enters a limiting amplifier and then a frequency doubler stage. Pulses derived from the leading and trailing edges of the F.M. signals are used to trigger a flip-flop which produces standard pulses at a repetition frequency depending on the input signal. Low pas filters remove all frequencies above 3 kHz) passband corresponding to modulation frequencies 0 . . . . 1.5 kHz before doubling). The average d.c. level of standard pulses is proportional to the instantaneous frequency of the original F.M. signal. A centre frequency compensation voltage is applied leaving a direct voltage

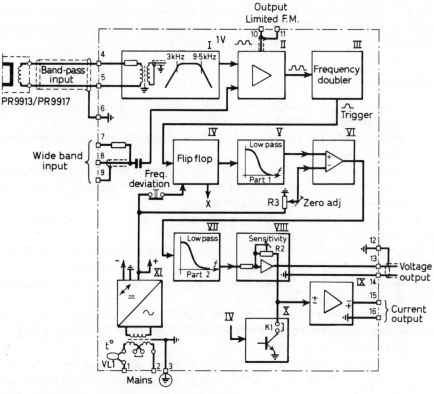

Figure 3.23 *Block diagram of receiving section*

corresponding to the modulation. This compensation voltage is adjustable, providing a zero offset adjustment.

The output signal is proportional to the measured torque and available as a $-1, \ldots, 0, \ldots, +1\,V$ signal corresponding to a frequency sweep of $-30, \ldots, 0, \ldots, +30\%$, and a second output of $-10, \ldots, 0, \ldots, +10\,mA$ is provided for feeding low impedance indicators or recorders. The actual limited F.M. signal may be of a magnetic tape.

3.15.3 Digital read-out radio telemetry

Frequency modulation is popular since the strain gauge output signal is converted to a constant-amplitude, variable-audio-frequency signal in a modulator. Deviations from some reference frequencies are proportional to amplitude. Data from a number of transducers can be transmitted simultaneously over one transmission channel by superimposing several discrete audio-frequency bands. At the receiver, the individual channels are separated, and the original information regained by use of discriminators, which convert frequency deviations into electrical signals analogous to the original strain amplitudes.

Recent developments in the telemetering field include systems employing pulse-duration modulation (PDM) and pulse-code modulation (PCM) which are consistent with the current trend towards digital data readout.

References

1 ——— (1975), *Non-dispersive X-ray analysis,* Applied Research Laboratories Ltd, Wingate Road, Luton LU4 8PU.

2 Bovankovitch, J.C. (1973), 'On-line corrosion monitoring', *Materials Performance*, **12**, No. 6.

3 Britton, C.F. (1974), 'Methods and equipment for electrochemical measurements and control in the field of corrosion', *Intl. Scandinavian Congress on Chemical Engineering,* Copenhagen January.

4 ——— (1974), *Proof Ring Transducers,* Sangamo Western Controls Ltd., North Bersted, Bognor Regis, Sussex, PO22 9BS.

5 ——— (1974), *Tourmaline Underwater Pressure Transducers,* The Meclec Company, No. 7 Unit, Star Lane, Great Wakering, Essex SS3 OPJ.

6 Bailey, E.E. (1974), 'Strain Gauges — what they are and their uses', Chartered Mech. Eng., September, 103—107.

7 ——— *Auer Detector Tubes for Auer Toximeter and Auer Gas Tester,* Auergesellschaft Gmbh, D 1000, Berlin 65 (also: Mine Safety Appliances Co Ltd., Coatbridge ML5 4TD)

8 ——— *Environmental Surveillance Products for Evaluating Potentially Hazardous Atmospheres,* Mine Safety Appliances Co Ltd., Coatbridge, ML5 4TD.

9 Golding K. (1974), 'Sub-ppm air monitoring by portable IR analyser', *Process Engineering,* August, 70—73.

10 ——— (1974), 'M-S-A Billion-aire trace gas analyser', Bulletin 0706-4, Mines Safety Appliances Co. Ltd., Coatbridge, ML5 4TD.

11 ——— (1972), 'Hagan probe type oxygen analysing system', Publication S 5092, Negretti & Zambra Ltd, Stocklake, Aylesbury, Bucks. HP20 1DR

12 Heusden, S.van (1974), 'Air pollution monitors based on chemiluminescence., *Philips Tech. Rev.*, **34**, No. 2/3, 73—81.

13 ——— (1975), 'Hand vibrograph system ASKANIA', Publication 1606-ELM6/73 Lehmann & Michels, 2000 Hamburg 54, Germany.

14 ——— (1975), 'Generator core monitor' *Electrical Times,* 14 March.

15 ——— (1973), *Water Quality Monitor,* Ionics Inc., Watertown Mass. 02172 (also: Tecmation Ltd, Edgeware, Middx. HA8 8TP)

16 Sage, P.K. (1968), 'Industrial measurement of liquid denisty', *Instrument Practice,* September, 767—772.

17 Latter, D.T. (1970), *Non-destructive Coating Thickness Measurement,* Fischer Instrumentation (GB) Ltd, Newbury, Berks.

18 ——— (1975), 'Materials sounder MS1', Inspection Instruments (NDT) Ltd, 32 Duncan Terrace, London N1 8BR

19 Rogers, L.M. (1971), 'The application of thermography to plant condition monitoring', Report TB/TH/71, British Steel Corporation (Tubes Division) Corporate Development Laboratory, Hoyle Street, Sheffield S3 7EY

20 Bolus, R.L., Fang, C.S. and Chia, S.N. (1973), 'Design of a thermal monitoring system' *M.T.S. Journal*, 7, No. 7, October/November.

21 Worthy, J.G.B. (1970), 'Radio telemetry of diesel engine piston temperatures' Central Electricity Research Laboratories (Leatherhead) Conference on Transmitting Signals from Rotating Plant, 25/26 June

22 Curwen, K.C. (1975), *Turbine Blade Pyrometer System in the Control of the Concorde Engine*, Kollsman Instruments Ltd, The Airport, Southampton SO9 3FR Hants

23 Barber, R. (1969), 'A Radiation Pyrometer designed for In-Flight Measurement of Turbine Blade Temperatures', SAE Paper 690432, April.

24 Belcher, P.R. and Wilson, R.W. (1966), 'Templugs' *The Engineer* 221, 305.

25 ——— (1972), 'The diagnosis of engineering failures', S. Af. Mech. Eng., November 22, No. 11.

26 ——— (1974), A. Levermore & Co Ltd, P.O. Box 654, 40–44 The Broadway, London SW19 1SQ

27 Swift, P. 'Real-time particulate measurement', Celesco Industries Inc, Costa Mesa, California 92626

28 Knudsen, R.K. (1973), 'Condition monitoring of marine propulsion machinery', *Veritas*, August 19, No. 75, 8–10.

29 ——— 'PM-1/B1 Proximity monitoring system' Vibrometer Limited, Newby Road, Hazel Grove, Stockport, Cheshire SK7 5EE.

30 Collacott, R.A. (1974), *Sonic Fault Diagnosis*, British Steel Corporation, Research Fellowship Report.

31 Develis, J.B. and Reynolds, G.O. (1967), *Theory and Applications of Holography*, Addison-Wesley Publishing Co.

32 Powell, R.L. and Stetson, K.A. (1965), 'Interferometric vibration analysis by wavefront reconstruction', *J. Opt. Soc. Am.*

33 Hockley, B.S. (1972), 'Measurement of vibration by holography', *Trans. I. Mar. E.*, 84 170–175.

34 Weise, P.R. and David, T.J. (1973), 'Collecting data off high speed rotors', *Engineering*, 27 December.

35 Drew, D.A. (1958), 'Developments in methods of measuring stresses in compressor and turbine blades on test bed and in flight' *Proc. I. Mech. E.*, 172 320–337.

36 Jones, D.H. (1963), 'A multi-channel contactless telemetry system for vibration studies on steam turbine blades', Intl. Telemetry Conference, I.E.E. London 62–81.

37 Adler, A.J. (1970), 'Wireless strain and temperature measurement with radio telemetry', CERL (Leatherhead) Conference 25/26 June.

38 Attwood, H.I.S. (1970), 'A simple telemetry system for acoustic strain gauges', CERL (Leatherhead) Conference 25/26 June.

4 Data processing and analysis

4.1 Introduction

Machinery produces messages or signals by its action on sensors. In most cases these signals are received as alternating electrical currents with a characteristic pattern of voltages comprising a mix of different magnitudes, frequencies and phase relationships. These signals are meaningful in condition diagnosis surveillance.

To unravel the significance of these mixes within the composite analogue signal the data requires to be analysed or further processed in order that the important characteristics can be determined.

While data analysis and processing techniques have been applied extensively to the interpretation of vibration signals it has also been necessary for some conditioning to be made to the analogue signals for other parameters.

4.1.1 Temperature processing

Signals received from the various temperature sensors and monitoring systems need to be processed beyond the recording of the raw signal. Thus with piston temperature radio telemetry [1], the signal needed further conditioning since it contained a high mean temperature with a small amplitude ripple component superimposed at a frequency of half the engine speed; this effect reached a peak when the fuel ignited. To resolve the ripple to less than $\frac{1}{2}°C$, corresponding to 0.2% of the overall signal; a.c. preamplification was used to improve resolution by processing engine order signals separately.

With the thermal discharge monitoring from a nuclear power station [2] each sensor was sampled and digitized by the data acquisition system and then recorded on magnetic tape providing 50 data, 3000 data points/sensor/hour, recorded on IBM compatible format and interfaced directly with an IBM 360 computer. In this computer, time constants were matched to the sampling interval to prevent the data from being 'aliased', Aliasing is the distortion which occurs from the sampling interval being too large for events with a cyclic nature or with rapid changes in randomly alternating directions (see Section 4.4.1).

With the Kollsman turbine blade pyrometer [3, 4] current generated at the silicon cell by the sensed radiation is amplified and sent to a remotely located signal process box. This box analyses the signal received and provides a digital readout of the maximum blade temperature, the average blade temperature, and the average peak blade temperature Figure 4.1. The output is read directly on three separate digital displays in °C or as a voltage proportional to the actual blade temperature values. At the same time, these outputs are available as a function to tape recorders,

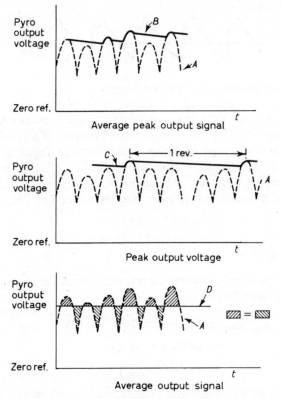

Figure 4.1 *Output signal wave forms*

oscilloscopes, fuel control systems, etc. The raw data output, (in linear or non-linear form) as supplied from the buffer amplifier to the detector, is also available as a coaxial output for blade profile analysis. The system bandwidth (150 KHz) is normally wide enough to respond to the temperature profile of each blade.

4.1.2 Vibration and sound processing

Signals from accelerometers and microphones have a complex of amplitudes, frequency, phase and time relationships. These signals need to be simplified by means of Fourier analysis to convert them into their component harmonics, digitized (to adapt them for computer applications) and then processed in order to identify and quantify the significant parameters.

4.2 Fourier Analysis

A typical analogue signal is a complex of alternation of amplitude with time such as that shown in Figure 4.2 [5]. On its own, such a signal can rarely provide much

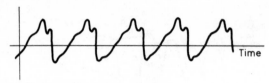

Figure 4.2 *Typical waveform as a periodic function*

worthwhile information until the complex harmonic is broken down into its constituent harmonics. This can be undertaken by means of a Fourier analysis to produce the constituent sine or cosine signals shown in Figure 4.3. These constituent harmonics are Fourier coefficients which can be derived mathematically from a given complex harmonic, and ideal exercise for computer processing, which in practical applications is performed by the process of Fast Fourier Transformation.

4.2.1 Harmonic analysis

Many signals obtained from monitors are periodic such that the displacement x may by written as a Fourier Series:

$$x = a_0 + a_1 \sin(\omega t + \alpha_1) + G_2 \sin(2\omega t + 2\alpha_2)$$
$$+ a_3 \sin(3\omega t + 3\alpha_3) +$$
$$+ a_n \sin(n\omega t + n\alpha_n) + \ldots \quad (4.1)$$

The terms of this series are described as follows:

$a_1 \sin(\omega t + \alpha_1)$ harmonic

$a_2 \sin(3\omega t + 2\alpha_2)$ octave or second harmonic

$a_3 \sin(3\omega t + 3\alpha_3)$ second octave or third harmonic

   . . .   etc

Considering the expansion of each of these terms based on the trigonometry of compound angles:

since, $a_n \sin(n\omega t + n\alpha_n) = a_n \cos \alpha_n \sin n\omega t + a_n \sin \alpha_n \cos n\omega t$

substituting $A = a_n \cos \alpha_n$

$B = a_n \sin \alpha_n$

$a_n \sin(n\omega t + n\alpha_n) = A \sin n\omega t + B \cos n\omega t \quad (4.2)$

hence the Fourier series can be written in the form:

$$x = B_0 + B_1 \cos \omega t + B_2 \cos 2\omega t + \ldots B_n \cos n\omega t$$
$$+ A_1 \sin \omega t + A_2 \sin 2\omega t + \ldots A_n \sin n\omega t$$

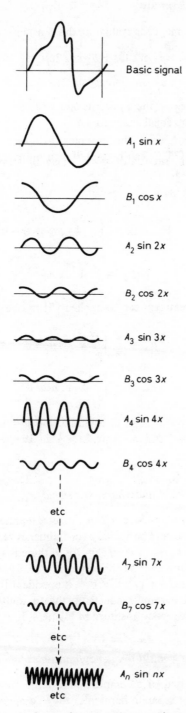

Figure 4.3 *Resolution of typical waveform into its constituent harmonics by Fourier analysis*

or in a form more convenient for evaluating the constants:

$$x = \tfrac{1}{2}b_0 + b_1 \cos \omega t + b_2 \cos 2\omega t + \ldots b_n \cos n\omega t$$

$$+ \, a_1 \sin \omega t + a_2 \sin 2\omega t + \ldots a_n \sin n\omega t \qquad (4.3)$$

To determine the values of the constants and since the integrals of the terms between limits 0 and 2π are equal and zero, i.e.

$$\int_0^{2\pi} \sin n\omega t \, dt = \int_0^{2\pi} \cos n\omega t \, dt = 0$$

integrating equation 4.3 gives

$$\int_0^{2\pi} x \, dx = \int_0^{2\pi} \tfrac{1}{2}b_0 \, dt + 0 + 0 \qquad (4.4)$$

thus

$$b_0 = \frac{1}{\pi} \int_0^{2\pi} x \, dx \qquad (4.5)$$

Similar mathematical treatment can be applied [5] to give

$$b_n = \frac{1}{\pi} \int_0^{2\pi} x \cos n\omega t \, dt \qquad (4.6)$$

$$a_n = \frac{1}{\pi} \int_0^{2\pi} x \sin n\omega t \, dt \qquad (4.7)$$

The manner in which the straight line relationship $y = x$ can be obtained from a summation series of $y = 2 (\sin x - \tfrac{1}{2} \sin 2x + \tfrac{1}{3} \sin 3x - \tfrac{1}{4} \sin 4x + \tfrac{1}{5} \sin 5x + \ldots)$ is shown in Figure 4.4.

4.2.2 Frequency spectrum – multi-harmonic signal

Each of the terms $b_1 \cos \omega t$, $b_2 \cos 2\omega t \ldots$, etc. represents harmonic motions of amplitude b_1, $b_2, \ldots$, etc. and periodicity (or frequency) ω, $2\omega, \ldots$, etc.

Since, for any angle θ, $\sin (\theta + \pi/2) = \cos \theta$, there is a $\pi/2$ i.e. $90°$ phase difference between the sine term and the cosine term.

Allowing for a constant $k = (a^2 + b^2)^{1/2}$, it is evident that in terms of frequency content and individual harmonic constants that a multiharmonic signal can be expressed in terms of frequency as shown in Figure 4.5.

4.2.3 Frequency spectra – multi-frequency/multi-harmonic signal

Most signals are composed of a complex mixture of different frequencies so that when resolved into their separate harmonics they comprise a medley of fundamentals and overtones, thus

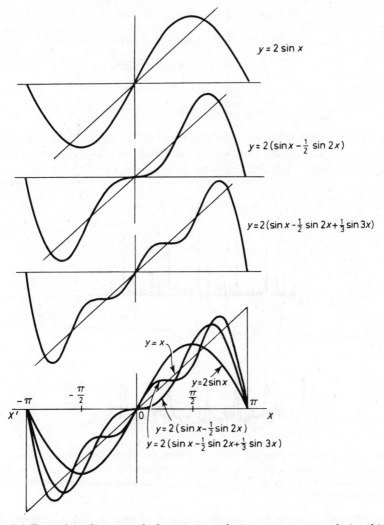

Figure 4.4 *Typical application of a harmonic analysis to represent a relationship*

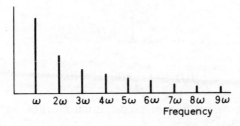

Figure 4.5 *Frequency spectrum for a multi-harmonic signal corresponding with the saw-tooth waveform (Figure 4.4)*

$$\omega_1, \quad 2\omega_1, \quad 3\omega_1, \quad \ldots, \text{etc}$$

$$\omega_2, \quad 2\omega_2, \quad 3\omega_2, \quad \ldots, \text{etc}$$

$$\omega_3, \quad 2\omega_3, \quad 3\omega_3, \quad \ldots, \text{etc}$$

$$\ldots, \text{etc}$$

$$\omega_n, \quad 2\omega_n, \quad 3\omega_n, \quad \ldots, \text{etc}$$

When plotted with appropriate constants on a basis of frequency, they appear as in Figure 4.6. In practice, allowing for the multiplicity of frequencies and inertias of the system, a frequency analysis plots out as an envelope as shown in (b).

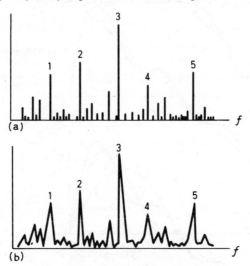

Figure 4.6 *Frequency spectrum for a multi-frequency, multi-harmonic signal showing (a) basic harmonics, (b) envelope of all harmonics such as would be plotted during an analysis. Peaks 1, 2, 3, 4, 5 are significant resonances corresponding to discrete frequencies*

For purposes of computer programming multi-frequency/multi-harmonic signals the Fourier coefficients of equations 4.6 and 4.7 are expressed as the following Fourier integral pair:

$$f(\omega) = \frac{1}{2\pi} \int_{-\infty}^{\infty} f(t) \, e^{-j\omega t} \, dt$$

$$f(t) = \int_{-\infty}^{\infty} F(\omega) \, e^{j\omega t} \, d\omega \tag{4.8}$$

which defines the complex frequency component $F(\omega)$ of the time function $f(t)$ for each angular frequency ω.

It is possible to write discrete equivalents of these equations as follows:

$$F(k) = \frac{1}{N} \sum_{n=0}^{N-1} f(n)\, e^{-jkn2\pi/N}$$

and (4.9)

$$f(n) = \sum_{k=0}^{N-1} F(k)\, e^{jkn2\pi/N}$$

which are known as the Discrete Fourier Transform (D.F.T.). This transform pair has very similar properties to (4.8) with the exception that it is discrete and periodic in both time and frequency domains.

4.2.4 Analogue to digital signals

When considering a signal obtained from a vibration transducer (the concept is relevant to any other form of sensor) the output is found to vary over some arbitrary but finite amplitude density range depending upon the force characteristics of the source vibration, the impedance of the transmission path to the transducer and the gain characteristics of the transducer system.

The instantaneous amplitude can have any value within the finite range and thus the signal is classified as an analogue. It is convenient to express this for processing by a computer in binary form in one of two forms

$$1 = \text{an impulse}$$

$$0 = \text{no impulse}$$

This is known as a digital signal, when the number of levels is two it is in binary form [6].

In general the amplitude range can be quantized into a number of discrete impulses or levels and at any one instant the signal can only be at one of these levels. This is referred to as an m-ary signal [7] and illustrated by Figure 4.7. Typical signals and their electrical characteristics are classified in the following table:

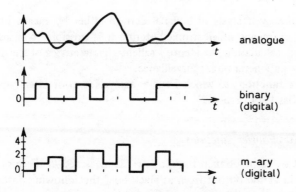

Figure 4.7 *Typical analogue and their equivalent digital signals*

Table 4.1

Type of signal	Classification of transducer output
Vibration	Analogue
Speech (telephony)	Analogue
Machine noises	Analogue
Teleprinter (telegraphy) Computer information (data)	Digital
Still pictures (facsimile)	Analogue
Moving pictures (television)	Analogue
Physical parameters (telemetry)	Analogue or digital
Navigational information (radar)	Analogue or digital
Remote control (telecontrol)	Analogue or digital

4.2.5 Boolean logic

In order to switch or gate signals, it is necessary to consider the logical possibilities of processing some signals particularly in relation to the control of functions. For such purposes use may be made of Boolean algebraic concepts.

4.2.6 Fast Fourier Transform (F.F.T.) – Computer program

The following program was given by Randall [8] as an appendix to the B and K application note.

Virtually all Fourier analysis is based on the fourier integral pair equations 4.8 and 4.9. To calculate all N coefficients $F(k)$ directly according to equation 4.9 requires N^2 complex operations, but the FFT algorithm has been devised to obtain the same values in $N \log_2 N$ operations; a considerable saving.

One implementation of the algorithm, written in Fortran IV uses the program NLOGN shown in Figure 4.8 with a further improved subroutine HALF given in Figure 4.9 to reduce the storage redundancy.

4.2.7 Fourier analysis – derived functions

From the frequency analysis of a signal derived either by means of a computer using FFT or by means of electronic filtering it is possible to introduce further computer-based calculations to identify salient characteristcs of the original signal and thus of the parameter under surveillance.

A list of the functions to which Fourier analysis is most often applied has been tabulated by Kiss [9] (Table 4.2).

4.2.8 'Black box' Fourier functions

Assuming that the time domain function $x(t)$ is measured at the input and that of $y(t)$ is measured at the output from a 'black box' the following observations were made by Kiss [9]:

Table 4.2 *Frequently-used functions in Fourier analysis applications*

Time Domain	Frequency Domain
Sampled time functions $x(t), y(t)$	Complex spectra $X(f), Y(f)$
	Power spectral estimates $G_{xx} = X(f) \cdot X^*(f)$ $G_{yy} = Y(f) \cdot Y^*(f)$
	Cross power spectral estimate $G_{yx} = Y(f) \cdot X^*(f)$
Autocorrelation functions $g_x(t), g_y(t)$	Ensemble averages of power spectra $\bar{G}_x, \bar{G}_y$
Cross correlation functions $g_{yx}(t)$	Ensemble average of cross spectra $\bar{G}_{yx}$
Impulse response	Transfer function $\bar{G}_{yx}/\bar{G}_{xx}$
	Coherence functions $|\bar{G}_{yx}|^2/\bar{G}_{xx} \cdot \bar{G}_{yy}$

(1) The Fourier transforms of these functions : $X(f)$ and $Y(f)$ are their complex spectra. Each spectrum line has two components: magnitude and phase, or real and imaginary components. Third octave, tenth octave, etc., spectra can be created by integrating the amplitude spectrum over suitable spectral values;

(2) Multiplying a complex spectrum by its own conjugate complex we get a power spectral estimate. Multiplying the complex spectrum of the output signal by the conjugate complex of the spectrum of the input signal by the conjugate complex of the spectrum of the input signal we get a cross power spectral estimate. Ensemble averaging of power spectral estimates is a very useful method of lifting out spectral components of coherent signals from uncorrelated noise;

(3) The inverse Fourier transforms of power spectra and cross power spectra are the auto and cross correlation functions of the original signals;

(4) The ratio of the cross and input power spectra is the transfer function of our 'black box'. Transfer functions are generally complex functions of frequency since the general black box has some gain and phase shift at each frequency of its pass band;

(5) The inverse Fourier transform of the transfer function is the impulse of the black box;

(6) Dividing the square of the cross power spectrum by both input and output power spectra results in a real, dimensionless function of frequency, called the coherence function. The values at each frequency component are between zero and one;

```fortran
      SUBROUTINE NLOGN(N,X,SIGN)
      DIMENSION M(25)
      DIMENSION X(2)
      COMPLEX X,WK HOLD,Q
      LX=2**N
      DO 1 I=1,N
      M(I)=2**(N-I)
      DO 4 L=1,N
      NBLOCK=2**(L-1)
      LBLOCK=LX/NBLOCK
      LBHALF=LBLOCK/2
      K=0
      DO 4 IBLOCK=1,NBLOCK
      FK=K
      FLX=LX
      V=SIGN*6.2831853*FK/FLX
      WK=CMPLX(COS(V),SIN(V))
      ISTART=LBLOCK*(IBLOCK-1)
      DO 2 I=1,LBHALF
      J=ISTART+I
      JH=J+LBHALF
      Q=X(JH)*WK
      X(JH)=X(J)-Q
      X(J)=X(J)+Q
    2 CONTINUE
      DO 3 I=2,N
      II=I
      IF(K.LT.M(I)) GO TO 4
    3 K=K-M(I)
    4 K=K+M(II)
      K=0
      DO 7 J=1,LX
      IF(K.LT.J) GO TO 5
      HOLD=X(J)
      X(J)=X(K+1)
      X(K+1)=HOLD
    5 DO 6 I=1,N
      II=I
      IF(K.LT.M(I)) GO TO 7
    6 K=K-M(I)
    7 K=K+M(II)
      IF (SIGN.GT.0.0) RETURN
      DO 8 I=1,LX
    8 X(I)=X(I)/FLX
      RETURN
      END
```

Figure 4.8 *Fast Fourier transform (FFT) algorithm NLOGN written in Fortran IV*

```
SUBROUTINE HALF(NS,X,SN)
DIMENSION X(2)
COMPLEX X,FI,A1,A2,W
FI=(0.0,0.5)
NC=2**NS
N2=NC/2
W=CEXP(2.0*3.14159*FI/FLOAT(NC))
CALL NLOGN(NS,X,SN)
C1=0.5*(REAL(X(1))+AIMAG(X(1)))
CN=0.5*(REAL(X(1))-AIMAG(X(1)))
X(1)=CMPLX(C1,0.0)
X(NC+1)=CMPLX(CN,0.0)
X(N2+1)=0.5*CONJG(X(N2+1))
DO 2 N=2,N2
MINUSN=NC+2-N
A1=0.5*(CONJG(X(MINUSN))+X(N))
A2=  FI*(CONJG(X(MINUSN))-X(N))
X(N)=A1
A2=A2*W**(-N+1)
A1=0.5*(X(N)+A2)
A2=0.5*(X(N)-A2)
X(N)=A1
X(MINUSN)=CONJG(A2)
RETURN
END
```

Figure 4.9 *Subroutine HALF used to remove inefficiencies from Figure 4.8*

(7) The coherence function is a very good measure of the casual relationship between the output and the input of the black box; the value between zero and one expresses the percentage of output caused by the input. It is also a very powerful indicator of the 'goodness' of the transfer function.

4.3 Frequency analysis technique

Reference was made to Fast Fourier Transform as an extension of the mathematical concept of periodic signal analysis but this is not the sole technique available. In practice, the methods by which a signal can be resolved into a frequency spectrum (sometimes called 'time-to-frequency domain conversion') resolves into a choice between the following types of analyser [10]:

(1) Fixed filter: In which a number of fixed electronic contigous filters together cover the whole of the frequency range of interest. A detector circuit successively measures the averaged output from each filter in turn. Generally only used for relatively broadband analysis such as 1/3 octave;

(2) Sweeping filter: An electronic filter circuit with a variable centre frequency which is swept over the frequency range of interest. There are two main types, (i) constant absolute bandwidth operating on the heterodyne principle, (ii) constant percentage bandwidth;

(3) High speed analysis: By recording the signal on a digital event recorder and playing it back through the analyser at considerably faster speed;

(4) Real-time analysis: In which the analysis is made so quickly that results are provided almost immediately. The methods may be (i) analogue parallel analysis, (ii) time compression, (iii) digital filters, (iv) FFT;

(5) Analogue parallel analysis: Contains a range of filters each with its separate detector for each filter so that the output levels from all filters are displayed simultaneously;

(6) Time compression: Similar to the high speed analysis method except that recording in the digital memory proceeds during the same time as playback and heterodyne frequency analysis;

(7) Digital filters: Similar to analogue filters in that the mathematical description of the analogue is converted to a numerical algorithm with which the input values from an analogue/digital converter are processed;

(8) F.F.T. (fast Fourier transform): By which the very efficient algorithm described in Section 4.2.6 calculates the discrete Fourier transform (DFT). It is possible to make rapid transformation (in either direction) between time and frequency domains while retaining phase information if desired.

4.4 Derived functions

Monitoring techniques may reasonably be performed by simply producing a frequency spectrum and recording trends in one or more resonant frequencies associated with the input from particular components. Experience suggests that greater sensitivity to change arising from deterioration can be achieved by further processing the signals in order to study changes in the derived functions such as those listed in Table 4.2. To undertake such calculations the use of computer-orientated hardware using FFT techniques offers the most practical method [11].

4.4.1 Aliasing

Consider a number of component sinusoidal signals as shown in Figure 4.10 with sample points (the crosses) uniformly spaced on the time axis. In Figure 4.10(b) shown, the period of the sampling is one-fifth that of the period of the wave; a sampling frequency of 5000 Hz = 5X frequency of the wave (1000). Thus it is not possible to pass a sinusoid through points of a lower frequency (longer period) than shown in (b). The waveform in Figure 4.10(c) sampled at a rate five-fourths the frequency of the waveform. We can draw another lower-frequency wave through the sample points. The waves have been drawn so that all 3 go through equivalent points.

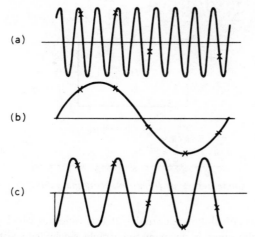

Figure 4.10 *Aliasing of component sinusoidal signals*

If the frequency of the middle wave were 1000 Hz, and the sampling frequency 5000 Hz, the other frequencies would be 4000 Hz and 9000 Hz. Other waves of frequencies 6000, 11 000, 14 000, 16 000, 19 000, 21 000 Hz, etc., of the same peak amplitude pass through the same points. Since these cannot be distinguished from one another by the selected set of points, they are called aliases.

The frequencies of components that are aliases are related by the equation:

$$\pm f_1 = f_2 \pm k f_s \qquad (4.10)$$

where f_1 and f_2 = alias frequencies,

k = an integer,

f_s = sampling frequency,

When sampled points are treated by digital processing they are usually assumed to be from a wave of the lowest frequency. If the sampling rate for an incoming signal is such that: sampling rate from an incoming signal > 2 × highest frequency of any component in the signal, some of the high-frequency components of the signal will be effectively translated down to be less than one-half the sampling rate. This translation may cause serious problems with interference of high and low frequency components.

4.4.2 Signal path impedance (transfer functions)

Forces generated by the components of a machine are transmitted through the enclosing structure. Condition monitoring based on transducers is influenced by the velocity through the transmission path depending upon the response of the system to different load and speed conditions. This response may be expressed in terms of the mechanical transfer function [12].

A transfer function may be regarded as the ratio between the input to a system and the resulting output.

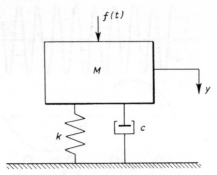

Figure 4.11 *Simple single-degree of freedom spring mass damper system*

A simple mathematical-dynamic model of this function which can be extended to relate forced vibrations is given in the simple single-degree-of-freedom system Figure 4.11 where

$$F = \text{input}$$

$$Y = \text{system response,}$$

$$k = \text{spring constant,}$$

$$c = \text{damping coefficient,}$$

$$m = \text{system mass,}$$

$$f(t) = \text{input force,}$$

$$y = \text{output displacement.}$$

Thus by Newton's law the following differential equation is obtained:

$$m \, d^2y/dt^2 + c \, dy/dt + ky = f(t) \tag{4.11}$$

and by Laplace transformation for zero initial conditions

$$(ms^2 + cs + k)Y(s) = F(s). \tag{4.12}$$

which, re-arranged gives:

$$\text{transfer function} = \frac{Y(s)}{F(s)} = \frac{1}{ms^2 + cs + k}$$

$$= \frac{K}{(s/\omega_n)^2 + (2\rho/\omega_n)s + 1} \tag{4.13}$$

where $K = 1/k = \text{static sensitivity,}$

$\omega_n = k/m = \text{natural frequency,}$

$= c/2(km)^{1/2} = \text{damping ratio.}$

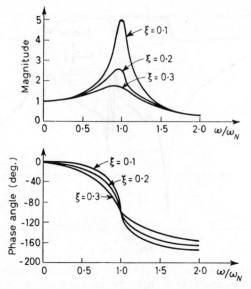

Figure 4.12 *Frequency response of second-order system*

Considering the case of a sinusoidal excitation force $f(t) = F \sin \omega t$ where $a = j\omega$ in which $j = (-1)^{1/2}$, the resulting amplitude ratio Y/F and phase angle for different amounts of damping ratio ξ are shown in Figure 4.12.

In a realistic structure the dynamic response to an input force is very much more complex than for the simple single-degree-of-freedom mode. There are several paths which a signal might follow to arrive at the output motion so that the frequency response plot of the simple second-order system must be extended. Most force inputs from gears and bearings are periodic and can be expressed as summations of sinusoidal terms through the Fourier series as follows:

$$f(t) = F_1 \sin (\omega_1 t + \alpha_1) + F_2 \sin (\omega_2 t + \alpha_2) + F_3 \sin (\omega_3 t + \alpha_3) + \ldots \text{etc} \quad (4.14)$$

$$Y(t) = Y_1 \sin (\omega_1 t + \alpha_1 + \alpha_1^1) + Y_2 \sin (\omega_2 t + \alpha_2 + \alpha_2^1)$$

$$+ Y_3 \sin (\omega_3 t + \alpha_3 + \alpha_3^1) + \ldots \text{etc} \quad (4.15)$$

where; Y_1/F_1 = amplitude ratio at frequency ω_1,

α_1 = phase angle at ω_1.

The amplitudes Y_1, Y_2, Y_3 depend strongly on the frequency response shape, if one of the frequencies ω_1, ω_2, ω_3 etc. approaches the fundamental a large amplification may occur (with lightly damped structures, this may exceed a hundred-fold).

Under conditions of varying speed the amplitude amplification as a consequence of the flexural path of the structure may become large and frequency normalization is inaccurate with resonance in the region of a source frequency.

A typical application is to compare the behaviour of an engine which operates

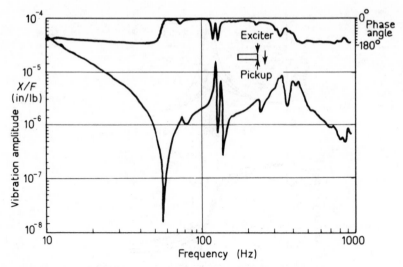

Figure 4.13 *Transfer function plot indicates specific frequencies of special weakness (highest points on chart) or strength. In this example, the structure was excited from 10–1000 Hz.*

smoothly on a test bed, with its behaviour when it is installed in a structure such as a ship, vehicle or aeroplane. New interactions arise involving the dynamic stiffness of the structural members under vibrating loads. Because a vibrating force can produce greater deflections than when applied statically, the lack of dynamic stiffness can be a serious problem. Transfer function analysis identifies the vibration frequencies at which the structure is weak (which might be at the resonance of a major engine vibration component).

Effectively, the technique compares the amplitude of the applied force to that of the resulting structured motion. Experimentally, the engine vibration is simulated by a vibration exciter which applies force to the structure at the same points and in the same direction as the actual operating forces. The structure's response to the excitation is picked up by sensors and plotted to show points of critical resonance, such as maximum deflection or bending. Tests are repeated in all important coordinate directions; axial, longitudinal, cross, torsional, etc. Results such as Figure 4.13 pin-points frequencies of special weakness [13].

Forces which arise from the intermesh of gear teeth are transmitted through the shafting and bearings as a consequence of both lateral and torsional vibrations of the shafting system. By using impulse testing with a force pulse applied at the gear mesh and appropriate data reduction associated with an assumed system linearity the transfer function can be calculated [14, 15, 16, 17].

Transient pulse excitation is a method of obtaining mechanical transfer functions [18]. The system is forced with a sharp pulse and the response measured at a remote location. Spectral density calculations involving the input force and output motion, using computer techniques, provide a value for the transfer function. The coherence function between the two signals enable values of system linearity and transfer function to be deduced. System resonance (but not transfer functions) can be determined from the amplitude peaks thrown up as a piece of rotating machinery is slowed down.

4.4.3 Transfer function analysis

The technique for determining transfer functions and tracking the frequency of an atomic reactor component in its fundamental mode was described by Macleod, Halliwell and Hale, [19].

4.4.4 Response testing

Chamberlain [20] observed that two signals from which a cross spectrum has been taken contain information of use for purposed of identification. Using the concept of a transfer function in a system (the ratio of the output to the input) this is also the point-by-point ratio of the square of the absolute value of the cross-spectral density function of the input and output divided by the auto-spectral density function of the input. The transfer function can be used to estimate the output of a system for any type of input wave. It is therefore possible to test a communication channel while it is being used to transmit normal signals, by means of the normal transmissions that provide the signals.

If an input is broadband, the cross-correlation of the input and output of a system is the impulse response of the system. Furthermore, this calculation is not hindered by the presence of noise or any other signal that may be applied to the system between input and output. This means that it is possible to obtain the impulse response of a system while it is in operation. An industrial process-control system can be tested without its operation being stopped or affected.

4.4.5 Correlation cross-correlation and autocorrelation

Correlation is a measure of the similarity of two wave-forms, it is a function of the *time displacement* between the two. If the wave-forms are distinct and separate the process is called cross-correlation [21, 22]. Autocorrelation relates to comparisons of the wave-form itself [23, 24]. Transfer function is the comparison of signal strengths whereas cross-spectrum is the measure of similarity of two wave-forms in the frequency domains.)

Correlation techniques are applied to location of the source of excitations, typical reports relating to:

(1) Location of noise sources by direction finding [25–27];

(2) Separation of aerodynamic noise of turbulent boundary layers from noise radiated by jet engines [28];

(3) Measuring panel transmission loss [29];

(4) Underwater echo ranging [30];

(5) Energy flow in structures [31].

A procedure proposed by Enesco Inc., Springfield, Virginia to diagnose the condition of gears uses two different techniques, one for use on bearings and the other for use on gears [32].

The autocorrelation function is used to detect bearing failure. By assessing the peaks which have time lags corresponding to the period associated with specific bearing components, a judgment can be made on each specific bearing failure. Stochastic correlation used in the computation of the correlation function can adjust the analysis to compensate for the effects of speed fluctuation. Bad parts baseline data is used to deduce limit values for the correlation amplitudes.

Gear deterioration detection is based on the increasingly rich harmonic content in worn gears. 'Good' and 'bad' baseline spectra must be prepared and the amplitude normalized by 'pre-whitening' the spectra. To test a gear a vibration signature is taken and a spectral correlation obtained with the baseline data. If the signals are related, peaks in the correlation function should be seen at frequency shifts corresponding to the frequency difference between harmonics of the gear mesh frequency in the signal. Discrimination between good and bad gears is obtained by the statistical measurement of the extent of correlation.

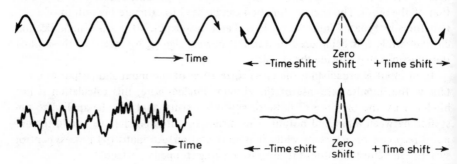

Figure 4.14 *Autocorrelation function of a periodic function, is itself periodic. The autocorrelation function of random noise is an impulse of duration inversely proportional to the bandwidth of the noise (a) Sinewave (b) Autocorrelation function (c) Broadband noise (d) Autocorrelation function.*

4.4.6 Correlation measurement techniques

Signals are multiplied to obtain a correlation. Multiplying one waveform point-by-point with an identical one and taking the average of these multiplied values over the full range of the wave the result is the mean square value, if there is no time displacement; it is usually termed, 'power', (although it may not be related to physical power). If the two waveforms are shifted in time with respect to each other, another averaged product can be obtained, and so on for many shifts in time. These averaged products as a function of the time displacement, or delay (usually designated by τ and also called 'lag'), are the autocorrelation function.

Cross-correlation may be obtained by multiplying two signals (rather than one signal alone, as with autocorrelation) and produces a resulting graph similar to that for autocorrelation shown in Figure 4.14.

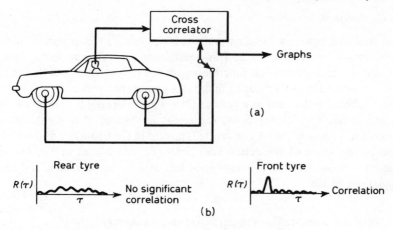

Figure 4.15 *Identification of noise source in a motor vehicle*

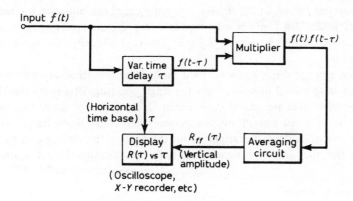

Figure 4.16 *Autocorrelation circuit*

4.4.7 Cross-correlation source identification

The use of cross-correlation in an automobile test is shown in Figure 4.15. In determining the noise source at the driver's ear the diagram shows only two of the many possible sources of noise. One graph (Figure 4.15b) shows cross-correlation of the noise (measured by microphone) with vibrations from the front axle; the other shows no correlation, so the rear axle does not contribute to this noise.

Implementation of autocorrelation can be accomplished by using a variable time delay, multiples and integrator (or averaging) circuit such as that shown in Figure 4.16. An almost identical circuit is used for cross-correlation analysis.

4.4.8 Convolution

Cross-correlation against a certain waveform is equivalent to 'convolution' with the same waveform reversed in time. This equivalence between waveforms does not mean that the cross-correlation function equals the output signal but that the process of cross-correlation is similar to that of the convolution process.

Convolution can be used as a general filter with independent control of amplitude and phase. The filter response can easily be changed. Any combination of selection and rejection can be constructed, limited in the frequency domain by the frequency resolution of the data blocks processed, and limited in dynamic range (or signal-to-noise ratio) by the number of bits used in the processing and analog-to digital conversion.

4.4.9 Signal spectrum, autospectrum (power spectral density)

A single signal when resolved into frequency components is a 'spectrum' of the signal. An autospectrum (also called 'power spectral density') is a derived spectrum in which the ordinates are (coefficients of the spectrum components $)^2$ [33].

A spectrum giving frequency information can be taken from an autocorrelation by calculating the Fourier transforms of the autocorrelation curve. The result gives a power spectral density of the signal, commonly called the 'auto-spectral function'. (The term 'power' used in this way implies a squaring and averaging operation and – need not be related to actual physical power). The symbol for power spectral density is G.

An auto-spectral display can be used more effectively than autocorrelation to find a sine wave buried in noise. Thus the single spectrum of a noisy signal would only show that there are several large amplitudes, any one of which may be either a noisy peak or a sine wave. If numerous spectra are averaged one frequency stands out since the noise-signal variations about the spectral line decrease as frames are averaged, but the sine-wave signal does not, thus distinguishing the sine wave.

4.4.10 Cross-spectrum

This is a measure of the similarity between two signals in the frequency domain. Cross-spectrum is the Fourier transform of the cross-correlation and expresses the similarity as a function of frequency [34].

A conjugate multiplication must be applied in order to calculate the cross-spectral density function from the Fourier transform of the two time series. This type of multiplication gives the product magnitude x difference of the two signals.

A Fourier transform of the cross-correlation function results in a cross-spectrum of the two signals involved. Only frequencies which are common to both waveforms are contained. For example, suppose that the flaps of an aeroplane wing can be vibrated. A vibration pickup at the end of the wing can detect the output waveform caused by the vibration generated by the flaps. A cross-spectrum of this waveform and the input will show the major frequency generated. This is a common application for almost every airframe manufacturer, which is often referred to as 'flutter testing' [35].

4.4.11 Coherence function (Y²)

This is a measure of the reliability of a transfer function estimate. The significant values are:

Coherence function = 0   When transfer function has no statistical validity;

Coherence function = 1   When transfer function estimate is not contaminated by interfering noise.

The value of the coherence function is given by the expression:

$$\text{coherence function } Y^2 = \frac{(\text{cross spectral density function})^2}{\left(\begin{array}{c}\text{power spectral density}\\ \text{signal 1}\end{array}\right) \times \left(\begin{array}{c}\text{power spectral density}\\ \text{signal 2}\end{array}\right)}$$

$$= \frac{(\bar{G}_{YX})^2}{\bar{G}_{XX} \times \bar{G}_{YY}}.$$

In addition to helping to correct for the major errors in the measurement of transfer functions arising from signal noise the coherence function can help to predict the number of averages of cross-spectra and auto-spectra that are required to define transfer function to a given accuracy.

When the coherence function $Y^2 = 1$, the spectral line at the monitored point is completely coherent with the measured source. A coherence function *of* $Y^2 = 0.5$ means that 50% of the power at the monitored point at any given frequency is coherent with the measured source.

Due to the way in which the coherence function is normalized across the measurement band the gain of the transducer and the transmission path is eliminated. The proportion of power at a monitored point arising from a given source may be compared with the noise coming from other sources on an absolute 0-to-1 scale without the need to compare transducers.

4.4.12 Composite exceedance

This is a statistical procedure [36] for the condition monitoring of any type of rotating machinery and is based upon the amplitude distribution of the peak values of acceleration of the system being monitored. It is hypothesized that in the case of normal operation, i.e., 'good' operating condition, the peak value amplitude distribution will exhibit Gaussian behaviour. However, when a fault occurs, the distribution tends to change shape. If the fault is not too severe, the distribution may still look Gaussian but will have larger means and standard deviations. If the fault is severe enough, the distribution may appear to have a non-Gaussian shape.

References

1 Worthy, J.G.B. (1970), 'Radio telemetry of diesel engine piston temperatures', Central Electricity Research Laboratories (Leatherhead) Conference on Transmitting Signals from Rotating Plant, 25/26 June.

2 Bolus, R.L., Fang, C.S. and Chia, S.N. (1973), 'Design of a thermal monitoring system', *MTS Journal,* 7, No. 7 Oct/Nov.

3 Curwen, K.C. 'Turbine blade pyrometer system in the control of the Concorde engine', Kollsman Instruments Ltd, The Airport, Southampton SO9 3FR, Hants.

4 Barber, R. (1969), 'A radiation pyrometer designed for in-flight measurement of turbine blade temperatures', SAE Paper 690432, April.

5 Collacott, R.A. (1975), 'Plant deterioration monitoring', *Engineering Materials & Design',* 19, No. 6, June, 17—19.

6 Gold, B. and Rader, C.M. (1965), *Digital Processing of Signals,* McGraw Hill, New York.

7 Betts, J.A. (1970), *Signal Processing, Modulation and Noise,* English Universities Press.

8 Randall, R.B. (1974), 'Cepstrum analysis and gear-box fault diagnosis', Bruel & Kjaer Application Note.

9 Kiss, A.Z. (1973), 'Real-time digital Fast Fourier Transform hardware', Audio Engineering Society, 45th. Convention, May, Los Angeles California (Time/Data Corporation, Palo Alto, California).

10 Randall, R.B. 'Frequency analysis of stationary non-stationary and transient signals', Bruel & Kjaer Application Note 14-165.

11 Oppenheim, A.V. (1969), *Papers on Digital Signal Processing,* M.I.T. Press, Cambridge, Mass.

12 Sloane, E.A. (1970), 'Measurement of transfer function and impedance', *General Radio Experimenter,* 44, Nos. 7—9, July—Sept.

13 Klosterman A.L. 'A combined experimental and analytical procedure for improving automotive system dynamics' (Structural Dynamics Research Corporation), SAE Paper No. 720093.

14 Wang, S.M. and Morse, I.E. (1971), 'Torsional response of a gear train system', Vibrations Conference, Toronto, Canada ASME Paper 71-VIBR-77, September.

15 Laskin I., Orcutt, F.K. and Shipley, E.E. (1968), 'Analysis of noise generated by UH-1 helicopter transmission', USAAVLABS Technical Report 68-41, Fort Eustis, Virginia, June, AD 675457.

16 Badgley, R.H. and Laskin, I. (1970), 'Program for helicopter gearbox noise detection and reduction', USAAVLABS Technical Report 70-12, Fort Eustis, Virgnia, March, AD 869882.

17 Badgley, R.H. and Chiang, T. (1972), 'Investigation of gearbox design; modifications for reducing helicopter gearbox noise', USAAMRDL Technical Report 72-6, Fort Eustis, Virginia, March, AD 742735.

18 Morse, I.E., Shapton, W.R., Brown, D.L. and Kuljanic, E. (1972), 'Application of pulse testing for determining dynamic characteristics of machine tools', *13th Int. Machine Tool Design and Research Conference,* University of Birmingham, September.

19 Macleod, I.D., Halliwell, G. and Hale, J.C. (9170), 'Signal analysis techniques for use in acoustic diagnostics', (UKAEA) Symposium, I.Mech.E. 20 Oct.

20 Chamberlain, L.J. 'A simple discussion of time-series analysis', Time/Data Corporation, 490 San Antonio Road, Palo Alto, California 94306.

21 Lathi, B.P. (1965), *Signals, Systems and Communication,* Wiley.

22 Anstey, N.A. (1966), 'Correlation techniques – a review', *J. Cam. Soc. Explor. Geophysics,* **2,** No. 1 Dec, 1–28.

23 Lee, Y.W. (1960), *Statistical Theory of Communication,* John Wiley, New York.

24 Heitzman, C.L. (1970) 'Signal analysis with digital time series analysers' *General Radio Experimenter,* **44,** Nos 7, 8, 9.

25 Goff, K.W. (1955), 'Application of correlation techniques to some acoustic measurements', *J. Acous. Soc. Am.,* **27,** No. 2, March, 236–246.

26 Faran, J.J. and Hill, R. (1952), 'Correlators for signal reception', Tech. Mem. 27, Sept, Acoustics Research Laboratory, Applied Science Division, Harvard University, Cambs, Mass.

27 Gilbreck, D.A. and Binder, R.C. (1956), 'Portable instrument for locating noise sources in mechanical equipment', *J. Acoust. Soc. Am.* **30,** No. 9 Sept, 842–6.

28 Bhat, W.V. (1971), 'Use of correlation techniques for estimating in-flight noise radiated by wing-mounted jet engines on a fuselage', *J. Sound & Vibration,* **17,** No. 3, 349–355.

29 Burd, A.N. (1968), 'The measurement of sound insulation in the presence of flanking paths', *J. Sound & Vibration,* **7,** No. 1, Jan, 13–26.;

30 Horton, C.W. (1969), 'Signal processing and underwater acoustic waves', US Government Printing Office, Washington DC 20402.

31 White, R.H. (1969), 'Cross-correlation in structural systems, dispersion and non-dispersion waves', *J. Acoust. Soc. Am,* **45,** No. 5, May, 1118–1128.

32 Demuth, H.P. and Rudd, T.J. (1972), 'Helicopter gearbox signature analysis', ENESCO Inc, August.

33 Blackman, R.B. and Tukey, J.W. (1958), *The Measurement of Power Spectra,* Dover Publications Inc, New York.

34 Lange, F.H. (1967), *Correlation Techniques,* Iliffe Press.

35 Wilby, J.F. and Gloyna, F.L. (1970). 'Correlation measurements of airplane fuselage vibrations', *J. Acoust. Soc. Am,* **48,** No. 1 (Part 1) July.;

36 Ziebarth, H. and Chang, J.D. 'Composite exceedance technique for the status monitoring of mechanical components', Airesearch Manufacturing Company, The Garrett Corporation, Torrance, California, U.S.A.

5   Vibration analysis

5.1 Introduction

Vibratory effects are sensed quite easily by most people, they are tiring and unpleasant, at the extreme they can be frightening. It is not surprising that vibration
analysis occupies a prominent place in machine diagnostics.

5.1.1 Vibration concepts – frequency

This is the time rate of repetition of a periodic phenomenon, measured in cycles per
second or the unit Hertz (hz) where

$$1 \text{ Hz} = 1 \text{ cycle s}^{-1} = \tfrac{1}{60} \text{ cycles min}^{-1} \tag{5.1}$$

Frequency (f) is the reciprocal of the periodic time (T_0)

$$f = \frac{1}{T_0} \tag{5.2}$$

5.1.2 Frequency octave bands

The numerous discrete frequencies of different amplitudes which make up a
vibration may be divided into octave bands each of which covers a 2 to 1 range of
frequencies. Such analysis in the frequency domain is said to yield a series of
'octave-band levels'. An octave is an interval between two frequencies such that
they have a ratio of 2. Thus, given a frequency f_1 the next frequency, f_2 would be

$$f_2 = 2 \times f_1 \tag{5.3}$$

The next octave frequency, f_3 would be given by

$$f_3 = 2 \times f_2 \tag{5.4}$$
$$= (2 \times 2)f_1$$
$$= (2)^2 f_1$$

Subsequently
$$f_4 = 2 \times f_3$$
$$= (2 \times 2 \times 2)f_1$$
$$= (2)^3 f_1$$
$$f_n = (2)^{n-1} f_1. \tag{5.5}$$

The mid-frequency or centre-frequency for each of these bands identifies the
octave and must bear the same relationship to the upper and lower frequencies.
Thus, if f_{a1} is the mid-frequency for the octave band between f_1 and f_2 then:

$$\frac{f_{a1}}{f_1} = \frac{f_2}{f_{a1}},$$

$$f_{a1} = (f_1 f_2)^{1/2} \tag{5.6}$$

Thus, for an octave band, since $f_n = (2)^{n-1} f_1$, the general expression for the mid-frequency fan in each octave up to the upper frequency f_n, i.e. between f_{n-1} and f_n is given by

$$f_{an} = [(2)^{n-1} f_1]^{1/2} \tag{5.7}$$

$$f_{a1} = (f_1 f_2)^{1/2}$$

$$= (f_1 2 f_1)^{1/2}$$

$$= (2)^{1/2} f_1 \tag{5.8}$$

$$f_{a2} = (f_2 f_3)^{1/2}$$

$$= (2 f_1 4 f_1)^{1/2}$$

$$= 2 \cdot (2)^{1/2} f_1$$

$$= 2 \cdot f_{a1} \tag{5.9}$$

From equations 5.8, 5.9 it is evident that the ratio of consecutive mid-frequencies

$$= \frac{f_{a2}}{f_{a1}} = 2 \tag{5.10}$$

It is convenient for the preferred range of mid-frequencies to include 1000 Hz (1 kHz), thus the typical octave frequencies are as given in the following table.

Table 5.1

Mid-frequencies (Hz)	31.5		63		125		250		500		1000	
Octave frequency (Hz)		44.25		88.5		177		345		707		1414

Mid-frequency (Hz)		2000		4000		8000		16000		32000	
Octave frequency (Hz)			2828		5656		11312		22524	

The number of frequencies between adjacent octave frequencies is known as the 'pass band'. Thus with a mid-frequency of 1000 Hz and octave frequencies of 707 and 1414 Hz the pass band would be $1414 - 707 = 707$ Hz.

5.1.3 One-third octave bands

For more detailed analysis it is convenient to split the pass band into three parts. A similar division of the mid-frequencies also occurs.

Thus f_x and f_y, the one-third octave frequencies for the octave frequencies 707 and 1414 Hz (with mid-frequency of 1000 Hz) would be such that

$$\frac{f_x}{707} = \frac{f_y}{f_x} = \frac{1414}{f_y}. \tag{5.11}$$

Taking the first and second, then the second and third equalities

$$(f_x)^2 = 707f_y$$

$$f_x f_y = 1414 \times 707,$$

hence

$$(f_x)^3 = 2(707)^3,$$

$$f_x = (2)^{1/3}707. \tag{5.12}$$

Thus, for an octave band, since $f_n = (2)^{n-1}f_1$, the general expression for the mid-value: 1.259921049). This can be applied to the values in Table 5.1 to give the following values.

Table 5.2 *One-third octave frequencies*

Mid-frequency (Hz)		795		1000		1260		1540	
One-third octave band frequency (Hz)	707		890		1122		1414		1780
Mid-frequency (Hz)	2000		2520		3180		4000		
One-third octave band frequency (Hz)		2244		2822		3560		4488	

5.1.4 Narrow bands

Frequency analysers for more precise investigation of the spectrum need to use narrower bands than one-third octave and include one-tenth, one-twelfth and one-fifteenth octave analysis.

As can be seen from equation 5.12) the general relationship involves $(2)^{1/M}$ where M = number of parts into which the octave is to be split. Typical values are given in Table 5.3.

Table 5.3 *Frequency ratios for narrow band octaves*

M	$(2)^{1/M}$ = *frequency ratio*
One-quarter octave (1/4)	1.189207114
One-fifth octave (1/5)	1.148698354
One-sixth octave (1/6)	1.122462048
One-seventh octave (1/7)	1.104089513
One-eighth octave (1/8)	1.090507732
One-ninth octave (1/9)	1.080059738
One-tenth octave (1/10)	1.071773462
One-eleventh octave (1/11)	1.065041089
One-twelfth octave (1/12)	1.059463094
One-thirteenth octave (1/13)	1.054766076
One-fourteenth octave (1/14)	1.050756638
One-fifteenth octave (1/15)	1.047294122

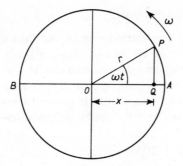

Figure 5.1 *Simple harmonic motion – polar concept*

5.2 Vibration – simple harmonic motion concept

A simple relationship between vibratory displacement, velocity and acceleration with frequency can be obtained by considering Figure 5.1. The relationship can be illustrated by considering a point (P) rotating at a radius (r) with uniform angular velocity (ω). The projection of P on AOB at Q is said to move with simple harmonic motion.

At any instant the horizontal component of vector, displacement x is given by

$$x = r \cos \omega t$$

The velocity of Q towards O is given by

$$v = \frac{\mathrm{d}x}{\mathrm{d}t}$$

$$= -\omega r \sin \omega t$$

$$= \omega r \cos (\omega t + \pi/2)^*$$

$$= \omega x. \tag{5.13}$$

The acceleration of Q is given by

$$a = \frac{\mathrm{d}^2 x}{\mathrm{d}t^2}$$

$$= -\omega^2 r \cos \omega t$$

$$= -\omega^2 x \tag{5.14}$$

The angular velocity ω is related to the periodic time T_0 by $2\pi = \omega T_0$ i.e. $\omega = 2\pi/T_0 = 2\pi f_0$ hence values for displacement, velocity and acceleration can be expressed in terms of frequency, f_0.

Structures rarely vibrate to simple harmonic motion but involve a complex of 'tones' or frequencies together with the effects of differing masses and damping combined with the effects of the forcing frequencies.

* The $\pi/2$ implies that velocity leads displacement by $\pi/2$.

5.2.1 Decibel vibration measurement

Amplitude, velocity and acceleration may be measured in terms of the S.I. units, millimetres (m), millimetres per second (mm s^{-1}) and millimetres per second^2 (mm s^{-2}). It is becoming increasingly common to use a logarithmic basis for measuring velocity and acceleration and the 'bel' is used internationally although the amplitudes of vibrations are measured in terms of various linear units. The bel is a large unit but for convenience the decibel abbreviation dB is normally used and is defined as:

(1) Acceleration $(a\text{dB}) = 20 \log_{10} a_1/a_2$ (5.14)

where $a_1 = $ the rms or peak measured acceleration (mm s^{-2})

and $a_2 = $ the predetermined reference level,
                                        normally $0\,a\text{dB} = 10^{-2}\,\text{mm s}^{-2}$

such that $a\text{dB} = 20 \log_{10} a/10^{-2}$ (5.15)

(2) Velocity $(v\text{dB}) = 20 \log_{10} v_1/v_2$ (5.16)

where $v_1 = $ the rms or peak measured velocity (mm s^{-1})

                              $v_2 = $ the predetermined reference level,
                                        normally $0\,v\text{dB} = 10^{-5}\,\text{mm s}^{-1}$

such that $v\text{dB} = 20 \log_{10} v/10^{-5}$ (5.17)

To avoid misinterpretation of results the reference level on which the measurements are based should always be given)(adB levels are easily converted to vdB's and vice versa should it be so required).

5.2.2 Conversion from acceleration to velocity in decibels

By considering the vector diagram, Figure 5.1 and the simple harmonic equations 5.13 and 5.14 with the frequency term f_0 in place of the angular velocity ω,

$$\text{velocity } (v) \text{ of } p \text{ in s.h.m.} = (2\pi f_0)x \tag{5.18}$$

$$\text{acceleration } (a) \text{ of } p \text{ in s.h.m.} = {}^* -(2\pi f_0)^2 x \tag{5.19}$$

Thus, from equations 5.18 and 5.19

$$v = \frac{a}{2\pi f_0}. \tag{5.20}$$

From equation 5.15 for acceleration decibels, $a\text{dB} = 20 \log_{10} a/10^{-2}$,

$$\log_{10} a/10^{-2} = \frac{a\text{dB}}{20},$$

$$a/10^{-2} = (10)^{a\text{dB}/20}$$

$$a = (10)^{(a\text{dB}/20)-2} \tag{5.21}$$

---

* The negative sign in equation 5.19 is a vector direction and not quantitative.

Substituting for a in equation 5.20

$$v = \frac{1}{2\pi f_0} \cdot (10)^{(a\,\mathrm{dB}/20)-2} \tag{5.22}$$

Substituting for v in the decibel equation 5.17

$$v\mathrm{dB} = 20 \log_{10} \left[\frac{1}{2\pi f_0} \cdot \frac{(10)^{(a\mathrm{dB}/20)-2}}{(10)^{-5}} \right],$$

$$= 20 \log_{10}(10)^{(a\mathrm{dB}/20)+3} - 20 \log_{10} 2\pi f_0,$$

$$= 20 \left(\frac{a\mathrm{dB}}{20} + 3 \right) - 20 \log_{10} 2\pi - 20 \log_{10} f_0,$$

$$= a\mathrm{dB} + 60 - 15.966 - 20 \log_{10} f_0,$$

$$= a\mathrm{dB} + 44.034 - 20 \log_{10} f_0,$$

$$= a\mathrm{dB} + K \tag{5.23}$$

where $K = 44.034 - 20 \log_{10} f_0$ and is thus related to frequency.
To convert from $a\mathrm{dB}$ to $v\mathrm{dB}$ equation 5.23 can be transferred,

$$a\mathrm{dB} = v\mathrm{dB} - K. \tag{5.24}$$

5.3 Vibration signature of active systems

Internal combustion engines, pumps, fluid valves, roller element bearings etc. are typical active systems which generate their own signatures during operation. According to Weichbrodt [1] it is possible to divide active systems into three groups:
   (1) Cyclic machinery (engines and transmissions);
   (2) Flow noise generators (pumps, boilers);
   (3) Single transient generators (switches, punches).
   Cyclic machinery produces a sound or vibration pattern which repeats with a certain time interval, this pattern or signature has been most successfully analysed for conditioning monitoring. Data recorded during the operation of engines, bearings and gear trains contains a large amount of background signal. Thus, if a piston becomes damaged or a fatigue crack appears in a bearing race, something changes in the signal, but the high background level makes it impossible to detect the change by simply measuring the overall level of the signal or some other obvious characteristics. It is necessary to develop a selective method which sorts out the background and clearly displays the part of the signal which is generated by the events under study.

5.3.1 Vibration monitoring precepts

The raw vibration signal in any mode from a single point on a machine is not a good indicator of the health or condition of a machine. Vibration is a vectorial parameter with three dimensions and requires to be measured at several carefully selected points and directions and the signal analysed into the basic components which make up the complex raw waveform. Selection of the location and direction of the sensor measurement is the single most critical factor in machinery vibration monitoring and analysis. If the raw signal does not contain the components that are representative of machinery condition, no amount of analysis will reveal that condition, as in the computer industry with respect to data input, 'garbage in — garbage out' [2].

Bearings are the best locations for measuring machinery vibration since this is where the basic dynamic loads and forces of the machine are applied and are a critical component of machinery condition. Vibration measurements should be made on the bearing cap of each bearing in a machine. If this is not feasible, the measurements should be made at a point as close as possible to the bearing with the minimum possible mechanical impedance between that point and the bearing. For a complete vibration signature of a machine, triaxial measurements should be made at each location with rotating machinery, although adequate information can usually be obtained from an axial and a radial measurement at each location. The different components of a machine vibrate at one or more discrete frequencies; different malfunctions cause vibrations at different discrete frequencies. The combination of these discrete frequency vibrations result in the complex vibration waveform at the measurement point. The measured signal is therefore analysed by reducing it to its discrete frequency components. The results of this type of analysis, presented as a plot of amplitude versus frequency, is referred to as the vibration signature of a machine.

The actual vibration signature of a machine contains frequency components which cannot be readily identified with a specific source of vibration. Some frequencies are caused by the mechanical resonance of various components the resonant frequency of which results from machine vibrations which are periodic but not sinusoidal. The spectrum of signals contain a component at a frequency equivalent to the basic period of vibration as well as many other components at harmonics of that frequency resulting from the non linear combination, in the machine, of signals at different discrete frequencies. This process (modulation) generates frequencies at the sum and difference of the frequencies being combined (sidebands). That every frequency component in a vibration signature cannot be traced to a specific source is no cause for concern. Attention should be paid to those frequencies that can be identified with a source and to changes in the overall signature with time.

5.3.2 Vibration measuring modes

There is general agreement that:

(1) Vibration severity at low frequencies is proportional to displacement;
(2) Vibration severity at mid-range frequencies is proportional to velocity;
(3) Vibration severity at very high frequencies is proportional to acceleration

There is less agreement as to the specific frequencies where crossover occurs. In the range of frequencies of interest for most machinery applications, vibratory velocity is the best indicator of machinery health. This is in accord with the intuitive feeling that the damage potential of vibration is proportional to the dynamic energy being dissipated which in turn is a function of velocity.

5.4 Vibration monitoring equipment

Sensors operating inductive pick-up displacement transducers, vibration pick-ups or strain gauges detect the vibration signal which has traversed the structure to be detected by the signal. This signal is then passed, either directly or through telemetry systems to a meter or an analyser which processes the signal data.

In its simplest form, the monitor may present a safe/unsafe indication of vibration severity. At a more advanced stage, the monitor may provide a frequency spectrum or signature by means of which the discrete frequency of the defective component can be used to identify the component which is failing. At its most sophisticated, the data can be processed in detail to reveal the type of defect within the faulty component itself.

5.4.1 Vibration analyser

Complex vibration wave-forms can be analysed continuously by means of variable narrow-band filters over the frequency range from 8 to 700 Hz. This enables maintenance engineers to make routine checks on rotating machinery to enable the vibration limits to be within specified limits. In a complex machine having parts rotating at various speeds, the individual vibration component frequencies can be measured or, alternatively, the overall vibration level can be measured with the filter switched out. The output of the analyser is available to trigger a stroboscope which can be used to give a stationary image of the vibrating part to which the analyser is tuned. The meter [3] indicates the acceleration, displacement or velocity of the part in absolute units.

5.4.2 Octave analyser

This instrument is primarily designed to measure the amplitude and frequency of the components of complex sound and vibration spectra using $1/3$ − octave and $1/10$ − octave bandwidths.

The high input impedance permits direct connection of piezoelectric transducers to measure sound pressures from 44 to 150 dB (re $N\,m^{-2}$) and acceleration from 0.0007 g to 100 g. Preamplifiers extend the full scale sensitivity by 20 dB allowing the use of long transducer extension cables. The analyser consists of a high impedance amplifier, a continuously tunable filter with a noise bandwidth of either $1/3$ or $1/10$ octave. The centre frequency of the filter is continuously adjustable.

5.4.3 Percentage bandwidth analyser

High resolution narrow bandwidth frequency analysis is performed by the use of this wave analyser which separates closely spaced frequencies using a constant percentage bandwidth technique. At low frequencies this analyser has a narrower bandwidth, better stability and more accurate calibration than fixed-bandwidth heterodyne wave analysers.

5.4.4 Narrow bandwidth analyser

This analyser has a narrow bandwidth in Hertz which is independent of the centre frequency. It is essentially a heterodyne type of voltmeter. The intermediate-frequency amplifier at 100 kHz includes a highly selective quartz-crystal filter whose bandwidth can be switched to 3, 10 and 50 Hz. The use of a heterodyne system makes it possible to vary the response frequency although the filter frequency is fixed. In one mode of operation the output is also heterodyned back to the original frequency. In another mode, the local oscillator beats with a 100 kHz quartz-crystal oscillator to function as a beat-frequency oscillator. These two outputs are also available at panel terminals as filtered input component and indicated frequency, respectively.

5.4.5 Real-time analyser

On-line frequency spectra analysis is provided by real-time instruments. In a typical instrument the input signal is applied to a set of analog filters (from 30 to 45 depending on requirements) that cover a frequency range from 3.15 Hz to 80 kHz. These filters include, as an option, individual attenuators to permit pre-whitening or other signal conditioning. The outputs of the filters are processed in another unit, the rms detector. It is unique in that it processes the signals from the filters digitally. Each channel is sampled, the sampled data converted to digital binary form, and the binary numbers are fed to a digital processor that computes root-mean-square levels.

The averaging method is true (linear) integration with a choice of nine accurate integration times from 1/8 seconds to 32 seconds. This scheme not only produces answers faster than the running-average circuits found in analog devices, but also make it possible to determine exactly what events in time have affected the answer. The computed band levels are stores in digital memory to be retrieved at a rate limited only by the output recording or storage device. The analyser simultaneously provides both digital and analog outputs.

Note: the term 'pre-whitening' used in connection with the pre-conditioning attenuators refers to the signal conditioning to eliminate unrequired 'noise'. The following terms are relevant:

(1) White Noise: Signals which have a band of random noise with a continuous distribution of frequency over each test band;

(2) Pink Noise: Signals containing random noise of equal power over each fraction of bandwith.

5.4.6 Time-series analysers

Programmable micro-computers have made it possible to develop a broad range of time series data programming analysis techniques can be performed with this instrument. Push buttons provide the selection of such operations as direct or inverse FFT, correlation, Auto Spectrum (PSD), transfer/coherence function, etc. with fast, error free and continuous measurements of input signals. The analysers combine the speed of a microprogrammed Fast Fourier-Transform processor with the flexibility of a digital controller.

These analysers permit real time continuous processing without any loss of data. Processing bandwidths are available up to 38 kHz (auto spectrum) directly from the panel. In addition, the full scale frequency range selection is in sequence steps of 1, 2, 2.5, 4, and 5 from 0.1 Hz to 50 kHz to allow maximum utilization of the bandwidth capability. Widest useful dynamic range is preserved by means of 16-bit words, double-precision calculations, and operator selected dynamic scaling. The pre-programmed pushbutton functions are:

DIRECT/INVERSE FFT
AUTO-/CROSS-SPECTRUM
TRANSFER/COHERENCE FUNCTION
AUTO-/CROSS-CORRELATION
AMPLITUDE HISTOGRAM
WAVEFORM AVERAGING.

5.4.7 Turbine vibration analyser

An electronic wave-form measuring technique described by Strickland [5] was specially developed for use in connection with turbine tests similar to those shown in Figure 5.2 in which a turbine blade is forcibly vibrated. The accelerometer

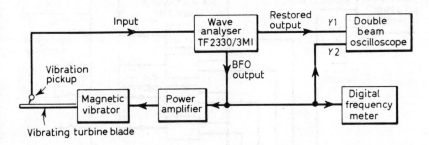

Figure 5.2 *Measurement of blade resonances*

vibration pick-up is placed in contact with the blade and the frequency varied. A resonance is visually and audibly observed. If the accelerometer is not at a nodal point it will give an output which will be indicated by the analyser both on the meter and on the oscilloscope. The analyser is then tuned for a maximum reading and the exact resonant frequency noted.

5.4.8 Turbine vibration (tracking) analyser

A modified version of the previous analyser known as the TM 8817 MI was developed for use in vibration tests on rotating machinery to track several different modes of vibration of the turbine regardless of the fact that the rotor speed may be changing. Vibration signals from strain gauges attached to the blades on the rotor are first recorded on a multichannel tape recorder, (Figure 5.3). The use of a variable speed tape recorder extends the range of measurement well outside the nominal 50 Hz to 5 kHz. After recording the channels are replayed and analysed in an arrangement similar to Figure 5.4.

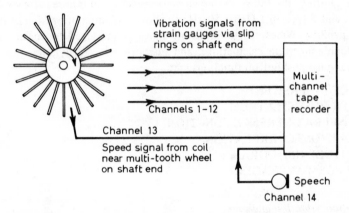

Figure 5.3 *All signals from the rotating blade are recorded during a speed run*

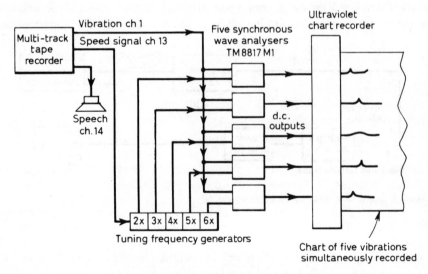

Figure 5.4 *Connection of the system for normal analysis of turbine blade vibrations during speed runs*

5.4.9 Modal plotters

Weaknesses of a structure under vibration which might be identified by transfer function plots may be analysed further by mapping the surfaces to locate the positions which would benefit most by stiffening.

The Modal Study is a logical progression from the Transfer Function Analysis. Plots are made at the various important resonances at the low impedance points disclosed by the previous step. A response pickup is moved to various locations over the structure and the responses physically marked. When the points of equal deformation or structural deflection are connected, the result is a true map, very similar to the familiar topographic terrain map as shown in Figure 5.5 . Points of bending and twisting are clearly shown.

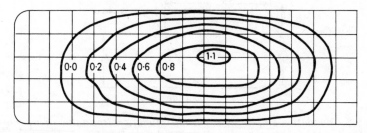

Figure 5.5 *Mode shape plot of a fabricated steel base for a turbine/compressor unit showing the actual deflection of the top plate as the structure is excited at the lowest resonant frequency*

5.4.10 Random noise analysis

Random noise is proving a useful test signal [6, 7]. The future behaviour of the signal cannot be predicted with precision yet it results from physical behaviour at the source [8]. The term 'noise' relates to signal interference and is characterised by white or pink noise mentioned in section 5.5.5.

When evaluating an unknown signal the longer the period over which it is observed, the more typical will be the average characteristics obtained. The statistical parameters of a signal usually measured define its average amplitude behaviour and average frequency characteristics.

The mean value of a signal is its average amplitude, analogous to its effective direct current level. A random signal fluctuates about its mean value. The mean-square value of a signal, which is proportional to its average power, takes into account both the DC component and the fluctuating component. These two signal parameters can be measured using standard meters, but they convey no information concerning the signal's variation throughout its range of amplitudes. The amplitude-probability-density function (p.d.f.) of a signal is the probability that its amplitude will be found within any particular very narrow band of amplitudes. For example Figure 5.6 shows the probability density function of Gaussian noise, as displayed on a statistical analyser by Garforth [9]. The probability of the signal's amplitude having a particular value, X_1 is given by p_1.

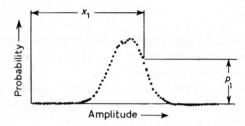

Figure 5.6 *Probability density function of Gaussian noise as displayed on a statistical analyser (the value of p_1 represents the probability that the signal will have an amplitude x_1).*

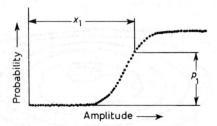

Figure 5.7 *Cumulative amplitude probability function (the value of p_1 represents the probability that the amplitude will be equal to or less than x_1)*

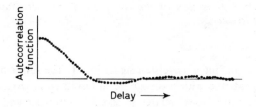

Figure 5.8 *Autocorrelation function (that this approaches zero with increasing time indicates a random signal)*

When the p.d.f. is integrated with respect to amplitude, the resulting function is the cumulative-amplitude-probability-distribution function (c.d.f.) of the signal. The c.d.f. expresses the probability that the signal's amplitude will be equal to or less than a given amplitude. For example, Figure 5.7 shows the cdf of the Gaussian noise signal presented in Figure 5.6. Thus, p_1 represents the probability that the signal will have an amplitude that is equal to or less than X_1.

The power-spectral density function describes the fraction of a signal's total power that is present at different frequencies. Direct measurement of the power spectrum of certain signals can be difficult, and it is more convenient to look at the correlation functions. Autocorrelation function measures the average similarity between a signal and a time-delayed version of itself, expressed as a function of delay.

An example of the autocorrelation function of band-limited white noise is shown in Figure 5.8. As the delay time increases, the value of the autocorrelation function tends toward zero, indicating that there is no similarity between the signal and its delayed version. An autocorrelation function that approaches a constant value is evidence that the signal is random.

The concept of correlation measurement can be extended to evaluate the similarity that exists between two signals. The cross-correlation function measures average similarity between two signals when one is delayed in time with respect to the other. Cross-correlation measurements constitute a powerful means for detecting delayed similarities between two signals, or for evaluating the dynamic response of systems. The new factor in the correlation field is the recent availability of instruments that are capable of performing simple on-line correlation measurements.

5.5 System monitors and vibration limit detectors

The use of on-line hardware or the implementation of an on-line data retrieval and centralized processing programme makes it possible to apply the instrumentation described to the condition monitoring of whole systems or single machines. These facilities range widely from sophisticated computer-controlled processing monitors down to simple inertia switches.

5.5.1 Multi-channel deviation monitor

Deviations of vibration severity from the norm can be established by the use of a deviation monitor which eliminates the need for the continuous logging of routine information and only activates the output device when the deviations exceed certain preset criteria. The schematic layout of the typical unit [10] is shown in Figure 5.9.

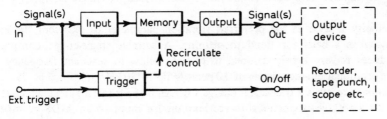

Figure 5.9 *Multi-channel deviation monitor. Block diagram*

The input unit converts the incoming signals into suitable form for the memory computer. Thus it samples the incoming signals and digitizes them into numbers which are easily stored. The number of signal channels to be monitored may range from 1 to 16. The sampling frequency needed to reproduce the signals with sufficient accuracy should be 3 or 4 times the required signal bandwidth. The accuracy required for reproduction is normally ± 1% with 8 bit resolution.

The trigger unit monitors the incoming signals and determines when a deviation from the norm has occurred. The unit may monitor only one channel or several channels simultaneously. A threshold may be set on absolute amplitude levels, rates and rise or fall, cumulative values, frequency of a signal or event and many others.

As the memory receives the new information it forgets the old, thus size determines how far back it can remember. Memory size is also governed by the signal parameters, namely number of channels, frequency response, and accuracy. A maximum memory size of 40 000 samples applies to a standard monitor. (Note: memory size = No. of channels × sampling rate × memory time). The output unit represents the signals in a suitable form to the output device. In many cases the output device will be a pen recorder: it may equally well be a paper tape punch, a digital printer, or a tape recorder.

5.5.2 Aircraft integrated data systems (AIDS)

Vibration is one of the several parameters monitored in a typical system integrated fault monitor such as the ASTROLOG installed in the BAC 1-11 aircraft operated by Americal Airlines, Tulsa, Oklahoma [11]. The data management unit receives 2 vibration signals (inlet case vertical and diffuser case horizontal) for each engine.

A deviation monitor initiates an alarm if the vibration signal from either source exceeds a prescribed value. All data is stored in a flight data recorder and at completion of a specified number of missions the data is analysed at the maintenance base to establish trends. From the information provided by these trends decisions are made concerning service condition and the need for maintance action.

5.5.3 Boeing 747 aircraft vibration monitors

Engine vibration monitoring systems (EVM) employing seismic mass, moving coil-type transducers have been employed in commercial jet engine aircraft for many years.

Velocity is used as the preferred vibration parameter to represent the engine vibration as a basis for trend monitoring. A suitable single-ended compression transducer is very lightly damped to operate below its resonant frequency such that operational characteristics of 10 pcmb/g up to 20 kHz are possible. The signal conditioner receives a transducer charge through a 165 ft (50 m) cable which it converts to a voltage-proportional-to-acceleration for input to an AIDS or other recorder.

5.5.4 'Vibraswitch'® malfunction detector

Machinery may be protected against damage by means of this acceleration-sensitive inertia device [12]. The arrangement of this detector is shown in Figure 5.10. The armature is constrained to respond to single direction motion by a flexure pivot.

When subjected to vibration perpendicular to the base, the inertia force, aided by the adjustable spring tends to force the armature away from the stop pin and the restraining force of the magnet. When the peak acceleration exceeds the set-point level the armature leaves the stop pin until it reaches the latch magnet, actuating the switch during its upward travel.

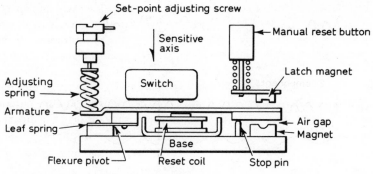

Figure 5.10 *Principle of a 'Vibraswitch'* ®

5.5.5 Vibration stethoscope

Failure prediction and preventive maintenance schemes can be assisted by means of small portable stethoscopes. A typical instrument comprises a: (a) vibration pick-up of piezo-electric type which can be electronically integrated to produce either velocity or displacement responses, with range from 0.01 to 100 g. The amplification circuit comprises high stability field effect circuits. (b) stethoset of the magnet binaural type.

The stethoscope facility increases the versatility of the instrument and enables identification of sharp cracking sounds that may occur in a bearing.

5.5.6 Vibration meter

A vibration meter provides overall vibration data in terms of acceleration, velocity, displacement or impulse (jerk) [13]. A typical vibration meter consists of an inertia-operated lead-zirconate-titanate ceramic pickup, to an adjustable attenuator, an amplifier, an indicating meter.

5.5.7 ASKANIA vibrograph

Practical waveform assessment to determine the more prominent vibratory displacements can be carried out by means of this hand-held vibrograph. By pressing the feeler of this instrument against a moving surface a mechanical linkage produces a trace of the vibration on a moving paper tape [14].

Any approximate evaluation of frequency may be based on the paper speed, provided there is no slipping. For accurate evaluation the recording length of one second is determined by means of the time marks, then the number of deflections — either the zero passages in one direction or the reversal points on one side — within this space is counted. The figure thus obtained represents the frequency in Hz. Ordinarily, the vibrations on a shorter length of tape are counted.

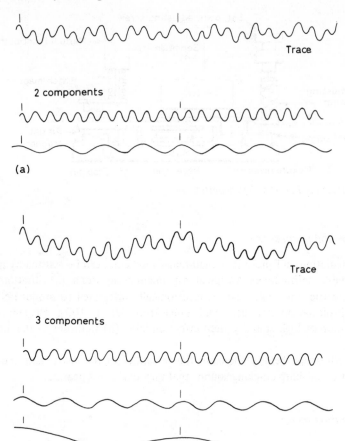

Figure 5.11 *Wave-traces and their components*

5.5.8 Waveform components estimation

ASKANIA tape records or U.V. recorder print-outs can be analysed by a manual estimation process to establish some of the larger components by graphical methods using the procedure described by Jackson [15] of Westland Aircraft Ltd.

Typical wave-traces are shown in Figure 5.11. In (a) there are two component vibrations, in (b) there are three component vibrations, wave trace of 2 components is made up of one frequency superimposed on a lower frequency, a trace with 3 components is made up in a similar manner. The trace in Figure 5.12 is made up of 3 components. As applied to a helicopter drive using the ASKANIA vibrograph, from the time base 4 cycles of 4 R (i.e. 1 rev of the main rotor) are established and this distance is used as unit time. The highest frequency is determined as a rotor order (i.e., if 11.3 cycles of this frequency occur in unit time then the rotor is

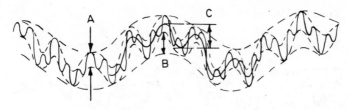

Figure 5.12 *Enveloped 3-component wave trace*

11.3 R. This rotor order is then enveloped and the double amplitude measured (i.e. the distance at A). The value of $\pm$ single amplitude is given by $\pm$ double amplitude/10.

The enveloping of the highest frequency reveals the next lowest frequency. This frequency is determined as a rotor order and the double amplitude is measured. In this case the double amplitude is the distance at C. On an actual vibrograph trace it is difficult to judge the line of this component because the trace is so small. As the high frequency envelope is parallel to the actual mean path of the lower frequency, the value of the double amplitude required, is taken as the value of the double amplitude of the high frequency envelope (i.e. the distance at B).

The process of enveloping and measuring the amplitudes of the rotor orders, continues until the lowest order is reached down to the frequency limitation of the recorder.

An example of a sequences of component estimation for the structural vibration in a helicopter is shown in Figure 5.13 (R = main rotor speed; T = tail rotor speed; E = engine speed). At 17 500 engine revs/min the rotor speed is 221.4 revs /min (3.69 Hz), with 4 rotor blades the significant forcing frequency is $4 \times 3.69 =$ 14.76 Hz (sometimes called the 4th main rotor order).

5.6 Vibration monitoring experience

Reference was made to the use of vibration monitoring systems in aircraft and experience in these techniques has developed in other industries. The rationale of monitoring vibration in process plant was reported by Downham and Woods [16] while other equipment such as compressors [17], pumps [18], diesel engines, steam and gas turbines have also been effectively maintained by the use of vibration monitoring.

5.6.1 Marine lubricating oil pump

The successful detection of a bearing failure was reported by Corben [19] in which a permanently installed vibration monitoring system on board a ship was provided with shore-based support.

The system was mounted on a main lubricating oil pump which was outwardly functioning perfectly satisfactorily. During sea-service the permanent vibration monitors indicated a failure condition and the suspect pump together with a similar

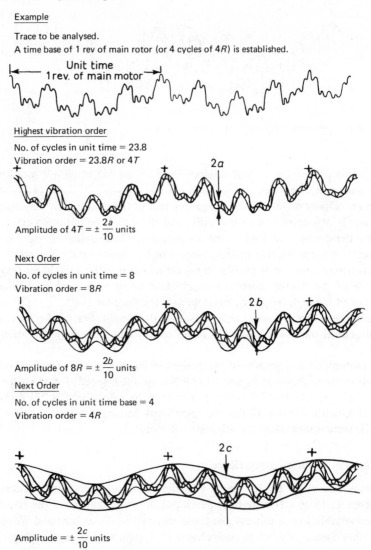

Figure 5.13 *Example of 3-component graphical resolution*

pump was accordingly tested during a routine harbour period. A one-third octave band analysis revealed:

(1) The levels were, for machine No. 1 within ± 11 dB of those recorded 2 years earlier, the levels of machine No. 2 were up to 40 dB higher than those recorded 2 years previously. The largest increases were at the high frequency end of the spectrum.

(2) Most levels measured on machine No. 2 exceeded those measured previously with the exception of the 63 Hz band, which remained constant.

It was decided that in view of these increases in vibration level, (which were indicative of a defective bearing), subject to the results of a further vibration test and a pump performance test, the motor bearings would be changed and any necessary repairs made to the pump.

Check vibration tests made 3 months later after the pump No. 2 had been operating satisfactory during sea-time showed that the vibration levels had increased over the previous values although the rate of increase was lower. It was decided to change the bearing of the pump motor.

Dismantling revealed that the drive-end bearing required immediate replacement. There was no grease left in the bearing apart from a thin, hard, dark brown layer in the non-contact areas and the outside diameter had suffered severe fretting corrosion due to creeping in its housing. On testing for freedom of movement the bearing produced a series of 'clicks' due to defects which had developed in the balls and/or raceways. The complete failure of the bearing was imminent and it was doubtful if it would have survived another patrol. When this bearing was cleaned and dismantled it was found that:

(1) The inner race was severely pitted;

(2) The outer race was showing slight corrosion;

(3) Most of the balls were heavily scored.

The non-drive bearing was in a reasonable condition; there was a quantity of grease in the end-cap although the grease in the bearing was very discoloured.

It was discovered that the probable sequence of events leading to failure was:

(1) Loss of the lubricant at a very early date when the machine overheated due to a restriction of cooling water. With the current practice of not greasing between refits, the lubricant was not replaced. (The drive-end bearing showed signs of overheating);

(2) After a long period of running without lubricant the drive-end bearing gradually deteriorated and was approaching failure point when the fault was located;

(3) Occasionally there was a temporary seizure of the bearing which caused the outer diameter to fret in the housing. (The markings on the outer diameter of the drive-end bearing were typical of severe fretting corrosion).

Lessons which can be learned from this investigation include:

(1) Bearing defects due to lubricant failure show up on the vibration monitoring system and may have a long warning period. In this instance it was many months which enabled the selection of a convenient refitting period and the avoidance of an embarrassing machine failure at sea.

(2) Vibration level increases of the order of 20 dB in the region above 3 kHz should be considered as a warning that the bearing probably requires regreasing. Ultimately the levels of the higher frequency bands increased by 30—40 dB but it would be unwise to assume that all bearings can be allowed to deteriorate to this extent because inspection showed the drive-end race to be near complete catastrophic failure.

(3) The measurement of acceleration up to frequencies of 10 kHz was justified. Had the system operated on measurements of velocity, or, not extended above 1 kHz the first indication of trouble would not have been seen until very much

later. A system operating on the total level of acceleration and extending up to 10 kHz would have detected the problem at the same time as the present installed system. Without the regular analyses, however, valuable advance information on the likely source of trouble and the degree of urgency which may be derived from the rate of increase in levels would not have been available.

5.6.2 Run-down vibration tests

Although on-load vibration measurements on a number of 500 MW turbo-alternators operated by the C.E.G.B. gave no fault indication it was known that cracks were occurring near mid-span in some of the low-pressure rotors. The Central Electricity Research Laboratories (C.E.R.L.), Leatherhead accordingly arranged recordings of vibration signatures during successive rundowns from the operating speed [20].

Rundown signatures were obtained from vibration transducers fitted on the main bearing pedestals, the signals being automatically tape recorded. Vibration analysis was carried out at the N.E. Regional Scientific Services Department, C.E.G.B., Harrogate, where the amplitude and phase of two components (harmonic and first overtone) were plotted against shaft rotational speed. A comparison between current analysed results and those from previous run-downs as is shown in Figure 5.14 shows the high increase in vibrational velocity level at about 1500 rpm which indicates the presence of a defect in the adjacent low-pressure rotor.

5.6.3 Power spectral density peak-to-peak ratio

Vibration data can be extensive and complex analysis occupy an unreasonably lengthy time. A simple procedure developed by the Northrop Corporation (Electronics Division, Hawthorne, California) plots the power spectral density output from a system. This was carried out on various helicopters with accelerometers mounted on the gear boxes and prepared for a number of 'good' systems as well as for those containing faulty parts.

In some frequency bands peak amplitudes increased when faulty parts were tested. These observations were based on a manual scan of the test data. Ratios of the peak amplitudes in the frequency bands were used to indicate the condition of a gear drive system. The PSD curve for a good gear is shown in Figure 5.15(a) where the ratio of peak amplitude in band X to that in band Y is 0.667. For a system with a faulty component as in Figure 5.15(b) the ratio is 3.2. It has been suggested that a critical threshold limit for this particular component should be 2.0 in excess of which the condition of this component would be considered to be deteriorating.

Values for threshold limits acquired for the UH-1H helicopter with a given set of failure modes are shown in Table 5.4.

Once the threshold ratios are determined a system can be used in which heterodyne filters are used to sweep through the peak frequencies of interest while peakhold networks pick off the peak amplitudes in each range, divide networks produce the ratios.

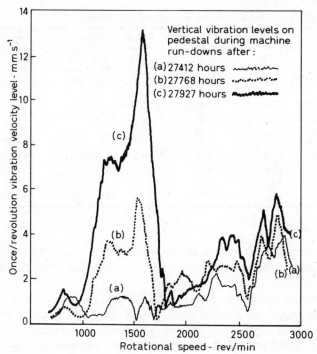

Figure 5.14 *Detection of a cracked rotor shaft from vibration signatures during 3 run-downs*

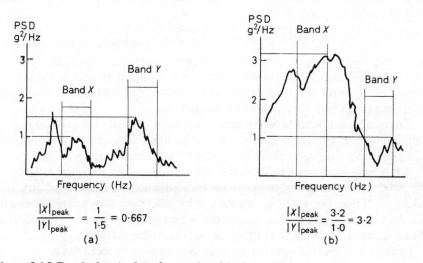

$$\frac{|X|_{peak}}{|Y|_{peak}} = \frac{1}{1\cdot5} = 0\cdot667$$

(a)

$$\frac{|X|_{peak}}{|Y|_{peak}} = \frac{3\cdot2}{1\cdot0} = 3\cdot2$$

(b)

Figure 5.15 *Test-bed ratio data for good and bad systems*

Table 5.4 *P.s.d. peak-to-peak ratios and threshold limits*

Component	Ratio	Threshold	Lower Band (Hz)	Upper Band (Hz)
Engine	B/A	1.0	(A) 605–695	(B) 755–875
	F/D	4.5	(D)1670–1890	(F) 2340–2660
	G/F	1.0	(F)2340–2660	(G)4000–4200
42° Gearbox	M/L	1.0	(L)2400–2700	(M)4600–5000
	J/K	2.0	(K)1400–1700	(J) 1300–1400
Transmission	Q/P	1.0	(P) 800–1000	(Q)3300–3600
90° Gearbox	θ/B	10.0	(B)1535–1581	(θ) 2046–2139
	n/B	4.0	(B)1535–1581	(n) 3023–3162
	θ/a	0.5	(a) 140–456	(θ) 2046–2139
	n/a	0.25	(a) 140–456	(n) 3023–3162

5.6.4 Likelihood index

A variation of the likelihood recognition technique as it is known in pattern recognition procedures can be applied to condition monitoring based on the statistical characteristics of the vibration signal.

The procedure reported by Kukel *et al.* [21] used a vibration signal ranging from 0 to 26 kHz which was divided into 128 partitions each having a bandwidth of 200 Hz. Such data was gathered both for 'good' engines and those with failures and an amplitude distribution produced for each of the 128 partitions of the analysis.

The amplitude for each of these 128 partition parameters was then found for an engine of unknown condition and the Likelihood Index then determined on the basis:

$$\text{Likelihood index} = \frac{\text{Probability of indication of a failure}}{\text{Probability of a 'good' condition}}. \quad (5.25)$$

A value of Likelihood Index greater than 1.0 shows a greater probability of failure than of a good condition.

Typical results for one parameter are shown in Figure 5.16. Any monotonic function of the two probabilities may be used, such as differences, sums, weighted ratios etc. The reference distributions are first acquired by means of a spectral analyser or bandpass filter, possibly aided in a total digital manner by the use of a computer. From this the total analysis can be automated with a bank of bandpass filters or a swept-frequency filter and logic circuitry.

Failure diagnosis is made on the basis of a value attained for the Likelihood Index with limit values set against each of the 128 partition parameters – if a number of parameters have excessive indices a diagnosis of failure can be made.

A screening procedure eliminates the effect of some of the statistical variations in the data. A number of index values for each parameter are calculated either by taking more than one set of sample data or by using more than one reference data set. An index value for a given parameter is considered to be an indicator only if all the values calculated using the various data sets exceeded the limits.

By the use of spectral windows the analysis is sensitive to speed variations.

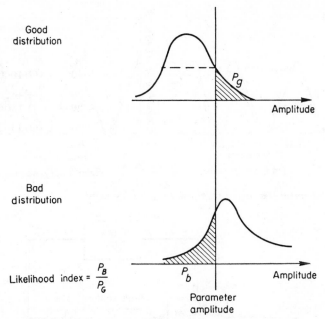

Figure 5.16 *Typical calculation of Likelihood index for one parameter*

5.6.5 Binary words frequency (FOBW)

This is a pattern recognition method developed by Houser and Drosjack [22] which uses sonic or vibration data by first defining it as an encoding algorithm with processing in a binary digital format. Thus an analogue-to-digital convertor is programmed to reduce the analog signal to a string of 1s and 0s. Polarity coding algorithms assign a 1 (positive) or 0 (negative) according to the scope of the signal — this encoding algorithm depends largely on the baseline data generated by the system. Once the coding is done the binary words must be defined.

The binary word analysis makes use of the significant frequency characteristics for purposes of identification; if for example a sinusoidal test were run with the frequency of occurrence of a given word as the amplitude the frequency response would be as shown in Figure 5.17.

Investigations show that there is a sensitivity to amplitude, i.e. if the amplitude of an incoming spectral component is large the frequency of the binary word will be large. Provided the signal to noise ratio is less than one an increase in the bit size of a word increases the amplitude sensitivity.

An analysis procedure can be implemented on the basis of these characters, binary words being chosen in order to use their frequency characteristics in a type of frequency analysis. Typical of this would be the detection of an inner race malfunction in which the binary word would be tuned to the inner race frequency, the choice of the encoding algorithm depending on base-line data.

The hardware for such an arrangement is shown in Figure 5.18 in which a

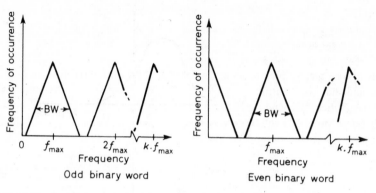

Odd binary word   Even binary word

Figure 5.17 *Typical frequency response for a binary word*

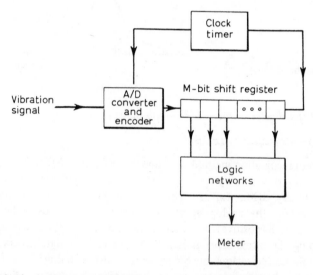

Figure 5.18 *Block diagram of FOBW analysis implementation*

programmable analog-digital converter puts the incoming data in the form of binary strings based on the sampling rate and encoding algorithm. These strings are fed into a shift register equal to the length of the words which are being detected, it is tied to the same clock mechanism as the convertor. Output from the register enters a logic circuitry which counts the frequency of occurrence of the given words. An alarm is triggered at previously determined limit levels.

Tests with the FOBW analysis procedure carried out by the General Electric Company, U.S.A. successfully discerned spalled bearings although some difficulties arose from spurious signals.

5.6.6 Gear fault diagnosis — 'manifestation pattern'

An attempt to determine the functional condition of gears was made by Loos and

Vanek [23] by producing what they termed a 'manifestation pattern' as an ideal representation of the mechanism producing the signal thereby making it possible to prepare a mathematical model.

The engagement of a pair of toothed wheels was considered, assuming both wheels had an ideally precise shape and that the wheel material was completely inelastic and that the transmission ratio was constant (a situation not completely fulfilled in practice). Depending on the inaccuracy of the tooth shape the periods of the different variations will be of varying duration – a cycle will be completed after the passage of such a number of revolutions so that the exact tooth pairing is repeated. One cycle compresses a sequence of elementary events which produces pseudostochastic signals of a statistical nature and line spectra form – and this discounts the effects of unbalance, bearings and other defects.

A typical experimental set-up is shown in Figure 5.19 with an acceleration

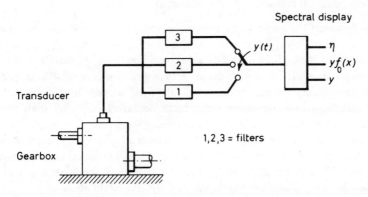

Figure 5.19 *Experimental set-up for manifestation pattern analysis*

transducer attached to the gearbox and filters split up the frequency range and the spectral display is produced by a form-factor circuit.

5.6.7 Gear deterioration – sideband analysis

As the condition of gears deteriorate increased side banding and modulation tend to occur. An analysis of this effect was conducted by Hogg [24] by means of a noise-corrected spectrum energy technique on the vibration produced by a power train. The parameters used to determine condition were:

(1) The spectral amplitudes of the gear meshing frequency;
(2) The first upper and lower sidebands of the mesh frequency;
(3) The first harmonic of the gear meshing frequency.

Amplitude levels of a given spectral frequency for a gear pair of the same mechanical condition are assumed to vary linearly with the extraneous noise levels. Using this assumption, the following normalization procedure was followed:

$$A_c(f) = C_1(1.5 - n) + A(f) \qquad (5.26)$$

where f = frequency of interest,

  $A_c(f)$ = corrected spectral amplitude db,

  C_1 = constant,

  n = noise level, dB,

  $A(f)$ = measured spectral amplitude, dB.

Once the amplitudes of the frequencies of interest for an individual gear mesh were found and the amplitudes corrected for noise, the amplitudes (in decibels) were added. This decibel amplitude was used as the discriminant. If the summed amplitudes exceeded limits set by the experiment, the gear was judged to be faulty.

5.6.8 Gear vibration sideband/signal modulation

Defects in which the mesh frequency is the carrier frequency and the shaft fre- (or one of its multiples) is the modulating frequency were shown by Shipley [25] and Provenzano *et al.* [26] to produce an infinite number of sidebands.

Amplitude modulation might occur with a gear mesh involving an eccentric gear, by periodically forcing the teeth into the mesh a cyclic loading pattern arises with a minimum and maximum mesh force occurring once per shaft rotation. Sideband amplitudes increase as eccentricity becomes larger.

Frequency modulation may result from errors in gear manufacture – with a cutting tool spacing error the tooth spacing on the gear would change as a function of rotation, i.e. a phase discrepance, hence a frequency modulation would arise. The phenomenon of sidebanding is induced by the modulation of a signal. This may be either (i) amplitude modulation and/or (ii) frequency modulation (fm). In the dynamics of geared systems either are common.

A mathematical description of sidebanding may be followed from the treatment of a sinusoidal signal represented by the equation

$$f(t) = A \cos (2\pi f_c t + \theta), \qquad (5.27)$$

where $f(t)$ = signal as a function of time,

  A = amplitude of the signal,

  f_c = carrier frequency, Hz,

  ϕ = phase angle, radians.

Amplitude modulation causes A to become a function of time such that

$$A(t) = K(1 + m \cos 2\pi f_m t), \qquad (5.28)$$

where K = constant,

  f_m = modulating frequency, Hz,

  m = fractional extent to which the amplitude is modulated.

Hence, for a typical amplitude modulated signal:

$$f(t) = K[1 + m \cos 2\pi f_m t, \cos (2\pi f_c t + \phi)]$$

$$= K \left\{ \cos (2\pi f_c t + \phi) + \frac{m}{2} \cos [2\pi(f_c - f_m) + \phi] + \frac{m}{2} \cos [2\pi(f_c + f_m) + \phi] \right\}$$

$$(5.29)$$

which shows that the spectra for a signal after modulation produces two sideband spikes as shown in Figure 5.20. This type of analysis (with trignometric expansion) for amplitude modulation may be performed for any modulating time function by

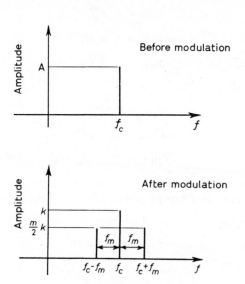

Figure 5.20 *Spectra of a typical sinusoidal signal before and after modulation*

simply expanding in a Fourier series — sidebands will occur at the carrier frequency ± modulation frequency. As a consequence of frequency modulation the phase angle ϕ becomes a function of time so that

$$\phi(t) = \beta \sin 2\pi f_m t \qquad (5.30)$$

where f_m = modulating frequency, Hz,

      β = modulation index (representing the maximum phase shift exhibited by the carrier frequency during a cycle).

Thus the equation for a frequency modulated signal will be

$$f(t) = A \cos (2\pi f_c t + \beta \sin 2\pi f_m t), \qquad (5.31)$$

which can be expanded by the use of Bessel functions.

5.6.9 Gear amplitude modulation

Sidebands due to wheel rotation effects produce full amplitude modulation about the primary (gear mesh) frequency, thus there are both upper and lower side-band frequencies.

Consider a pinion (wheel 1) rotating at frequency f_1 and engaging with a gear (wheel 2) of frequency f_2 at a mesh frequency f_c. Defects will be present in the pinion if the sidebands have spacings f_1 since the frequencies of amplitude modulated side-bands for this (wheel 1) are:

Upper side-bands

$$\text{first} = f_c + f_1$$
$$\text{second} = f_c + 2f_1$$
$$\text{third} = f_c + 3f_1 \quad \text{etc.}$$

Lower side-bands

$$\text{first} = f_c - f_1$$
$$\text{second} = f_c - 2f_1$$
$$\text{third} = f_c - 3f_1 \quad \text{etc.}$$

Defects will be present in the gear if the side-bands have spacings f_2 since the frequencies of amplitude modulated side bands for this (wheel 2) are:

Upper side-bands

$$\text{first} = f_c + f_2$$
$$\text{second} = f_c + 2f_2$$
$$\text{third} = f_c + 3f_2 \quad \text{etc.}$$

Lower side-bands

$$\text{first} = f_c - f_2$$
$$\text{second} = f_c - 2f_2$$
$$\text{third} = f_c - 3f_2.$$

5.6.10 Gear wheel eccentricity – lower side-bands

The side-effects associated with an eccentric wheel are to produce lower side bands.

Thus, for the pinion, an analysis with no upper side bands but an extensive display at $f_c - f_1$, $f_c - 2f_1$, etc. implies pinion eccentricity. Similar for the gear wheel an exclusive display of the lower side-bands $f_c - f_2$, $f_c - 2f_2$ etc. indicates gear eccentricity. In practice some upper side-banding may appear but the dominant modulation would be at the lower frequencies.

5.6.11 Gear discriminant analysis

Deviations from pure sinusoidal mesh action due to gear faults have led to the use of 3 gear discriminant ratios. Thus, these discriminants cover the effects that if a failure is present the harmonics of the mesh frequency together with sidebands of both the mesh frequency and its harmonics will monitor these effects. The three discriminant values used are:

$$\text{Discriminant (1) ratio} = \frac{\text{peak value of vibration signal}}{\text{average value}}$$

$$\text{Discriminant (2) ratio} = \frac{\text{total energy at gear mesh frequency's harmonics}}{\text{total energy at fundamental frequency}}$$

$$\text{Discriminant (3) ratio} = \frac{\text{total energy contained in sideboards}}{\text{total energy at the mesh frequency}}$$

schematically in Figure 5.21.

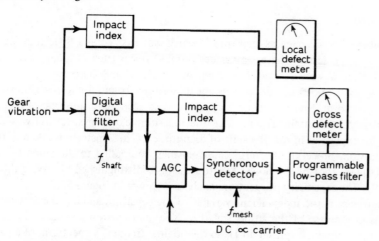

Figure 5.21 *Block diagram of gear discriminant analyser*

A digital comb filter tuned to the mesh frequency provides the discriminant (2). The characteristics of this enables all the energy at the harmonic frequency to be passed through the filter, it is accordingly possible to use a band pass filter at the meshing frequency and to use a dividing network to produce the harmonic ratio. With a normal gear, discriminant (2) is slightly less than 1.0 this ratio increases as the malfunction becomes more severe; it is generally sensitive to discrete failures on a small portion of the surface area of the gear tooth as distinguished from gross failures.

With the digital comb filter set at the shaft frequency and passing the output through an automatic gain control (AGC) to a synchronous detector set to the gear mesh frequency, a signal is produced which is proportional to the ratio of total sideband energy to that of the mesh frequency carry, discriminant (3). This device has the effect of folding the signal about the gear mesh frequency and then translates the gear mesh frequency to zero where it appears as a DC bias in the signal. This causes the modulation energy to appear at shaft frequency intervals. Gross defects are monitored by discriminant (3).

5.6.12 Cepstrum gear side-band analysis

Periodicity in a frequency spectrum can be detected by cepstrum analysis. Previous applications have been mainly in speech analysis for voice pitch determination and related questions [27, 28], periodicity in the spectrum being given by the many harmonics of the fundamental voice frequency. Cepstrum analysis detects the presence of sidebands spaced at equal intervals around one or a number of carrier frequencies. In gear box vibrations the sidebands are grouped around the tooth-meshing and its harmonics, spaced at multiples of the modulating frequencies [29, 30].

5.6.13 Cepstrum terms

A number of terms such as 'cepstrum' (which was first coined 'spectrum' by a reversal of some letters) have similar derivations. Terms such as 'quefrency' 'saphe' and 'rahmonics' were derived from 'frequency', 'phase' and 'harmonics'.

Cepstrum is normally defined as the power spectrum of the logarithm of the power spectrum. Since absolute calibration is of secondary importance (provided consistency is maintained) and since the logarithmic power spectrum would normally be expressed in dB, the unit of amplitude of the cepstrum is taken to $(dB)^2$. On occasion, however, the term cepstrum may be applied to the amplitude spectrum (square root of the power spectrum), this is distinguished by having the units dB [29]. The sensitivity of spectrum analysis is shown in Figure 5.22.

Quefrency is the independent variable of the cepstrum and has the dimension of time as in the case of the autocorrelation. The quefrency in seconds is the reciprocal of the frequency spacing in Hz in the original frequency spectrum, of a particular periodically repeating component. Just as the frequency in a normal spectrum says nothing about absolute time, but only about repeated time intervals (the periodic time), the quefrency only gives information about frequency spacings and not about absolute frequency.

5.6.14 Cepstrum determination techniques

Six methods of applying Bruel and Kjaer instrumentation to produce cepstrum data were described by Randall [29]. Until recently, digital techniques were used exclusively. Use of the same algorithm for both the original spectrum and the cepstrum is feasible with operation in real time. Other possibilities range from complete analogue set-ups to arrangements where the spectrum analysis is carried out by analogue methods with spectrum output in digital form for cepstrum discrimination.

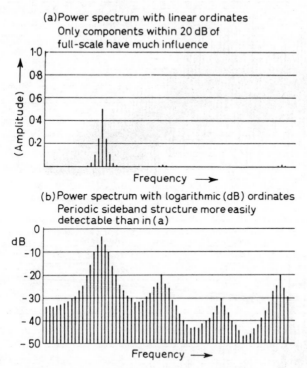

Figure 5.22 *Detection of periodicity in a spectrum using cepstrum analysis compared with the less sensitive frequency spectrum analysis*

5.6.15 Gearbox-cepstrum condition analysis

Vibration spectra and cepstra from a large slow speed gearbox driving a cement mill were obtained by Randall [29]. The signals taken 'before repair' show the result of many years' operation. At this time it was found necessary to replace a bearing, and the machine was then started up in the reverse direction, which could be expected to give the same effect as a new gearbox. Figures 5.23(a) and (b) show the spectra 'before repair' and 'after repair' respectively. They result from 4 K FFT transforms giving 2 K frequency points up to the Nyquist frequency and thus 1 K points up to the lowpass filter cutoff of the analyser.

Figure 5.24(a) and (b) show the two corresponding cepstra (note the difference in amplitude scaling). The dominant spacing in the spectrum 'after repair', at 92 Hz (Figure 5.23(b) is a major component in the cepstrum (Figure 5.24 (b) a 'ghost' component resulting from the gear cutting machine [30] and could be representative of the normal state of affairs, since it is present at about the same level in the cepstrum 'before repair'.

(a)

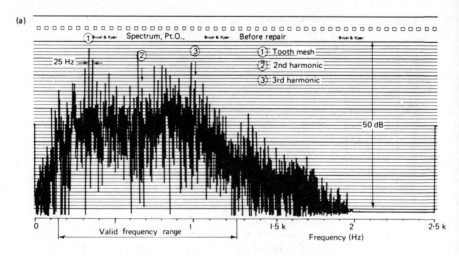

(b)

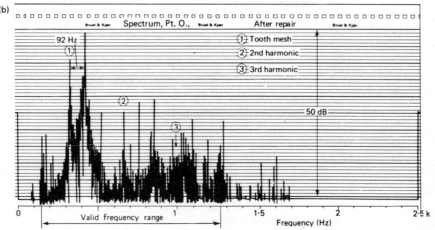

Figure 5.23 *Low-speed gearbox vibration spectra (a) before repair, (b) after repair*

The major component at 24 Hz in the cepstrum 'after repair' (Figure 5.24(b)) is the 3rd harmonic of the input shaft speed. This frequency is also known to result from a 3-times-per-revolution variation in the pinion tooth cutting process. This 3rd harmonic component is approximately $2\frac{1}{2}$ times larger in the cepstrum 'before repair' and is the dominant alteration in the spectrum.

5.7 Critical vibration levels

Threshold vibration levels in excess of which a machine can be regarded as in a bad condition have been recommended by various bodies from empirical data but are seldom based on scientific analysis and prediction. It has been stated that the failure of a machine is preceded by an increase in its vibration level in more than

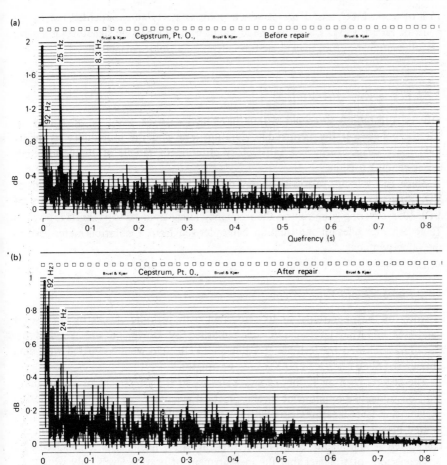

Figure 5.24 *Cepstra corresponding to Figure 5.23, (a) before repair, (b) after repair*

90% of the cases [31]. All machines vibrate regardless of how well they are de-signed and assembled, and it has been found in industrial practice that good cor-relation exists between the characteristic vibration signatures of machines and their relative condition [32].

A practical method for judging vibration severity is to establish baseline sig-natures for a machine known to be in good operating condition (this is not necess-arily a new machine as some machines 'wear in' to their normal operating levels) and to monitor changes in these signatures with time [2].

In deciding upon the significant magnitude of change in a signature component the Canadian Navy determined that an increase in vibration level is not significant unless it doubles. As important as the absolute level of change is the rate of change. The Canadian Navy has data which indicates that the mean level of a signature component as a function of time is a straight line with a slight positive slope for

75% of a machine' useful life at which point it starts an exponential rise to the point of failure [33]. Therefore, trend monitoring of vibration signatures is a more useful maintenance tool than a one-time survey of absolute magnitudes.

5.7.1 VDI severity criteria

Efforts to establish threshold criteria have tended to be based on an energy relationship between the permissible vibration level and the probability of failure [34] shear strain energy is a satisfactory criterion of fatigue of ductile materials under bending torsion.

Since rms velocity relates to energy this has been adopted in such standards as VDI 2056, October (1964) and is the basis of BS 4675:1971. A summary of VDI 2056 covering vibrations in the frequency range from 10 to 1000 Hz is given in Table 5.5, the relevant information being derived from a vibration meter used in the velocity mode.

5.7.2 Rathbone and Yates criteria

A study of the operational limits associated with periodic checks by the use of hand-held vibration level meters was published by Downham and Woods [16]. Criteria established by Rathbone, Yates and VDI 2056 were compared with data obtained at Shell Chemicals (U.K.) Ltd., Carrington.

The criteria used by Rathbone and Yates based the assessment of acceptability both on subjective opinions of several practical engineers and inspectors, followed by vibration level measurements with relatively crude instruments. Rathbone [35] limited his assessments to machinery running at speeds less than 5000 rev/min and considered it reasonable to extrapolate his curves for higher rotational speeds. He was conscious of the fact that the measurement of bearing levels only would not necessarily produce valid criteria for all machines.

Yates [36] produced his criterion by numerous tests on marine geared turbine installations. The fact that in this case the machinery was installed in relatively flexible steel shells, as opposed to the more massive foundations of Rathbone's machines, would account in part for the differences between the two criteria. For example, at 22 Hz 'too rough' on the Rathbone curve is classified as only slightly rough on the Yates curve. The line classifying an unpleasant level of vibration compares closely with the acceptable boundary levels on machines.

Values in the VDI 2056 specification for group G machines for power generation and process machinery on heavy rigid foundations were taken from Rathbone's work, although his curves were approximated to straight lines and the maximum rotational speed extended to 12 000 rev/min. Shell Chemicals (U.K.) Ltd., produced their own guidelines for acceptable vibration levels by modifying VDI, Rathbone and Yates criteria to suit their own particular machinery in the light of their own experience.

Table 5.5 *Summary of VDI 2056/1964 vibration severity criteria*

Vibration severity ranges		Examples of evaluation stages for individual groups of machines			
Range classification	Effective velocity in $mm\ s^{-1}$ rms at the range	Group K	Group M	Group G	Group T
0.28	0.28	Good			
0.45	0.45		Good		
0.71	0.71				
1.12	1.12	Usable		Good	
1.8	1.8		Usable		Good
2.8	2.8	Still acceptable		Usable	
4.5	4.5		Still acceptable		Usable
7.1	7.1			Still acceptable	
11.2	11.2				Still acceptable
18	18	Not acceptable			
28	28		Not acceptable	Not acceptable	Not acceptable
45	45				

Group K = Individual drive units of prime movers and processing machines which are rigidly fixed to the entire machine in the operating condition; particularly mass-produced electric motors up to about 15 kW.

Group M = Medium size machines, particularly electric motors from 15 to 75 kW capacity, without special foundations: also rigidly mounted drive components and machines (up to about 300 kW) with rotating parts only on special foundations.

Group G = Larger machines, prime movers or processing machines with rotating parts only, mounted on rigid or heavy foundations with a high natural frequency of vibration.

Group T = Larger prime movers and processing machines with rotating masses only, mounted on foundations with a low natural frequency of vibration, e.g. turbine groups, particularly those on foundations built on light construction principles.

5.7.3 IRD Mechananalysis severity limits

Another aid to critical vibration limit decision making is used by IRD Mechanalysis Inc. Colombus, Ohio and quoted by Nicholls [37]. The General Machinery Vibration Severity Chart (Figure 5.25) applies to machinery where vibration does not directly influence the quality of a finished product, can be used in establishing vibration tolerances. The left-hand vertical scale on the chart is graduated in mils peak-to-peak vibration displacement. The horizontal scale is graduated in cycles per minute (CPM). The diagonal lines on the chart divide the chart into regions of

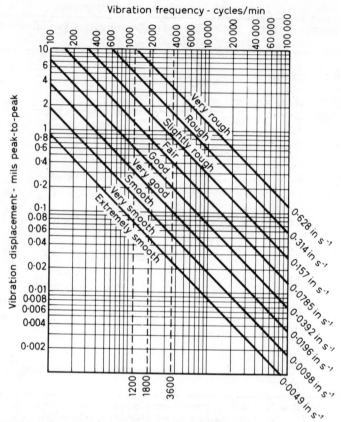

Figure 5.25 *General machinery vibration severity chart*

degrees of vibration severity. These degrees of vibration severity are based upon information and historical records of vibration readings taken on many machines. A classification of vibration severity is not a vibration tolerance. This means that the value for each machine on the list has to be selected. A reasonable rule to follow in setting the limit of vibration is one and a half to two times the normal vibration level – assuming of course, the normal level is acceptable.

5.7.4 G.M.C. Allison diesel severity limits

Allison 501 engine vibration limits were originally established by destructive test in terms of mils displacement, Figure 5.26 is a plot of these limits in terms of displacement, velocity, and acceleration. It will be noted that the limit of velocity is almost constant over a wide frequency range.

Based on such a review, the Detroit Diesel Allison Division used velocity measurements on all new development engines. The velocity signature in terms of velocity level versus 1/3 octave band frequency was used very successfully to

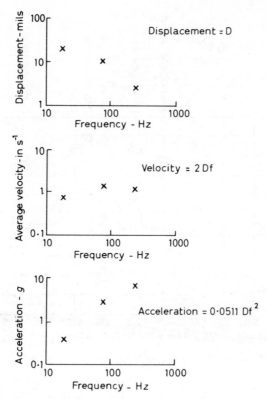

Figure 5.26 *Allison 501 vibration criteria based on destruction tests*

diagnose engine malfunction. In the majority of cases, it was not necessary to use narrow band analysis to accurately identify a malfunctioning engine component. A typical jet engine velocity signature with identified malfunction is illustrated by Figure 5.27. General Motors concluded that wider use of vibratory velocity signature could result in improved engine condition analysis and simplification of data acquisition systems. Figure 5.28 shows a general scheme for such a system which could be set up in any of several combinations of on-board, on-line or base-analysis systems.

5.7.5 Vibration limits – psychological criteria

Vibration levels which may be structurally acceptable may in fact be uncomfortable, annoying or even dangerous to persons subjected to their influences.

Vibration levels of 120 to 150 dB at certain resonant frequencies of the body structure can produce noticeable symptomatic reactions which have been comprehensively reviewed by Goldman and von Gierke [38], Guignard [39] and Parks [40]. Comfort criteria and tolerance criteria are listed in SAE J6a 'Ride and

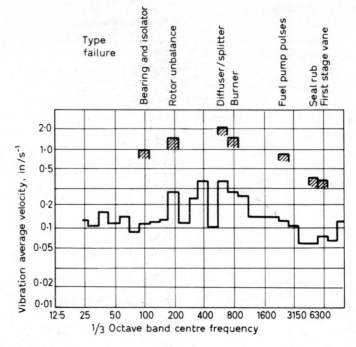

Figure 5.27 *GMC jet engine 1/3 octave limits*

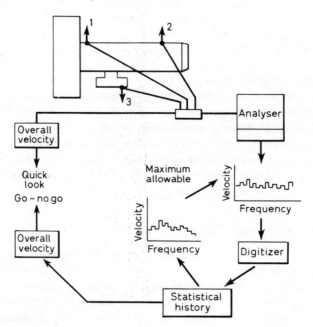

Figure 5.28 *GMC vibration limit control system*

vibration data manual' Society of Automotive Engineers (1965) in connection with which Janeway has prepared a chart giving recommended limits of vertical vibration for passenger comfort in automobiles using data obtained for vertical sinusoidal vibration at a single frequency range.

In the low-frequency range from 1 to 6 Hz the recommended limit is a fixed value of jerk. The corresponding maximum comfortable displacement at any frequency between 1 and 6 Hz is 2 divided by the frequency cubed (f^3). Over the frequency range from 6 to 20 Hz the recommended limit is a constant acceleration. The corresponding displacement is $1/3f^2$. From 20 to 60 Hz the recommended limit is a constant velocity, and the corresponding displacement is $1/60f$. In each instance, the amplitude calculated from these formulae is the maximum displacement from the static position, expressed in inches.

Resonance effects of the internal organs and their supports, and the upper torso and the shoulder-girdle structures, probably account for the marked sensitivity to vibration in the range from 4 to 10 Hz.

References

1. Weichbrodt, B. (1968), 'Mechanical signature analysis; a new tool for product assurance and early fault detection', GEC(USA) Schenectady, R & D Report 68-C-197.

2 Bowes, C.A. (1973), 'Shipboard vibration monitoring as a diagnostic/maintenance tool', *Int. Symp. Marine Engineering Society of Japan,* November.

3 ——— 'Vibration analyser type 1461 BV and 1461 BV(M)' Dawe Instruments Ltd., London W3 OSD.

4 ——— 'Sound and Vibration', Publication JN 497-393, General Radio Corp, Concord, Mass 01742.

5 Strickland W.J. (1966), 'Vibration analysis', *Marconi Instrumentation,* **10,** No. 6 December.

6 Tsien H.S. (1954), *Engineering Cybernetics,* McGraw Hill Book Co., New York.

7 Locos, H.R. (1973), 'Machine diagnosis without dismantling – a novel measuring technique', *Vibrometer News* No.5 23–33, Vibrometer Corporation, Fribourg, Switzerland.

8 Mjansikow, I. (1946), 'Sprachlaute und ihre objektive Ermittlung', *Jahrbuch L.G.U.,* **3,** (original in Russian).

9 Garforth, P.J. (1970), 'The use of correlation and statistical analysis in instrumentation', *Electronic Instrument Digest,* **6,** No. 8 7–13.

10 ——— 'Deviation monitor' Engineering System Developments, Allweather House, High Street, Edgeware, Middx.

11 Kruckenberg, H.D. (1972), 'Design and testing of the American Airlines Prototype B-747 AID system', *J. Aircraft* (U.S.A.) **9,**No. 4, April.

12 ——— 'Vibraswitch (R) malfunction detectors' S.K. Instruments Ltd, Skelmersdale WN89SB.

13 Anon. (1967), 'Sound and Vibration', No JN 497-393, General Radio Corp., Concorde, Mass. 01742.

14 ———— 'Hand vibrograph system *ASKANIA*', Publication 1606-E LM6/73 Lehmann & Michels, 2000 Hamburg 54, Germany.

15 Jackson, C.E.P. 'Diagnosis of faults by vibration analysis', Report G.604 (1966) Westland Aircraft Ltd, Yeovil.

16 Downham, E. and Woods, R. (1971), 'The rationale of monitoring vibration on rotating machinery in continuously operating process plant', *Trans. A.S.M.E.*, 71-Vibr-96, Sept.

17 ———— (1969), 'Ultrasonic vibration tests extend time between machinery overhauls', *J. Am. Acous. Soc.*, **44**.

18 Tatge, R.B. (1968), 'Acoustic techniques for machinery diagnostics', *J. Am. Acoust. Soc.*

19 Corben, F. (1972), 'Detection of defective bearings using the vibration monitoring system', Report LC/501/3909/1A, Electrical Laboratory, Admiralty Engineering Laboratory, West Drayton, Middx.

20 Beatson, C. (1975), 'Sounding out defects in power stations', *The Engineer*, 11 September, 43–45.

20a Lewis, D.L. and Walken, A.E. (1973), 'Mechanical aspects of comutation in traction motors', GEC(UK) *J. Science & Technology,* **40**.

21 Kukel, J., Ziebarth, H.K., Chang, J.D., Minnear, J.E. and Lau, J.L. (1971), 'Turbine engine diagnostic checkout system', Technical Report AFAPL-TR-71-33, Air Force Aero Propulsion Laboratory, Wright Patterson Air Force Base, Ohio, June.

22. Houser, D.R. and Drosjack, M.J. (1973), 'Vibration signal analysis techniques', Ohio State University USAAMRDL, Fort Eustis, Report TR-73-101, December.

23 Loos, M.R. and Vaněk, J. (1968), 'Kypernetische methoden für montagelose maschinendiagnostik', Stavba automolilič, regelmassige vierteljahrige Beilage der Zeitschrift Automobil, Prague III, (in Czech) (also:Vibrometer Coporation, Fribourg, Switzerland).

24 Hogg, G.W. (1972), 'Evaluation of the usefulness of using sonic data to diagnose the mechanical condition of Army helicopter power train components', USAAMRDL Technical Report 72-30, May Fort Eustis, Virginia.

25 Shipley, E. (1967), 'Gear failures', *Machine Design*, December 7.

26 Provenzano, J., Wyrostek, A., Ostheimer, A. and Yound, J. (1972), 'UH-1H Aidaps test bed program final technical report', USAAVSCOM Tech. Report 72-18, US Army Aviation Systems Command, St. Louis, Missouri, September.

27 Noll, A.M. (1967), 'Cepstrum pitch determination', *J.A.S.A.*, **41**, No. 2 293–309.

28 Luck, J.E. (1969), 'Automotive speaker verification using cepstral measurements', *J.A.S.A.*, **46**, No. 4, 1026–1032.

29 Randall, R.B. (1974), 'Cepstrum analysis and gearbox fault diagnosis', Bruel & Kjaer Application Note.

30 Mitchell, L.D. and Lynch, G.A. (1969), 'Origins of noise', *Machine Design,* May 1.

31 Lundgaard, B. (1971), 'The relationship between machinery vibration levels and machinery deterioration and failure', SNAME, Pacific Northwest Section.

32 Sankar, T.S. and Xistris, G.D. (1971), 'Failure prediction through the theory of stochastic excursions of extreme vibration amplitudes', ASME Paper 71-VIBR-60.

33 Glew, A.W. and Watson, D.C. (1968), 'Vibration analysis as a maintance tool', *Trans. I. Mar. E.* (Canadian Division) Supplement 32.

34 Monk, R. (1972), 'Vibration measurement gives early warning of mechanical faults', *Process Engineering,* November, 135–137.

35 Rathbone, T.C. (1963), 'A proposal for standard vibration limits', *Production Engineering,* **34**, March 68.

36 Yates, H.G. (1949), 'Vibration diagnosis in marine geared turbines', *Trans. N.E. Coast Inst. Eng. Shipbuilders,* **65**, 225.

37 Nicholls, C. (1970), 'Preventive maintenance using vibration analysis' Publication No. 1146, IRD Mechanalysis Inc, 6150 Huntley Road, Columbus, Ohio 43229.

38 Goldman, D.E. and Von Gerke, H.E. (1961), 'The effects of shock and vibration on man' *Am. National Standards Inst.* S3-W-39.

39 Guignard J.C. (1971), 'Human sensitivity to vibration' *J. Sound and Vibration,* **15**, No. 1 March, 11–16.

40 Parks, D.L. (1962), 'Defining human reaction to whole-body vibration', *Human Factors,* October, 147–159.

6 Sound monitoring

6.1 Introduction

Direct listening is intuitively used by experienced operators as a means of detecting malfunction in moving machinery. A mechanic may listen to an engine to determine its condition; or he might hold a screwdriver against a bearing housing to detect a malfunction in that particular bearing. Even entirely passive structures can be mechanically excited to produce a sound signature which is characteristic for its internal condition.

Such methods provide an intuitive evaluation which is more an art than a science their success currently lies in the inherent perceptiveness of the individual person. It requires a skill, often of a very high order, on the part of the listener and the result is not amenable to quantitative definition, nor is the skill easily acquired.

Quantifiable diagnostic techniques using sound signatures and data processing have been designed. That sound analysis is not used so widely as vibration analysis is probably the result of greater difficulty in the interpretation of sound signals. Certainly the instrumentation techniques are similar, the difficulties lie in the fundamental differences between sound and vibration transmission.

6.1.1 Sound

This term is discussed in relation to the monitoring of air-borne pressure waves. In the broadest meaning of the word it is a mechanical vibration normally transmitted through a fluid which, when acting upon the ear, produces a sensation on the brain known as hearing.

Relevant terms of 'noise' and 'acoustics' have other connotations in diagnostic and electronic terminology which may sometimes cause confusion. Pressure waves applied to the ear at a pleasing tonal frequency are known as sounds. Tonal sounds are musical and involve an expression of melody and harmony ('melody' = an agreeable succesion of single musical sounds) ('harmony' = a component whose frequency is an integral multiple of the fundamental frequency). Noise is a random 'sound of any sort' and is not necessarily a musical sound, it involves a combination of different frequencies with different pressure levels which may vary in an irregular manner. Acoustics is the term used to describe the science of sound and its phenomena, also the phenomena of hearing. In technological terms, the word 'acoustics' has also been extended to include solid-borne stress-wave emissions and thus embraces a much broader aspect of the phenomenon than air-borne sound.

Sound monitoring, as herein described, refers to fluid-borne pressure-wave phenomena. In general, this is simply air-borne and may not be either melodious or harmonic in the tonal sense.

6.1.2 Sound and vibration

Primarily, sound is a scalar quantity with no directional parameter. Vibration is essentially a vectorial quantity with both magnitude and direction. In practice, the influence of position and direction of the sources of sound in relation to its sensors makes nonsense of the contention that it is only scalar. In fact, positional influences are possibly the largest problem in the diagnostic monitoring of air-bourne sound.

6.1.3 Sound waves

Waves are classified according to the nature of the forces causing them, with sound waves it is necessary for the propagating media to have the properties of inertia and elasticity. Displacement of the medium creates a restoring force due to its elasticity or compressibility, inertia of the medium causes the motion to 'overshoot' the equilibrium position and thus produce a harmonic motion.

A particular sound wave motion can be considered in terms of a simple harmonic motion of particles or elements of the medium which execute identical orbits but with each element a little later than that nearest the source, i.e. out of phase. Such a longitudinal wave is shown in Figure 6.1.

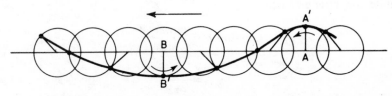

Figure 6.1 *Wave travelling from right to left, due to particles describing circular orbits anticlockwise, e.g. wave on the surface water.*

By considering the displacement characteristics of any individual particle or element while under simple harmonic motion as in Figure 6.2 it is possible to understand the relationship between displacement, density and particle velocity. There is physically a time lag for the transfer of momentum between the particles or elements (molecules) of the medium which accounts for the speed of the propagation and therefore its frequency. The speed of propagation is a physical property of the medium and in air is approximately $344\,\mathrm{m\,s^{-1}}$ ($1127\,\mathrm{ft\,s^{-1}}$) at $20°\mathrm{C}\cdot(68°\mathrm{F})$.

In a gaseous medium such as air the propagation of mechanical vibrations takes place in the form of density variations in the direction of propagation. The most common method of measuring these density variations is to measure the associated variations in pressure.

If the sound propagates in one direction only it is said to propagate as a plane, free progressive wave or shortly as a plane wave. Except for transmission losses in the medium (and dispersion) the root mean square value of the sound pressure

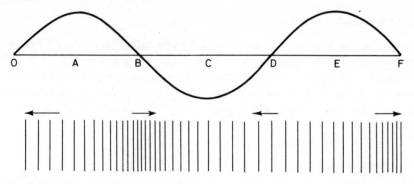

Figure 6.2 *Relation between displacement, density and particle velocity*

would then be the same everywhere along the direction of propagation. An approximation to this simple wave is obtained very far away from a sound source (a situation which is seldom met in normal circumstances).

Another type of wave which is found in practice, is a free spherical progressive wave, or just spherical wave, Such waves propagate radially away from a small 'pulsating' sphere (a 'point'-source), see Figure 6.3. For such a wave the emitted noise power is W, this spreads out from the source in the form of spheres of increasing radius r, Figure 6.4. Accordingly, if the average noise intensity (I) is defined as the noise power per unit area:

$$\text{intensity } I = \frac{\text{power } (W)}{\text{area}}$$

$$= \frac{W}{4\pi r^2}. \tag{6.1}$$

This is the inverse-square law applied to relate sound intensity to the distance from a source.

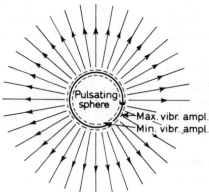

Figure 6.3 *Illustration of how sound waves propagate away from a 'pulsating' (vibrating) sphere*

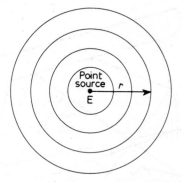

Figure 6.4 *Sketch illustrating the concept of spherical wave 'fronts' (see also Figure 6.3)*

6.1.4 Wave reflection

As soon as one or more reflecting objects are introduced into the sound wave field, there will exist at some point in the field, both the progressive source wave and the reflected wave. Source wave and reflected wave travel in roughly opposite directions. The sound pressure at a particular point in a field is the combination of the pressure at a particular point in a field is the combination of the pressure due to the original wave and the pressure(s) due to the reflected wave(s).

If the dimensions of the obstacle are large compared with the wavelength of the sound a reflected wave moves back from the obstacle, with the formation of a sound 'shadow' behind the obstacle, and a certain amount of diffraction or bending around its edges, the curvature being such that a convex surface will increase convexity, a concave surface diminish convexity, and a plane surface leaves the curvature unaltered.

6.1.5 Reverberation

Reflected waves interfere with source waves at particular points in a field. The amount of reflection also depends upon the sound absorbing properties of the object. Thus, both the physical dimensions and the sound absorption of an obstacle affects its reflecting properties.

Reverberation is the persistence of sound in an enclosed space as a result of multiple reflections after the sound source has stopped. It may also be regarded as the repeated reflections of residual sound in an enclosure and described in terms of the transient behaviour of the modes of vibration of the medium bounded by the enclosure.

It is possible to assess reverberation for a room in quantitative terms by measuring the reverberation time. This is the time taken for the mean squared sound pressure level (in a steady state) to decrease by an amount of 60 dB after the source of sound has stopped.

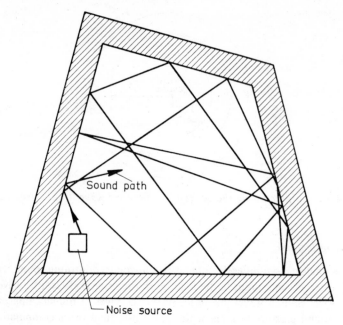

Figure 6.5 *Illustration of how sound 'diffuseness' is obtained*

Reflections of the original source sound within an enclosure produces a so-called diffuse sound field. A diffuse sound field is a field where a great number of reflected waves, from all directions, combine in such a way that the average sound energy density is uniform everywhere in the field. An approximation of this kind of field exists in large, reverberant rooms, and the way in which it is achieved is demonstrated in Figure 6.5.

6.2 Sound frequencies

Apart from the interference effects due to wave interaction the actual frequency of sounds and the wave-length of their propagations influence the acoustic pattern within an enclosure. Such frequencies are directly related to the source conditions and the associated discrete frequencies of the components of a machine in the same manner as for vibration monitoring. Additionally, interaction between frequencies produces tonal effects which are the basic distinction between noise and sound. While this may be environmentally undesirable, its effect in acoustic monitoring is to complicate the significance of the recorded frequency spectral peaks in attributing changes, whether to enclosure-reflection influences or to source deterioration.

6.2.1 Speed of sound

It can be proved that the velocity of propagation of a sound wave is given by:

$$c^2 = \frac{\mathrm{d}p}{\mathrm{d}\rho} \tag{6.2}$$

and that for a perfect gas

$$c = \sqrt{(\gamma R T)^{1/2}} \tag{6.3}$$

where c = sonic velocity

p = pressure

ρ = density

γ = ratio of specific heats, c_p/c_v

R = gas constant

T = absolute temperature.

While air is the normal medium for sound transmission it is possible in industrial monitoring conditions for other media to be involved. The speed of sound in various gases and vapours is given in Table 6.1.

Table 6.1 *Sonic velocity $(m\,s^{-1} \times 10^{-3})$ for various media*

Media	$t\,^{\circ}C$	c	Media	$t\,^{\circ}C$	c
Air	0	0.31	Methane	0	0.430
	20	0.343	Nitric oxide	10	0.324
Ammonia gas	0	0.415	Nitrogen	0	0.334
Carbon dioxide	0	0.259		20	0.351
Carbon monoxide	0	0.333	Nitrous oxide	0	0.260
Chlorine	0	0.206	Oxygen	0	0.316
Ethane	10	0.308		20	0.328
Ethylene	0	0.317	Sulphur dioxide	0	0.213
Hydrogen	0	0.128	Water vapour	0	0.101
Hydrogen chloride	0	0.296		100	0.105
Hydrogen sulphide	0	0.289			

6.2.2 Sonic velocity, frequency and wavelength

Since the wavelength (λ) and frequency (f) of a wave propagation are related to the velocity (c) by the relationship;

$$\text{velocity} = \text{wave-length} \times \text{frequency}, \tag{6.4}$$

then, considering air at 20°C for which $c = 343\,\mathrm{m\,s^{-1}}$, certain values apply (Table 6.2).

The effect of increasing the temperature of the medium is to increase the sonic velocity, thus as a temperature of 20°C, $T = 273 + 20 = 293$ while at 200°C, $T = 473$, thus the sonic velocity at 200°C will be $\sqrt{(473/273)^{1/2}} = 1.315$ times that at 20°C, i.e. 451 m s^{-1} and the frequency − wavelength relationships alter accordingly (Table 6.3).

Table 6.2 *Frequency (Hz) and wavelength (m) for air at 20°C*

Frequency	λ	Frequency	λ
20	17.15	500	0.686
50	6.86	1000	0.343
100	3.343	5000	0.0686
200	1.715	10000	0.0343

Table 6.3 *Frequency (Hz) and wavelength (m) for air at 200°C*

Frequency	λ	Frequency	λ
20	22.6	500	0.904
50	9.04	1000	0.452
100	4.52	5000	0.0904
200	2.26	10000	0.0452

6.2.3 Frequency and 'pitch' of a sound

Sound waves which act on the ear as pure sinusoidal impulses produce aurally pleasant sensations. An octave scale which is accepted musically has been termed the 'pitch' of each sinusoidal musical note and a letter assigned to each note related to its particular frequency (Table 6.4).

Table 6.4 *Frequency/pitch relationships on the musical scale*

Frequency (Hz)	Pitch
256	Middle C
288	D
320	E
342	F
384	G
426	A
480	B
512	Top C

A series of these notes A, B, C, . . . , G is known as a 'musical scale'. It should be noted that 'pitch' is a physiologically pleasing quantity whereas frequency is a physical quantity.

The fact that such a musical scale is acceptable arises from the non-linear response characteristics of the human ear. This non-linearity produces aural tones: all the summation frequencies, difference frequencies, harmonic frequencies as well as the impressed frequencies produce nerve stimulation. When the fundamental and the first few overtones are eliminated from the external tone, they are again introduced by the ear mechanism as 'subjective' tones, although with quite different intensities.

Combination tones are produced when two strong tones are sounded together and thus produce a third tone which can be distinctly heard. Apart from the summation beat tones $f_1 + f_2$, which are known as 'subjective' tones there are also tones due to sideband frequencies as previously described in connection with vibration monitoring.

The significance of the existence of combination tones in the acoustic (airborne sound) monitoring of machines was first established by the author in relation to the fatigue testing of materials [1, 2]. Experiments showed that fatigue tests based on the forcing cyclic frequency may be unreliable since the overtones and combination tones of flexure set up fatigue conditions of severity and repetition dependent more on these higher frequencies.

6.2.4 Machine sounds

The fact that sounds from machines are not in any way 'musical' arises from the differences in characteristics of the source and the influence of the path through which the source displacement and the reverberation of the enclosure on the final sound pattern at the observation point. For this reason, words commonly used to describe a sound may be as follows:

Table 6.5 *Words commonly used to describe sounds*

BANG	CLUCK	HUM	RING	SWOOSH
BARK	CLUNK	JINGLE	RIPPLING	TAP
BEEP	CRACK	JANGLE	ROAR	TATTOO
BELLOW	CRACKLE	KACHUNK	RUMBLE	TEARING
BLARE	CRASH	KNOCK	RUSHING	THROB
BLAST	CREAK	MEW	RUSTLE	THUD
BLAT	DINGDONG	MOAN	SCREAM	THUMP
BLEAT	DRIP	MOO	SCREECH	THUNDER
BONG	DRUMMING	MURMUR	SCRUNCH	TICK
BOOM	FIZZ	NEIGH	SHRIEK	TICK-TOCK
BRAY	GLUG	PATTER	SIZZLE	TINKLE
BUZZ	GNASHING	PEAL	SLAM	TOOT
CACKLE	GOBBLE	PEEP	SNAP	TRILL
CHEEP	GRATING	PING	SNARL	TWANG
CHIME	GRINDING	POP	SNORT	TWITTER
CHIRP	GROAN	POW	SPLASH	WAIL
CLACK	GROWL	POUNDING	SPLUTTER	WHEEZE
CLANG	GRUMBLE	PULSING	SQUAWK	WHINE
CLANK	GRUNT	PURR	SQUEAK	WHIR
CLAP	GURGLE	PUT-PUT	SQUEAL	WHISPER
CLATTER	HISS	RAP	SQUISH	WHISTLE
CLICK	HOOT	RAT-A-TAT	STAMP	YAP
CLINK	HOWL	RATTLE	SWISH	YELP
				ZAP

6.3 Sound Loudness Measurement

Apart from the wave and frequency characteristics of sound, the loudness, which is a function of the sound pressure may be measured subjectively in relation to frequency as a loudness level in phases or directly in terms of pressure as bels or decibels (dB).

A 'phon' is the loudness level equal to the median sound pressure level in decibels, relative to 0.002 microbar, of a free progressive wave of frequency 1000 Hz presented to listeners facing the source and which in a number of trials is judged by the listeners to be equally loud. A sone is another such subjective unit such that a simple tone of frequency 1000 Hz which is 40 dB above a listener's threshold produces a loudness of 1 sone.

A decibel is one-tenth of a bel. It is a unit of loudness level when the base of the logarithm is the tenth root of ten. The quantities involved are power. Parameters which qualify as power are: (sound pressure)^2, (particle velocity)^2, sound intensity, sound energy density, and (voltage)^2. Through the definition of the decibel, for a value of x^2, $\log_{10}(x^2)^{1/10} = 10 \log_{10}(x)^2 = 20 \log_{10}x$ which is the equation normally used for defining the decibel.

6.3.1 Decibel

A direct measurement of sound pressure in terms of microbar (μbar) is inconvenient because of the wide dynamic range involved as shown in Table 6.6 [3].

Since $1 \, \mathrm{N m^{-2}}$ is about 10-millionth of normal atmospheric pressure the sound pressures involved in monitoring ranges from perhaps $0.001 \, \mathrm{N m^{-2}}$ (14.5×10^{-8} lbf in^{-2}) to $20 \, \mathrm{N m^{-2}}$ (2900×10^{-2}). The pressure values are miniscule but the pressure range is $20\,000:1$.

For convenience, the logarithmic scale makes sound levels more easily comparable. The decibel is a term borrowed from telecommunications engineering to cover the wide range with easily manageable numbers, thus the number (N) of decibels within the pressure range p_0 to p is a measure of the sound power and thus of the sound pressure squared, hence

$$N = 10 \log_{10} \left(\frac{p}{p_0}\right)^2$$

$$= 20 \log_{10} \left(\frac{p}{p_0}\right) \tag{6.5}$$

This expresses in terms of decibels, the well-known relation that the effective (rms) value of a number of random electrical or acoustic signals is the square root of the sum of the squares of the individual voltages or sound pressures. It may be noted in passing that for the very special case of adding two voltages or sound pressures which have the identical frequency and phase, the increase in overall level is 6 dB.

Since the decibel value implies a ratio, the reference level p_0 must always be stated or at least clearly understood. For sound pressures, the internationally agreed

Table 6.6 *Typical noises and their rating*

Noise	Decibels, dB	Relative energy	Sound pressure ratio, Nm^{-2}	Typical examples
	120	1,000,000,000,000	20	Jet aircraft at 150 m (500 ft). Inside boiler-making factory. 'Pop' music horn at 5 m (16 ft).
DEAFENING	110	100,000,000,000		
	100	10,000,000,000	2	
				Inside tube train. Busy street. Workshop. Small car at 7.5 m (24 ft).
VERY LOUD	90	1,000,000,000		
	80	100,000,000	0.2	
				Noisy office. Inside small car. Large shop. Radio set full volume.
LOUD	70	10,000,000		
	60	1,000,000	0.02	
				Normal conversation at 1 m (3 ft). Urban house. Quiet office. Rural house.
MODERATE	50	100,000		
	40	10,000	0.002	
				Public library. Quite conversation. Rustle of Paper. Whisper.
FAINT	30	1,000		
	20	100	0.0002	
				Quiet church. Still night in the country. Sound-proof room. Threshold of hearing.
VERY FAINT	10	10		
	0	1	0.00002	

reference level is 20 micropascals ($20 \mu Pa = 20 \mu N\,m^{-2} = 0.0002$ dyne cm^{-2}) which is in the region of the threshold of audibility for a 1000 Hz pure tone. Sound pressures referred to this level are defined as 'sound pressure levels'. Thus:

$$\text{Sound pressure level} = 20 \log_{10} \frac{\rho}{0.00002} \text{ dB} \qquad (6.6)$$

6.3.2 Combining decibels

Sound pressure levels, expressed in decibels, cannot be added arithmetically, for example 80 dB + 80 dB does not equal 160 dB. In fact two equal sound pressure levels give an increase of 3 dB. Thus 80 dB + 80 dB gives a total sound pressure level of 83 dB.

A multiple-logarithmic equation needs to be used for the general decibel combinations, although a chart based on one developed by R. Musa and introduced by Peterson and Gross [4] simplifies the combination.

6.4 Acoustic power

This is the total sound energy radiated by a source in unit time. The power in the sound emitted from a source can range from the voice at conversational level (10^{-5} W) to a turbo-jet engine (10 kW) to that of a Saturn rocket (25 to 40 MW).

To calculate the power level (PWL) an equation is again involved. Thus, using a reference $W_0 = 10^{-12}$ watts,

$$PWL = 10 \log_{10} \frac{W}{W_0}$$

$$= 10 \log_{10} W + 120 \qquad (6.7)$$

The value for PWL is expressed in decibels (dB) since it is a power ratio. Thus a 0.02 watt source would have a power level of

$$PWL = \log_{10}(0.02) + 120,$$

$$= -17 + 120,$$

$$= 103 \, dB$$

6.5 Sound measurement

Objective sound measurement is based on the use of a microphone, amplifier and a meter or data-processing equipment.

6.5.1 Microphones

There are three common types of microphone, namely the moving-coil, piezo-electric and condenser type.

Moving-coil microphones operate when sound waves strike the diaphragm and cause the attached moving coil to vibrate axially in the field of the permanent magnet, so generating a voltage proportional to the oscillatory velocity. Leakage fields from electrical machines produce spurious signals.

Piezo-electric microphones depend on the piezo-electric properties of certain crystals and ceramics by which they generate small voltages when they are stressed mechanically.

Condenser microphones (which tend to be the most expensive) have diaphragms which are made to form one plate of a condenser. Vibration of the diaphragm by the sound varies the capacitance of the condenser and produces corresponding currents through a high resistance from a high-voltage polarizing d.c. supply.

6.5.2 Microphone selection and use

Environmental and technical factors influence the choice of the most suitable microphone; in all of these the condenser microphone shows superior performance to the piezo-electric type. For monitoring purposes where changes in dB level are used to provide trends, absolute measurement calibration is possibly less critical but stability of calibration important. The piezoelectric microphone is only slightly inferior to the condenser microphone in this respect and accordingly may be acceptable for some purposes. In view of the importance of higher reliability of the monitoring equipment than the system being surveyed and in view of the comparatively low cost of the microphone, only the best should be employed.

If several microphones are used for multi-stereophonic analysis, each microphone should at all times have identical characteristics to the others. Under such circumstances the condenser microphone has good linear frequency response but the piezo-electric microphone is more reliable in responding to phase shift.

6.6 Magnetic tape recorders.

Signals from a transducer (in this instance a microphone, but accelerometers, strain gauges and thermocouples are equally recordable in this manner) are stored as variations in the magnetic particles on the tape. The time scale then becomes a length scale on the tape. Signals recorded and stored in this manner are then available for later reproduction and analysis.

6.6.1 Tape recorder requirements

When noise is recorded on magnetic tape the recorder must be absolutely dependable and have a flat frequency response, wide dynamic range and a minimum of wow and flutter. In noise measurement, two recording techniques are in widespread use, (1) direct recording (high frequency bias) and (2) frequency modulation. When noise is to be recorded for later ordinary spectrum analysis of single samples the single track, direct recording technique is simplest; and if the stored data are to be analysed with respect to time coincidence (impulse type measurements, cross power spectrum analysis, correlation analysis etc.) i.e. in cases where phase preservation is important and/or the signal contains necessary DC information, frequency modulation is superior.

6.6.2 Tape loops

Signals made from a recorder may be selected, made into a loop and played through an analyser until the whole range has been analysed.

Some time compression or scaling can be achieved by the use of this technique. If the speed at which the tape loop is different from the original recording speed, all the signal components are translated in frequency by the speed ratio, and the repetition period of the loop is changed by the inverse of this ratio. Thus, with a one-second loop recorded at $45.4\,\mathrm{mm\,s^{-1}}$ and played back at $(1\frac{7}{8}\,\mathrm{in\,s^{-1}})$ i.e. at $8 \times$ times the speed, the loop will repeat every $\frac{1}{8}$th of a second, and a $1000\,\mathrm{Hz}$ component will become an $8000\,\mathrm{Hz}$ component. The loop when played back at its original speed repeats every second. The output signal will then have components that are spaced 1 Hz apart. The speeded-up loop will have components with 8 Hz spacing.

If the speeded-up loop is analysed in third-octave bands, the band at $8000\,\mathrm{Hz}$ is used to find the value of the components in the original signal in the $1000\,\mathrm{Hz}$ band. The band at $8000\,\mathrm{Hz}$ is 8 times as wide as the one at $1000\,\mathrm{Hz}$, and the response of its filter is correspondingly 8 times as fast, thus the signal is processed 8 times as rapidly as at the original speed.

If the signal is converted into digital form, it can be sorted in a circulating digital memory rather than in a tape loop. The speed-up that is then possible is many times greater, being 1000:1 or even more.

6.6.3 Shock (single impulse) recording

Impulse-type noises such as those produced by punch presses or drop hammers, the sonic boom from aircraft or projectiles, or the impact click from defective bearings involve a fast pressure rise and slow decay of single impulses.

Each single impulse may produce a time-domain filter response such as that shown in Figure 6.6 [5] and a wave-front distribution such as that for a point-source shown in Figure 6.7.

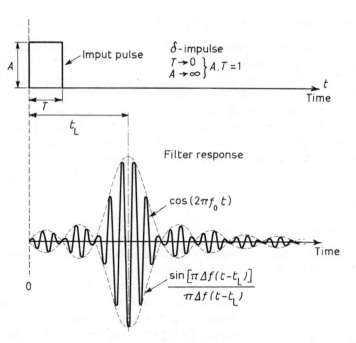

Figure 6.6 *Response of the 'ideal' filter to a unit-impulse*

Tape-loop repetion under such conditions repeats the impulses to obtain a periodic signal for line spectrum measurement. Figure 6.8 shows the result of an actual frequency analysis made from a tape loop [5]. The bandwidth of the analysing filter was 3.16 Hz, and the repetition frequency of the pulse was 4.2 Hz. To obtain a pulse repetition frequency of 4.2 Hz use was made of a tape recorder running at a speed of $1.524 \, \text{m s}^{-1}$ ($60 \, \text{in s}^{-1}$) and a tape loop of length 362 mm.

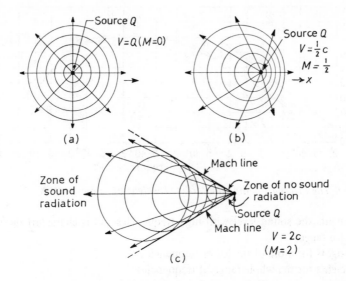

Figure 6.7 *Sound propagation from a 'point source' (a) Stationary 'point-source'. (b) Source moving at a speed equal to half that of the local speed of sound (v = ½c). (c) Source moving at super-sonic speed(v = 2c).*

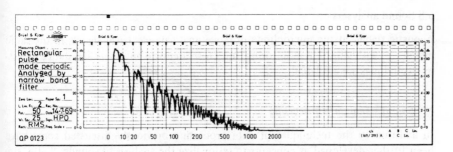

Figure 6.8 *Recording of the line spectrum of a periodically repeated pulse*

6.7 Sound level meters

Sound could be easily measured by determining its overall sound pressure level but this would neither indicate the frequency distribution nor the significance of human perception (particularly when used for noise-environment testing). Sound level meters are therefore provided with a set of frequency weighting networks termed A, B, C and D.

Figure 6.9 shows the frequency response of meters with such weightings as specified by ANSI-SI.4-1971 [63] and B.S. 4197:1967.

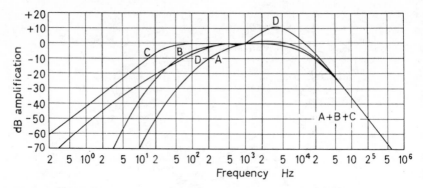

Figure 6.9 *The internationally standardized weighting curves meters*

A: weights the sound level with high frequencies and is characteristic of human voice range,
B: weights for more of the lower frequencies,
C: weights for the whole range of frequencies,
D: weights for specifically aircraft noise measurements.

Readings obtained by the use of such sound level meters should indicate the weighting. This is usually done by ascribing the units as dB(A), dB(B), dB(C), or dB(D) as the case may be.

6.8 Sound analysers

Acoustic signals at the microphone produce analogue electrical signals in the same manner as with vibration monitoring, therefore identical analysers and data processing techniques are available.

6.9 Sound signal data processing

In addition to the techniques of data processing previously discussed some have been further developed as concepts to rationalize the analysis of sounds. 'False' signals are more likely to arise in air-borne sound monitoring due to air-movement, density gradients, reverberation, absorbtion etc. which not only influence the frequency characteristics but may also affect the wave-front pattern from the sound source(s) to the microphone.

The concept of acoustic spectrum moments [7] is one such concept leading to the evaluation of such properties as:

Second moment: known as the 'variance' and a measure of the compactness of the spectrum
Third moment: related to the distribution symmetry
Fourth moment: known as 'kurtosis' and used to measure the peakiness of the spectrum distribution

Evaluations of the usefulness of the moments theory have been made by Thomas and Wilkins [8] and the use of filter networks to measure kurtosis have been proposed by Mills and Robinson [9] and Hillquist [10].

Amplitude analysis has been suggested as a more sensitive detection technique than frequency analysis, it consists of counting the number of times the amplitude of a signal exceeds a prescribed value.

6.10 Sound monitoring

Cyclic machinery presents a high potential for sound monitoring. Examples of installations which have been monitored by the analysis of their air-borne sounds were quoted by Lavoie [11] to include the following:

(1) High precision gear trains;
(2) Radar antenna bearings;
(3) Diesel engine structures;
(4) Jet engine malfunctions;
(5) Machinery in submarines;
(6) Cavitation in pumps;
(7) Wear in hydraulic valves.

Most of the data processing equipment is based solely on power spectral densities, autocorrelations or other phase-insensitive characterisations. Timing and phase relationships among the various amplitudes creating the overall acoustical input waveforms are generally lost or suppressed.

Nonlinear analog processing can be of great benefit prior to digital-type processing. For example, the key factors which depict the status of the hydraulic pump system may be carried largely in the envelope of the nongaussian signals emitted by the pump system. In this case, it would be appropriate to full-wave rectify and partially filter the signals before making a power spectral density or correlation analysis. In addition, other non-linear techniques can be employed which recognize a random train of impulses in the presence of other nongaussian noise [12].

6.10.1 Diagnostic sound data

Trend analysis to monitor a machine or system involves measurement of the sound (or other parameter) under normal conditions and an assessment of the significance of differences. An essential first step is therefore a knowledge of what is the normal sound produced by the plant or, more usually, the range of sounds which may be produced under normal running conditions.

One system developed at the U.K. Atomic Energy Authority uses the memory of a digital computer to store the information on the normal sound of plant; recognition techniques are used to detect faults. The system is shown schematically in Figure 6.10. In addition to serving as a memory for the collection of experience this computer also corrects for such things as the frequency response, the sensitivity

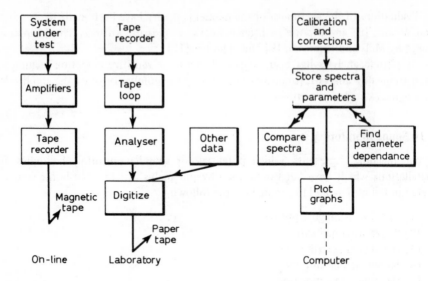

Figure 6.10 *System for fault diagnosis comparative studies*

of the detectors, the performance of amplifiers and tape recorders which may have been used and the bandwidth of the analyser. The computer output is therefore nearer to a true measure of the sound or vibration being studied than the electrical signal input. This function is of value where measurements are taken with different sets of equipment.

In this system the output of the analyser is digitized and a series of pairs of values produced to define the frequency response curve. The sampling rate is matched to the analyser sweep rate so that there is at least one, normally two, measurements digitized in each bandwidth. The data is punched on paper tape for transfer to the computer which processes the data and stores it either on magnetic tape or discs. The computer also produces graphs of amplitude or power spectral density against frequency and can produce comparative graphs of any set of data from its memory.

In addition to the logging and presentation of data the computer can be programmed to examine sets of data which should be the same, (e.g. periodic recordings from a machine) and draw the users' attention to any significant difference of a new set from the normal for that group. The criterion for significance has of course to be built into the programme.

6.10.2 Sound source identification

Assessment tests by the author [13] to determine the possibility of monitoring gross failure such as the machining action of machine-tools produced the illustrations, Figure 6.11(a)(b)(c)(d) showing the different time-domain waveforms for a bandsaw cutting materials and a hacksaw at different times before cut-off of the stock material (failure).

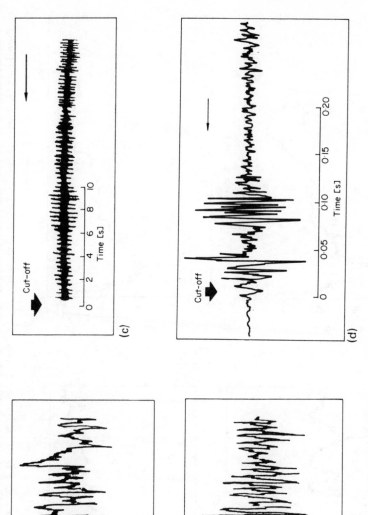

Figure 6.11 *Time-domain wave-forms for machine tools: (a) A bandsaw cutting wood emits an irregular waveform; (b) A bandsaw cutting aluminium emits a regular rising and falling waveform; (c) A slow print-out of the waveform from a hacksaw is generally regular; (d) The last split second of sawing before cut-off has a markedly different acoustic pattern*

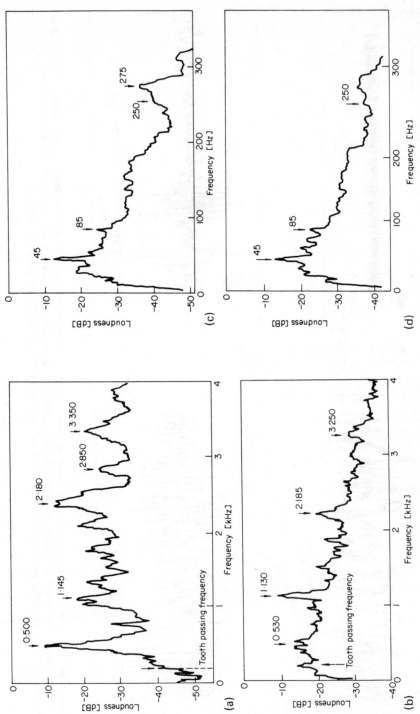

Figure 6.12 Frequency spectrum of machine tools: (a) Acoustic spectrum of a bandsaw cutting aluminium; (b) Acoustic spectrum of a bandsaw cutting wood; (c) Sound signature at start of sawing (hacksaw) (d) Sound signature at end of sawing (hacksaw)

The related sound spectra are shown in Figure 6.12(a)(b)(c)(d). The peak frequencies are the discrete frequencies of components of the machine/stock material combination and it is accordingly possible to identify sound sources by discrete frequency analysis in the same manner as with vibration monitoring.

6.10.3 Multiple-source identification

Since most machines operate in enclosures among other machines it is important to be able to identify individual machines and the components within these machines.

An investigation by the author [14] studied the sound spectra from six machine tools in a workshop when operating individually (solo) and when operating together (simultaneously). Studies are made of the respective time-domain wave-forms, decibel–frequency spectra (sound signatures) and power spectral density–frequency spectra. A comparison was also made between the resulting power spectral density spectra when all machines are working together and that for the computer-summated spectra of the machines when running individually.

Power spectral density/frequency spectra produced the greatest characterization of individual machines as shown in Figure 6.13 in which the spectra for individual machine tools are compared with that for all tools operating simultaneously.

6.10.4 Curtiss–Wright sonic analyser diagnostic system

This analyser [15] and related technology has been primarily utilized in diagnosing the mechanical condition of military engines and associated power train systems.

Inputs from a number of microphones are normalized to the level of a predetermined band of noise, characteristic of a given engine, which provides a base for component condition limits and analyzer system calibration. In addition to data input, one or more of the microphones monitor engine RPM by sending a discrete signal produced by the engine into a phase-locked filter, the output of which is fed into a frequency reference generator, thereby performing a tracking function and precise control and placement of a filter centre frequency.

When a part or component begins to wear or go through some other physical change, the character of the acoustic signal is altered. By monitoring these characteristics it is possible to detect changes in mechanical condition and to pinpoint the individual component which is deteriorating.

6.10.5 Helicopter components – rejection criteria

A Curtiss-Wright Sonic Analyser CWEA-4 was used by the U.S. Army Air Mobility Research and Development Laboratory, Fort Eustis, Virginia in conjunction with the Bell Helicopter Company and the U.S. Army Aeronautical Depot Maintenance Centre, Corpus Christi, Texas to investigate the possibility of using sound emissions

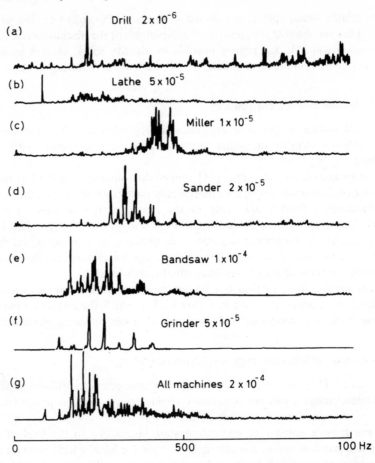

Figure 6.13 *Power spectral density spectrum, (a–f), solo machines and (g) all machines*

to determine the condition of an aircraft's mechanical components. As a result of these tests [16] changes were recommended in the rejection criteria for a number of components. In the tests, the microphones were located as follows:

Microphone No. 1 – the vertical centre line of the transmission
Microphone No. 2 – a point between the engine and the accessory drive.
Microphone No. 3 – at the gearbox.

The analyser itself receives a signal from one of the 3 microphones and passes it through a narrow band-pass filter, then compares the amplitude of the signal with a pre-determined critical amplitude. The normalized signal is then read on the condition-level meter.

Narrow-band spectrograms were produced for all recordings with a bandwidth of 10 Hz, from which a frequency had normally been selected to represent each

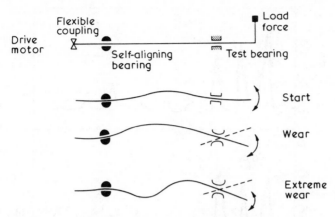

Figure 6.14 *Changes in system dynamics as wear progresses*

component under analysis. An amplitude was nominated above which the component' condition was considered to be faulty.

For bearing analysis, this frequency/amplitude limit technique proved to be satisfactory. For the acoustic monitoring of gear trains this method was less satisfactory. No particular harmonic had a particular sensitivity to faults.

6.10.7 Sound 'shrieks' monitoring

That sonic peak counting may provide a rational basis tor monitoring, under potential seizure conditions was demonstrated in the author's tests on plain bearings subject to seizure [17]. There is evidence to suggest that the flexure of a shaft in plain bearings may change somewhat like that shown in Figure 6.14 as a consequence of wear both of the bearing and the journal.

Spectral tests showed that the second harmonic $(3 \times f)$ again dominated as shown in Figure 6.15 and that the fifth harmonic $(7 \times f)$ also developed considerably during the progress of the test. Likewise, from a time-domain analysis of the behaviour near seizure the metallic bonding produced regular periodic load noises. The development of the frequency of occurrence of these shrieks is shown in Figure 6.16.

6.10.7 Sound recognition

Human observers are capable of recognizing the sounds from different engines installed in different vehicle bodies under differing conditions. Thomas and Wilkins [7, 8] set out to investigate two features of automobile recognition from their air-borne sounds:

(1) The identification of a particular engine in a variety of bodyworks and in a variety of environments, compared with other engines also in different bodyworks and environments.

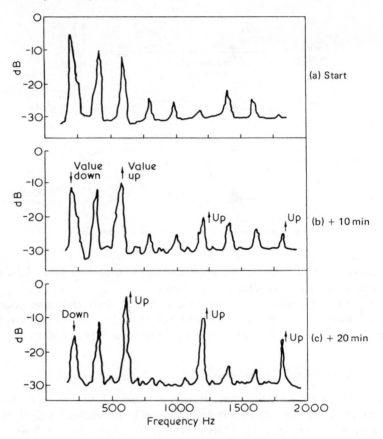

Figure 6.15 *Changes in amount of sound pressure at various frequencies as the test progresses*

(2) Distinguishing between different types of vehicle (e.g. tracked and wheeled, petrol and diesel etc.).

The factors which contribute to these noises are shown diagramatically in Figure 6.17, indicating the need for such techniques as spectrum moment analysis to select the relevant identifying sounds from the total waveform.

Although presented with some caution, Wilkins and Thomas showed in a plot of the second and third moments, Figure 6.18 that the Ford and Perkins engines could be distinguished from each other but that for each engine two of the vehicles are difficult to separate – which supported subjective information that the Ford engine has a similar sound when in different vehicles.

6.10.8 Engine firing rate cepstrum analysis

Autocorrelation analysis was shown by Wilkins and Thomas [18] to be unsatisfactory as a means of obtaining a signal periodicity from a signal obscured by noise

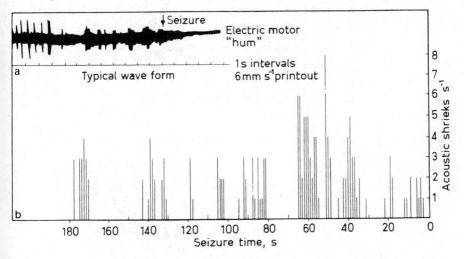

Figure 6.16 *Development of the number of shrieks during the final 180 s of a life test and including an outline of a typical wave-form print-out on which the counts were made*

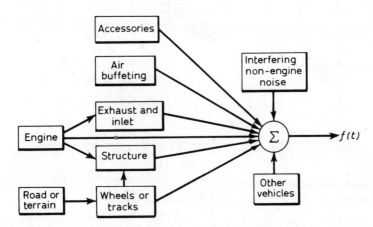

Figure 6.17 *Factors contributing to the total acoustic waveform obtained from a vehicle*

and other irrelevant components even by means of pre-filtering. Some success was reported by using the 'cepstrum' of the signal and the independent time variable, 'quefrequency'.

The cepstrum obtained from a moving Fordson tractor shown in Figure 6.19 shows a clear peak at the point corresponding to the firing point with additional peaks showing the harmonics. The cepstrum from rain falling through trees as shown in Figure 6.20 contains no firing rate indication, so that the use of cepstrum enables the presence of a vehicle to be detected even under inclement weather conditions by locating a cepstrum peak.

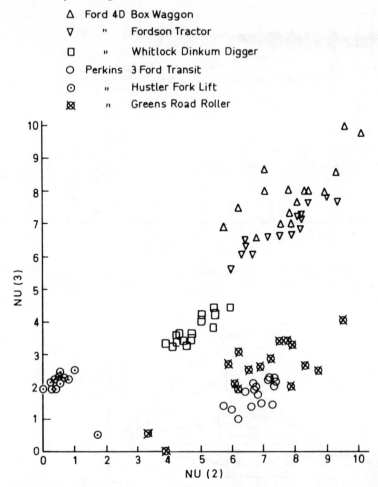

Figure 6.18 *Spectrum moments analysis, normalized variance NU(2) versus normalized symmetry NU(3) − stationary engines 'ticking over'*

6.10.9 Subjective diesel engine sound identification

To discover whether the identification of diesel engine sounds depends more on the low or the high frequency components in the sound spectra Webster, Woodhead and Carpenter [19] gave naval ratings the task of identifying the sounds of 3 lorry diesel engines at running speeds of 1000, 2000 and 3000 rev/min. The tests showed that engines types were better recognized than speeds, and that when the names of the engines were given the identification accuracy increased. The accuracy of engine identification decreased as the noise masking the engine increased.

The investigators indicated that the sounds of internal combustion engines have many of the same acoustical characteristics as two other classes of complex sounds

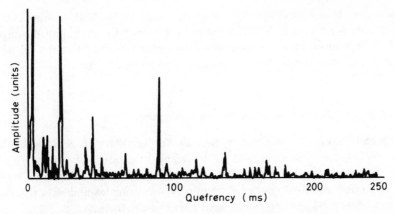

Figure 6.19 *Cepstrum obtained from a moving vehicle. The peak at a quefrency of 95 ms indicates the firing rate of 10.5 Hz*

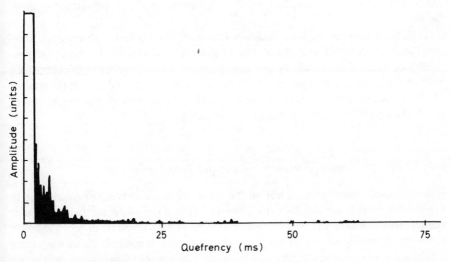

Figure 6.20 *Cepstrum obtained from very heavy rain falling through trees*

— musical instruments and vowels — in all cases the basic source is a complex tone consisting of a fundamental frequency and a series of harmonics or partials.

Engines typically have crank-shaft speeds of from 500 to 5000 rev/min (i.e. from 8.33 to 83.3 Hz). The fundamental frequency of the periodic explosions depends upon rev/min, the number of cylinders and whether the cylinders fire on a two- or four-stroke cycle. Relative to voice and musical sounds, the frequencies of the fundamentals of engines tend to be an order of magnitude lower. But for the purpose of identifying them, the acoustic correlates of engines, musical and voiced sounds are similar. In the case of engines the resonant formats tend to lie above 200 Hz and the unique harmonics are usually distinct below about 200 Hz. A third clue for their identification could be the rhythmic cycling of the engine, which is dependent upon rev/min. With increasing rev/min, resonant frequency format regions that do not change in frequency lie above 200 Hz.

6.10.10 Bissett–Berman pattern recognition programme

Cortina and Engel from the Bissett–Berman Corporation with Scott of the U.S. Army Tank-Automotive Command [20] examined the possibility of applying the signal classification techniques of the Bayes' Processor to the design of a statistically optimized automatic failure detection system using learning data from tests on a 4-cylinder spark ignition engine from a M151 army truck.

The Bissett–Berman Matrix Analysis Program is a computer implementation of a particular pattern recognition method. Its purpose is to design prediction filters that have the property of maximizing the ratio of interclass to intraclass separations and to develop compact descriptors for each fault in terms of the output of these filters.

Typical measurements are, vibration, sound, or fluid pressure waveforms. Thus measurement vector dimensions of 100 or more are conceivable implying the processing of tens of thousands of numbers per class. Bayes' Processor computations depend solely on the dimensions of the measurement vector and not on the quality of the measurements, so that by reducing the dimensions of the measurement vector the necessary computations may be reduced.

6.10.11 Bissett–Berman/Bayes processor engine defect parameters

From physical considerations, not all the measurement points are of equal value in determining which conditions exists. For example, a piston engine radiates sound energy from subsonic through ultrasonic regions (10 Hz to 50 Hz or greater) and it is not unreasonable to expect that most of this spectrum contains useful diagnostic information [21]. However, any one fault, such as a worn main bearing, will affect some parts of the spectrum significantly, others only slightly, and many parts not at all. If this is generally true for all the faults then substantial data compression can be achieved by defining each fault by those combinations of the measurement vector elements which are useful in discriminating that fault from others.

Experimental engine data was obtained prior to and subsequent to the introduction of known faults in (a) main bearings, (b) connecting rod bearings, (c) wrist pins. Typical waveforms contained an upper trace in the frequency band between 4–25 kHz, a middle trace in the band from DC to 5 KHz. A lower trace contains timing pulses when a particular cylinder was at top dead centre – with a signal attenuation by a signal from the sparking plug to distinguish the compression and exhaust strokes.

5.10.12 Bisset—Berman data analysis

Data processing was used to form measurement vector sets to be used subsequently by the Matrix Analysis Program; a Hewlett—Packard 2115A mini-computer equipped with an analog—digital converter and digital tape unit was programmed for this task, the measurement vector being based on frequency spectra.

The sets of measurement vectors were recorded on a digital tape unit with each set containing several hundred samples of the vector and identified by a code number that describes the particular engine condition used to generate the data. The input to the Matrix Analysis Programme was this data tape which contains samples of all the conditions, since the filters designed by the programme depended on the type and number of classes to be detected.

6.10.13 Bayes' processor decision technique

The technique is based on Baysian statistical procedure [22] and the elements if the Bayes Processor as applied to diagnostics is illustrated in Figure 6.21. The measurement set is denoted as a column vector, M, comprising N discrete values, m, (which may represent such measurements as average pressure drop in a manifold, peak pressures in a cylinder, or sound, vibration or pressure samples in a time or frequency domain).

This set of measurements, N is used to determine the presence (or absence) of faults F_1, F_2, F_3, . . .F_R, to recognize a fault its symptoms must be known in

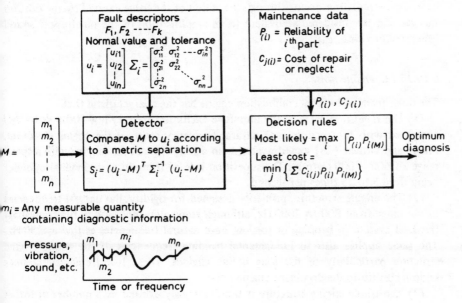

Figure 6.21 *Bayes' Processor*

terms of the measurement set. Thus, for example, in the case of a single number, such as the compression pressure, the manufacturer provides a nominal value and a tolerance. Values within the acceptable range indicate normal operation while values below the acceptable range may indicate leaky rings or defective valves. With more extensive measurements, we need the nominal values of these measurements and their expected deviation about the nominal. These descriptions of the measurements can be represented by two types of matrices, (higher order moments may be required to describe the probability distributions – an involved feat).

6.10.14 Sounds of solid particles in gas flow

During experiments to meter the solid particle mass flow rate during high speed gas-solid flows in pipes and convergent-divergent nozzles, Mobbs, Riches and Cole [23] reported that at a fixed air flow rate, changes in the solids flow rate resulted in audible variations in the level of noise issuing from the flow. The main noise source lay in the diffuser, located between the test section and the cyclone used to separate the solid particles from the flow. The cause was believed to be separation of the flow at entry to the diffuser, giving rise to a region of intense turbulence.

The basic theory of the aerodynamic generation of sound is contained in two papers by Lighthill [24, 25], who considered turbulence as a source of sound and showed that the sound field is that which would be produced by a distribution of quadrupole sources. Crighton and Ffowcs Williams, extended Lighthill's theory of the consideration of sound generated by two-phase turbulent flow. Water flow with air bubbles and suspensions of dust particles in a gas were investigated in the latter case, dipole radiation occurs through the action of the force exerted by the dust on the gas and is considered equivalent to an amplification of the quadrupole sound generated in a clean gas.

6.10.15 I.C. engine noise

The noise from an internal combustion engine has the characteristic that:

(1) The repetition rates of the impulsive excitation are very low (from $\frac{1}{20}$ to $\frac{1}{100}$) compared with the predominant natural frequency range of the engine structure. Priede and Grover [26] experimented with a four stroke cycle engine with a speed range of 600–3000 rev/min. The repetition rate of firing in an individual cylinder being from 5 to 25 times per second.

(2) The engine structure, primarily designed for rigidity, has natural frequencies in the range from 800 to 2000 Hz, although some elements such as the crankshaft flywheel system in bending or torsion, have natural frequencies as low as 250 Hz. The same applies also to fundamental bending frequencies of the whole engine structure particularly of the long in-line engines. These low frequency sources seldom constitute the dominant engine noise.

(3) The whole engine structure is simultaneously excited by a number of forces of widely different characteristics namely:

(a) Primary exciting force which is a gas force in the cylinder resulting from combustion.

(b) Secondary exciting forces of considerably different characteristics which are generated by the operation of the crank mechanism but related to primary gas force in some non-linear manner resulting in piston impacts, impacts in bearings, impacts in timing gears, etc.

(c) Forces produced in accessories (usually not related to primary gas force) — valve gear system, fuel injection system, etc.

6.10.16 Combustion induced noise

Of all sources that induced by combustion predominates. The characteristic 'knock' of the diesel combustion process was first heard and recognized in the Augsburg laboratories of M.A.N. in 1894. This harsh noise is characteristic of the diesel combustion process compared with the smoother combustion of the spark-ignition petrol engine. Some abnormal combustion processes in the spark-ignition petrol engines produce their own characteristic noises such as:

'Pinking': the noise of detonation arising from the spontaneous ignition of the air/vapour mixture;
'Rumble'; extremely fast burning due to ignition whithin the mixture by hot particles.

6.10.17 I.C. engine dynamics

The primary gas force within an engine acts through a piston/crank mechanism as shown in Figure 6.22. .

Secondary forces such as the normal force, N, and the crank torque force, T, arises from the primary gas force. These secondary forces accelerate various parts such as the piston and bearing journals across the running clearance. The major effect of the impulsive action of these secondary forces is to cause contact when parts are forced across the running clearance — and because of the mass of the moving part they also impart kinetic energy.

The acoustic significance of these forces are:

(1) The geometric configuration changes continuously with crankshaft revolution;

(2) At the top and bottom dead centre positions the connecting rod force alone applies a direct bending load to the crankshaft;

(3) On either side of dead centre a horizontal force component is additionally applied to the crankshaft;

(4) On either side of dead centre the crankshaft is also subjected to torque;

(5) With multi-cylinder engines the action of the gas forces successively changes from one piston to another and thus the load is successively applied to different parts of the engine structure;

(6) As a result of (5) the forces in each cylinder will excite essentially different

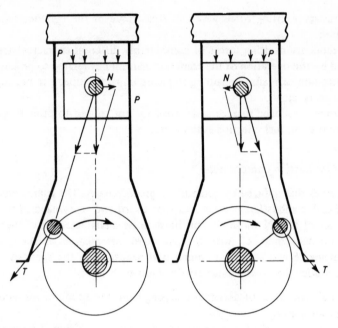

Figure 6.22 *Engine forces*

modes of engine structure vibration even if the exciting forces of every cylinder are identical;

(7) As the engine structure is not symmetrical no two cylinders will cause identical effects – Priede and Grover stated that the front end cylinder produces marked axial vibration of the crankshaft while the rear end cylinder produces pronounced flywheel tilting vibrations;

(8) The noise measurements in (7) taken at the front and rear end of a 6-cylinder diesel engine consist of individual impulses coinciding with the firing in individual cylinders of the engine;

(9) At the front end the noise is a high frequency impulse at around 1200 Hz with maximum impulses produced by the firing in the 1st, 2nd and 3rd cylinders – the minimum impulse being produced by the 6th cylinder firing;

(10) At the rear, the noise impulses are at the considerable lower frequency of 250 Hz resulting from flywheel tilting vibration while the maximum impulse results from the 6th cylinder firing.

6.10.18 Engine noise assessment

Priede assessed the relationship between the exciting forces in an engine and the emitted noise [27]. Extremes in cylinder pressure development in a diesel engine were considered, namely (a) an extremely abrupt pressure rise, (b) a smooth pressure rise.

Frequency analyses showed marked differences in gas spectra; (a) was louder by 15 to 20 dB in the frequency range between 800 and 2000 Hz. The whole frequency range being controlled by the gas force.

For the smooth pressure diagram rise the dominant force depends upon the frequency range, thus:

(1) Over the low and middle frequency range (up to 800 Hz) the gas force predominates;

(2) Above 800 Hz the major noise excitation is from the piston impacts. The effect of large piston clearances and the resulting 'piston slap' noise is not influenced by gas or piston impact forces.

The relationship between the forces and noise for various engine components is shown in Figure 6.23 where N = engine speed.

6.10.19 Timing gear noise

There appear to be three well-defined regions within an an engine in which timing gear effects can be distinguished:

(1) From 1000 to 1500 rev/min where the frequency spectrum has a slope of 20 dB per 10-fold increase of speed — typical of timing gears where the loads are applied suddenly by the meshing of the individual teeth, suddenness generally accentuated by backlash;

(2) From 1500 to 2000 rev/min frequency spectrum slope is 30 dB per 10-fold speed increase which suggests that the combustion noise is predominant;

(3) Above 2000 rev/min the tooth meshing frequency of the timing gears is sufficiently high to coincide with major frequencies of the engine structure and causes a rapid increase of noise to arise due to dynamic magnification.

6.10.20 Fuel injection equipment

Vibration of the solid structures of the pump, high pressure pipes and injectors arise primarily from hydraulic forces developed within the system. Investigations by Priede [28] show that vibrations in the pump are induced by rapid pressure rises in the pump chamber followed by a rapid drop in pressure on termination of injection. Both transients produce impulsive noise similar to that resulting from rapid pressure rise in the engine combustion chamber. The spectrum frequency has a slope of 20 dB per tenfold increase of speed. Injection pump mechanical malfunctioning can be detected by a simple overall noise measurement in dB(A).

6.10.21 Rotating auxilliary machinery noise

Additional auxilliary machinery fitted to an internal combustion engine includes a cooling fan, dynamo or alternator, pressure chargers and on two stroke cycle engines a Roots blower to produce scavenging. These machines produce noise entirely different from the impulsive characteristics of the engine itself.

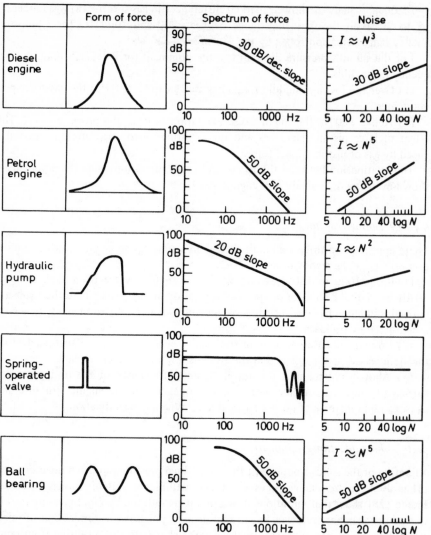

Figure 6.23 *Relation between force–time diagram, force spectrum and noise*

Cooling fan noise can be detected by narrow band frequency analysis. The fan speed generally is not a direct multiple of the engine speed and thus harmonics of the fan rotation can be readily distinguished from the engine harmonics.

Turbocharger noise generally falls in the high frequency range of the spectrum. Unlike fan noise the speed of the turbocharger is not directly related to engine speed as it is governed by exhaust gas flow velocity, therefore it cannot be identified from a single narrow band spectrum. Any progressive change of the factors affecting exhaust flow velocity such as load or speed will affect the components of turbocharger noise.

6.10.22 Exhaust and inlet noise

Exhaust noise is produced by the sudden release of gas into the exhaust system by the opening of the exhaust valve; the closing of the valve produces only a very minor effect.

Inlet noise is produced both by the opening and closing of the inlet valve. At opening, the pressure in the cylinder is generally above atmospheric pressure and a sharp positive pulse sets the air in the inlet passage into oscillation at the natural frequency of the column. This oscillation is rapidly damped by the change of volume of the system caused by the downward movement of the piston. During this period, high-frequency noise is generated by high-velocity air-flow across the valve seat. Inlet noise is markedly affected by the timing of both valves and the flow properties of the exhaust valve and exhaust system; late inlet valve timing, late exhaust valve closing, and large exhaust valve flow area tend to reduce inlet noise.

6.10.23 Comparison of noise emission from Diesel and petrol engines

Noise measurements on a large number of automotive Diesel engines (with inlet and exhaust silenced) show a striking similarity in shape of noise spectrum. All the spectra show a broad peak in the frequency range 800–2000 Hz.

The sound frequency spectrum of the petrol engine, is different. The components in the frequency range 800–2000 Hz are of lower intensity and the peaks of greatest intensity are in the frequency range 400–600 Hz. There are differences in the excitation due to cylinder pressure between petrol and Diesel engines.

6.10.24 Valve signatures

A feasibility study using mechanical signatures to detect incipient malfunctions in combustion engines was reported by Passanti [29] following an earlier study of a V-12 engine for Frankfort Arsenal.

The relationship between valve and tappet action and the mechanical signature showed that during each cycle two impacts take place:

(1) When the tappet hits the valve (valve opens);

(2) When the valve hits the seat (valve closes).

In both instances valve and tappet are in the same position and the impact speed was the same in both instances.

(1) Tappet–valve impact occures with an imposed oil film and is transmitted to the engine structure through tappet/camshaft/camshaft bearing. Oil films which separate these components damp the impact. Magnitude of the impact depends upon the quantity of interposed oil; 'splash' lubrication is inconsistent and the signature is accordingly variable.

(2) Valve-seat impact occurs between dry surfaces and is transmitted directly to the structure of the engine. The resulting vibrations are therefore strong and consistent.

With defective tappet action a high velocity on the tappet return takes place and spring resonance occurs at high engine speeds.

A bent valve stem, a misaligned valve rotator or an uneven spring force may cause the valve to act off-centre with respect to the seat. The signature of such valve when closing comprises two impacts spaced by a crank angle corresponding to the vertical displacement(s). The eccentricity varies with the position of the valve and accordingly the signature is also changeable.

6.10.25 Fault diagnosis in large diesel engines

The first attempts to synthesize a Diesel engine spectrum were probably those of Mercy [30], although the general practicability of fault diagnosis by frequency analysis may have been demonstrated earlier [31]. The work of Mercy was later extended by Williamson [32] who was apparently able to detect quite detailed changes in rail traction Diesel engine during idling.

Early attempts to apply a practical preventive maintenance technique arose with the availability of frequency spectrum analysers [33]. Tests by Bertodo and Worsfold [34], made on a 268 mm ($10\frac{1}{2}$ in) bore, 6-cylinder diesel driven alternator, showed that it was possible to detect the defects by both vibration and sound monitoring ring gear failures recorded at vibrations levels in the range 24.9–3.1 mm s^{-1} (0.98–1.5 in s^{-1} r.m.s.). Tests on a 362 mm ($14\frac{1}{4}$ inch) bore engine included 4 camshaft bearing and cambox holding down studs failures coupled with severe fretting at 47.5 mm s^{-1} (1.87 in s^{-1}) r.m.s. with near field noise levels of 123 dB. Tests on a 317 mm ($12\frac{1}{2}$ in) bore engine included fuel pump and feed-line fractures at noise levels of 118 dB and vibration levels of 31.0 mm s^{-1} (1.22 in s^{-1}) r.m.s.

These tests were inconclusive, due, it was felt, to the fact that the failures chosen for monitoring did not provide signals capable of easy recognition.

References

1   Collacott, R.A. (1974), 'Sonic fault diagnosis', British Steel Corporation, Research Fellowship Report.

2   Collacott, R.A. (1976), 'Monitoring to determine the dynamics of fatigue testing', ASTM, *Journal of Testing & Evaluation, May.*

3   Richings, W.V. 'Noise measurement techniques', Dawe Instruments Ltd, Concorde Road, Western Avenue, London, W3 OSD.

4   Peterson, A.P.G. and Gross, E.E. *Handbook of Noise Measurement,* General Radio Co, Concord, Mass, USA.

5   Broch, J.T. and Olesen, H.P. (1970), 'On the frequency analysis of mechanical shocks and single implulses', *B & K Technical Review,* 3,

6   ———— 'Specification ANSI-SI-4-1971 sound level meters', American National Standards Institute (ANSI), 1430 Broadway, New York, NY 10018.

7   Thomas, D.W. and Wilkins, B.R. (1970), 'Steps towards the automatic recognition of vehicle sounds', *Proc. Symposium I. Mech. E,* October.

8   Thomas, D.W. and Wilkins, B.E. (1972), 'The analysis of vehicle sounds for recognition', *Pattern Recognition,* 4, 379–389.

9   Mills, C.H.G. and Robinson, D.W. (1964), 'The subjective rating of motor vehicle noise', Noise-Final Report, H.M.S.O. 173.

10 Hillquist, R.K. (1967), 'Objective and subjective measurement of truck noise', *Sound & Vibration,* April 8.

11 Lavoie, F.J. (1969), 'Signature analysis; product early-warning system', *Machine Design,* January 23.

12 Bridges J.E. (1961) 'Pseudo-rectification and detection by simple non-linear resistors', *Proc. I.R.E.* February, 469–478.

13 Collacott, R.A. (1975), 'Condition monitoring by sound analysis', *Non-Destructive Testing,* October, 245–248.

14 Collacott, R.A. (1974), 'Sonic fault diagnosis', British Steel Corporation Research Fellowship Report.

15 Zabriskie, C. 'CW-2200 sonic analyser diagnostic system', Curtiss-Wright Corporation, 1 Passiac Street, Wood Ridge, New Jersey, 07075.

16 Hogg, G.W. (1972), 'Evaluation of the effectiveness of using sonic data to diagnose the mechanical condition of Army helicopter power train components', USAAMRDL Technical Report 72-30, May, Fort Eustis, Virginia.

17 Collacott, R.A. (1975), 'Sonic monitoring of plain bearings subject to seizure', *Tribology,* June.

18 Thomas, D.W. and Wilkins, B.R. (1970), 'Determination of engine firing rate from the acoustic wave-form', *Electronic Letters,* 6, 193.

19 Webster, J.C., Woodhead, M.M. and Carpenter, A. (1969), 'Identifying diesel engine sounds', *J. Sound Vib.,* 9, (2) 241–246.

20 Cortina, E., Engel, H.L. and Scott, W.K. (1970), 'Pattern recognition techniques applied to diagnostics', *SAE* Paper 700497, May.

21 Griffiths, W.J. and Skorechi, J. (1964), 'Some aspects of vibration of a single cylinder diesel engine', *J. Sound & Vibration,* 1, 345–364.

22 Anderson, T.W. (1958), *An Introduction to Multi-variate Statistical Analysis',* John Wiley & Sons Inc, New York.

23 Mobbs, F.R., Riches, D.M. and Cole, B.N. (1970), 'An investigation of noise emission from a high-speed gas-solid flow as a means of metering the solid phase', *Symp., I. Mech. E.,* 20 October.

24 Lighthill, M.J. (1952), *Proc. Roy. Soc.* (A) 211, 564.

25 Lighthill, M.J. (1954), *Proc. Roy. Soc.* (A) 222, 1.

26 Priede, T. and Grover, E.C. (1970), 'Application of acoustic diagnosis to internal combustion engines and associated machinery', *Symp. I. Mech. E.,* 20 October.

27 Priede, T. (1966), 'Some studies into origins of automotive diesel engine noise and its control', *11th International Auto. Tech. Congress,* June, Paper C12, FISITA.

28 Priede, T. (1967), 'Noise of diesel engine injection equipment', *J. Sound & Vibration,* 6.

29 Passanti, F.A. (1972), 'Mechanical signature analysis for diagnosis of reciprocating engines', General Electric Company (U.S.A.) Report 72-CRD-105, March.
30 Mercy, K.R. (1955), 'Analysis of the basic noise source in the Diesel engine', ASME Paper 55-OGP-4.
31 Bruel, P.V. (1957), 'Sound analysis in industrial processes and production', *Bruel & Kjaer Technical Review,* No. 2.
32 Williamson, K. (1967), 'Sonic analysis', *Design News Report,* 11 October.
33 Bowen, K.A. and Graham, T.S. (1967), 'Analysis of machinery noise as a technique of preventive maintenance', ASME Paper 67-VIBR-33.
34 Bertodo, R. and Worsfold, J.H. (1968/9), 'Medium speed diesel engine noise', *Proc. I. Mech. E.,* **183**, Pt. 1 129–151.

7 Discrete frequencies

7.1 Introduction

Sources of vibration or air-borne sound may be identified in the appropriate spectrum by their resonant peaks. To ascertain precisely which component or part of a system creates each peak it is necessary to evaluate the discrete (or 'natural') frequency which each component is able to produce.

The procedure for calculating such frequencies develops from the equilibrium balance of forces, couples, momenta and energy. As a technique it is an elaborate form of applied mathematics coupled with an adequate knowledge of the dynamics within the system combined with a high degree of aptitude in pure mathematics.

As a basis for health monitoring it is necessary to be able to make some relevant calculations for which reason some of the more elementary concepts are presented together with practical examples. The comprehensive list of further references should assist readers who require more detailed information.

7.1.1 Forced vibrations

A system responds to an imposed periodic force according to the mass, stiffness (or flexibility) and damping of the system. Imposed forces can be complex in nature but reduce by Fourier transformation to a number of identifiable harmonics. For this reason the inter-relationship between the discrete (or 'natural') frequency of a system and the forcing frequency may be simplified by considering the effect of a simple sinusoidally varying force.

7.1.2 Forced vibrations – simple linear system

A body of mass m in an elastic system of stiffness, s, when subjected to the influence of a sinusoidal excitation force, $F \sin \omega t$, involves the equation

$$\Sigma \text{ inertia force} + \text{stiffness} = \Sigma \text{ excitation forces}$$

i.e.
$$m(\mathrm{d}^2 x/\mathrm{d}t^2) + sx = F \sin \omega t,$$

$$\mathrm{d}^2 x/\mathrm{d}t^2 + (\omega_1^2)x = F_1 \sin \omega t \qquad (7.1)$$

where ω_1 = periodic frequency of the system rads/s = s/m,

$\omega_1/2\pi$ = system frequency, Hz,

$F_1 = F/m$.

The solution to this differential has two parts, namely the complementary function and the particular integral such that

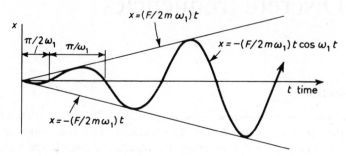

Figure 7.1 *Forced vibration waveform*

$$x = A \sin \omega t + B \cos \omega t \qquad (7.2)$$

which yields

$$A = F_1/(\omega_1^2 - \omega^2), \quad B = 0.$$

This shows that the amplitude increases with time in the manner shown in Figure 7.1 with the envelopes $x = \pm\frac{1}{2}(F/m\omega_1)t$ end the wave-form $x = -\frac{1}{2}(F/m\omega_1)t$ $\times \cos \omega_1 t$ for the critical condition of $\omega = \omega_1$ for an infinite value of amplitude, this being the resonant condition. Thus the critical condition for an undamped vibration corresponds to the natural frequency of the system.

7.1.3 Forced vibrations – phase relationships

The vibratory motion and driving force are in phase for values of $\omega < \omega_1$ i.e. at frequencies below resonance. For values of $\omega > \omega_1$ there is a phase difference of π radians (180°) between the driving force and the resultant displacement of the mass.

7.1.4 Damped forced vibrations

Assuming damping to be a means of dissipating energy which involves a specific force proportional to velocity (a simplification of the practical situation), let this damping force be given by $2K \, dx/dt$, then the force balance equation becomes:

$$\Sigma \text{ inertia force} + \text{damping force} + \text{stiffness} = \Sigma \text{ excitation forces}$$

$$m \, d^2x/dt^2 + 2K \, dx/dt + sx = F \sin \omega t,$$

$$d^2x/dt^2 + 2K/m \, dx/dt + \omega_1^2 x = F_1 \sin \omega t, \qquad (7.3)$$

where $\omega_1^2 = s/m$ (periodic angular frequency)^2,

$F_1 = F/m$.

This equation solves to give

$$x = (A^2 + B^2)^{1/2} \sin \{\omega t + \Psi\} \qquad (7.4)$$

where $A = (F/m), \phi^2(\phi^4 + 4K^2\omega^2),$

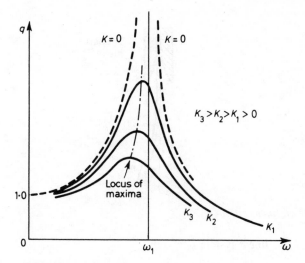

Figure 7.2 *Effect of damping on resonance*

$$B = -2K\omega A/\phi^2 = -(2K\omega)(F/m)(\phi^4 + 4K^2\omega^2),$$

$$\phi^2 = \omega_1^2 - \omega^2,$$

$$\tan \Psi = B/A = 2K\omega/(\omega^2 - \omega_1^2).$$

The effect of the coefficient K is for the resonant amplitude to have a limiting value which becomes more limited, i.e. of smaller value, as the amount of damping increases. This is shown in Figure 7.2 for progressively increasing values of $K = 0$, K_1, K_2, K_3.

At the same time, damping reduces the resonant frequency below that of the natural frequency for the system. By substituting terms for A and B in equation 7.4 and differentiating with respect to x to obtain the value for ω for maximum deflection under damped vibrations it can be shown that

$$\omega^2 = \omega_1^2 - 2K^2. \tag{7.5}$$

Most discrete frequency calculations are satisfactorily evaluated in terms of ω_1, i.e. without damping, but for components operating in viscous fluids or where they are likely to be lubricated excessively, allowance for the damping effect may be necessary.

7.1.5 Non-linear systems (one degree of freedom)

A system is regarded as being linear if the characteristic force (ϕ) due to elastic or equivalent effects is proportional to the displacement. The system is non-linear if the characteristic force is proportional to some power of the displacement. Thus, a typical solution [13] may be obtained as follows:

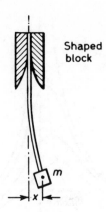

Figure 7.3 *Non-linear stiffness*

$$\text{linear system } \rho = s \cdot x \tag{7.6}$$

$$\text{linear system stiffness} = d\rho/dx = s \tag{7.7}$$

$$\text{non-linear system } \rho = s(x)^N \tag{7.8}$$

$$\text{non-linear system stiffness} = d\rho/dx = Ns(x)^{N-1}. \tag{7.9}$$

In a typical non-linear system such as that shown in Figure 7.3 the stiffness is partly due to the effective length of the cantilever, partly due to the shape of the clamps, thus

$$\text{stiffness} = d\rho/dx = s_1 + 3s_3 x^2,$$

$$\text{integrating: } \rho = s_0 + s_1 x + s_3 x^3$$

in which $s_0 = 0$, i.e.:

$$\rho = s_1 x + s_3 x^3. \tag{7.10}$$

The equation of motion for the mass thus becomes

$$m\ddot{x} + s_1 x + s_3 x^3 = 0 \tag{7.11}$$

which can be written as

$$\ddot{x} + \alpha x + \beta x^3 = 0 \tag{7.12}$$

where $\alpha = s_1/m,$

$\beta = s_3/m.$

This equation can be solved exactly by means of an elliptic integral or approximately by iteration, perturbation or Fourier series.

7.1.6 Whirling

As a form of shaft vibration this phenomenon was first discussed in a paper by Rankine [2] and interpreted by Jeffcott [3] in the form known today. The phenomenon is a vibration arising from the excitation of the natural frequencies of a shaft by a periodic source.

If a rotating shaft is disturbed from its equilibrium position the restoring forces due to shaft stiffness will be opposed by the centrifugal forces of the displaced masses. At a critical speed the two forces will be equal and proportional to the deflection of the shaft, this is a condition of instability which will cause the phenomenon of 'whirl'.

For a uniform shaft, if x is taken as an axial dimension and y the transverse,

$$EI\, d^4y/dx^4 = myw^2, \tag{7.13}$$

where m = mass/unit length of shaft, this produces the general equation

$$y = A \sin \omega x + B \cos \omega x + C \sinh \omega x + D \cosh \omega x \tag{7.14}$$

where A, B, C, D = constants in terms of E, I, m. Consequently, for short bearings equivalent in end conditions to simple supports

$$\omega = (\pi/l)^2 (EI/m)^{1/2} \tag{7.15}$$

for long bearings equivalent in end conditions to fixed ends

$$\omega = \tfrac{9}{4}(\pi/l)^2 (EI/m)^{1/2} \tag{7.16}$$

The subject of whirling is discussed further in Section 7.7.

7.2 Simple vibrations

The following frequency calculations relate to simple systems which provide relatively easy discrete frequency evaluations. The motion is assumed to be simple harmonic and therefore the acceleration of the vibrating mass is always towards the middle point of the path and is directly proportional to the displacement from that point.

7.2.1 Simple spring

Consider a simple spring of stiffness, carrying a mass, M to be displaced a distance from the equilibrium position and then released to vibrate freely Figure 7.4. When at a distance (x) from the equilibrium position the inertia force will equal the spring force (since at the equilibrium position the tension in the spring equals the gravitational force on the mass) hence

$$M\, d^2x/dt^2 + sx = 0 \tag{7.17}$$

i.e.

$$d^2x/dt^2 + \frac{s}{M}x = 0$$

$$x = a \sin(\omega t + \phi)$$

where, $\omega^2 = s/M,$

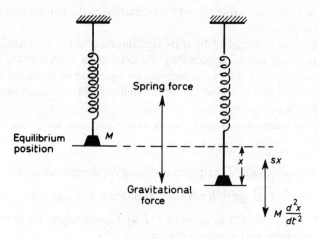

Figure 7.4 *Force equilibrium, simple spring*

$$\text{frequency } (f) = \frac{1}{2\pi}\sqrt{(s/M)}$$

$$= \frac{1}{2\pi}\sqrt{(\text{stiffness}/\text{mass})}. \tag{7.18}$$

7.2.2 Simple pendulum

The restoring force on a simple pendulum with a small angular displacement, produce the equation:

$$d^2x/dt^2 + (g/l)x = 0 \tag{7.19}$$

giving

$$\omega^2 = g/l \tag{7.20}$$

hence

$$\text{frequency } (f) = \frac{1}{2\pi}\sqrt{(g/l)}. \tag{7.21}$$

7.2.3 Compound pendulum

Treating this as an angular oscillation to allow for the motion of the constituent inertia

where: I_0 = mass moment of inertia above the centre of suspension,

I_g = mass moment of inertia above the centre of gravity,

h = distance from the centre of gravity to the pivot,

K = radius of gyration about the pivot,

the equation of motion therefore becomes

$$d^2\theta/dt^2 + (gh/K^2)\cdot\theta = 0 \tag{7.22}$$

giving
$$\omega^2 = gh/K^2$$

and from the relationship

$$I_0 = I_g + Mh^2, \quad K^2 = k^2 + h^2 \tag{7.23}$$

$$L = (k^2 + h^2)/h = \text{length of the equivalent simple pendulum}$$

$$\text{frequency } (f) = \frac{1}{2\pi}\sqrt{[gh/(k^2 + h^2)]}$$

$$= \frac{1}{2\pi}\sqrt{(g/L)} \tag{7.24}$$

7.2.4 Bifilar and trifilar suspension

By further applications of the force and couples balances in conjunction with simple harmonic motion, the following relationships can be established:

Bifilar suspension: $\quad$ frequency $(f) = [(1/4\pi)(k/d)]\sqrt{(g/l)}$ $\tag{7.25}$

Trifilar suspension: $\quad$ frequency $(f) = [(1/2\pi)(k/r)]\sqrt{(g/l)}$ $\tag{7.26}$

where $\quad k$ = radius of gyration about the centre of gravity,

$\qquad d$ = distance between the supports (Bifiliar suspension),

$\qquad l$ = length of supporting thread,

$\qquad r$ = radius of thread positions about the centre of gravity (Trifilar suspension).

7.2.5 Double pendulum

For a double pendulum as shown in Figure 7.5

$$\text{Horizontal restoring force (Lower pendulum)} = T_2 \sin\theta_2 = T_2\frac{(y-x)}{l_2}$$

hence
$$M_2\omega^2 y = -M_2 g\frac{(y-x)}{l_2} \tag{7.27}$$

Horizontal restoring force (Upper pendulum)

$$= T_1 \sin\theta_1 - T_2 \sin\theta_2 = T_1\cdot\frac{(x)}{l_1} - T_2\frac{(y-x)}{l_2}$$

hence
$$M_1\omega^2 y = -(M_1 + M_2)g\frac{x}{l_1} + M_2 g\frac{(y-x)}{l_2} \tag{7.28}$$

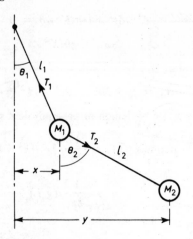

Figure 7.5 *Force equilibrium, double pendulum*

Combining equations 7.27 and 7.28 with $R = M_2/M_1$

$$\left[\omega^2 - (1 + R)\frac{g}{l_1} - R\frac{g}{l_2}\right]x + R\frac{g}{l_2}y = 0 \qquad (7.29)$$

$$\frac{g}{l_2}x + \left(\omega^2 - \frac{g}{l_2}\right)y = 0 \qquad (7.30)$$

Eliminating x and y from the simultaneous equations 7.29 and 7.30 gives the frequency equation

$$(\omega^2)^2 - (1 + R)\left(\frac{g}{l_1} + \frac{g}{l_2}\right)\omega^2 + (1 + R)\frac{g}{l_1}\frac{g}{l_2} = 0. \qquad (7.31)$$

For the simple case in which $M_1 = M_2$, $l_1 = l_2$

$$\omega^2 = \frac{g}{l}(2 \pm \sqrt{2}) \qquad (7.32)$$

hence $$\text{frequency} = \frac{1}{2\pi}[(2 \pm (2)^{1/2})g/l]^{1/2}. \qquad (7.33)$$

7.2.6 Two spring suspension system

A load supported on two springs and free to vibrate in a vertical plane will involve both linear and angular vibration as shown in Figure 7.6. Thus while the centre of gravity undergoes a linear oscillation the load will perform an angular oscillation about the centre of gravity.

Thus the equation of linear motion is

$$M\frac{d^2x}{dt^2} = -s_1(x + l_1\theta) - s_2(x - l_2\theta)$$

$$= -(s_1 + s_2)x - (s_1l_1 - s_2l_2)\theta \qquad (7.34)$$

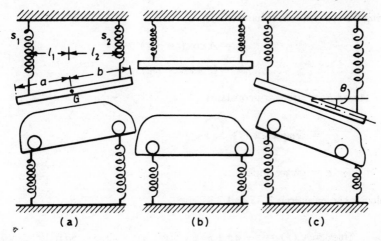

Figure 7.6 *Twin suspension of a single bar (with automobile suspension equivalent)*

which may be expressed as

$$\frac{d^2x}{dt^2} + ax + b\theta = 0 \tag{7.35}$$

The equation of angular motion is

$$I\frac{d^2\theta}{dt^2} = -s_1(x + l_1\theta)l_1 + s_2(x - l_2\theta)l_2$$

$$= -(s_1l_1 - s_2l_2)x - (s_1l_1^2 + s_2l_2^2)\theta. \tag{7.36}$$

These equations may be simplified to

$$\frac{d^2\theta}{dt^2} + dx + g\theta = 0 \tag{7.37}$$

where: $a = -\dfrac{(s_1 + s_2)}{M}$,

$b = -\dfrac{(s_1l_1 - s_2l_2)}{M}$,

$c = -\dfrac{(s_1l_1^2 + s_2l_2^2)}{M}$,

$d = \dfrac{b}{k^2}$,

$g = \dfrac{c}{k^2}$,

k = radius of gyration.

Assuming that the motion is harmonic,

$$x = A \cos (\omega t + e)$$
$$\theta = B \cos (\omega t + e)$$

(7.38)

where A and B = arbitrary constants,

$$\frac{\omega}{2\pi} = \text{frequency } (f),$$

e = phase angle.

The solutions to equations 7.35 and 7.37 give

$$\text{frequency } (f) = \frac{1}{2\pi} \{(g + a) \pm \sqrt{[(g + a)^2 - 4(ag - bd)]}\}^{1/2}. \quad (7.39)$$

7.3 Transverse vibrations of bars – approximate frequency calculations

A close approximation to the frequency of both loaded and unloaded bars can be derived from the application of their laws of deflection [4].

This method consists in equating the strain energy of the bar in its static deflection position to the kinetic energy the system would have in passing through its mean or undeflected position. The bar is assumed to vibrate throughout its length with the same frequency and with the same deflection profile – an assumption which is not totally correct but sufficient for a simple calculation [5].

7.3.1 Uniformly distributed load – general equation

$$\text{Strain energy} = \int_0^l m \frac{y}{2} \, dx \quad (7.40)$$

$$\text{Kinetic energy} = \int_0^l \frac{m}{2} (\omega y)^2 \, dx \quad (7.41)$$

where m = uniformly distributed local mass per unit length,

ω = periodic frequency, angular velocity, whereby

$$f = \frac{\omega}{2\pi}.$$

Equating 7.40 and 7.41,

$$\frac{1}{2} m \omega^2 \int_0^l y^2 \, dx = \frac{1}{2} m \int_0^l y \, dx,$$

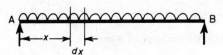

Figure 7.7 *Uniformly distributed loaded beam simply supported at ends*

$$\omega^2 = \frac{\int_0^l y \, dx}{\int_0^l y^2 \, dx}.$$

(7.42)

7.3.2 Uniformly distributed load – simply supported beam (Figure 7.7)

From the general bending equation $M = EI \, (d^2 y/dx^2)$ and evaluating the constants of integration in terms of known values for y and dy/dx the static deflection curve becomes

$$y = \frac{m}{24EI}(x^4 - 2lx^3 + l^3 x)$$

(7.43)

hence

$$\omega^2 = \frac{3024}{31} \frac{EI}{ml^4}$$

(7.44)

$$f = \frac{\omega}{2\pi} = \frac{1.572}{l^2} \sqrt{\left(\frac{EI}{m}\right)} \, (Hz).$$

(7.45)

7.3.3 Uniformly distributed load – bar fixed at both ends

The static deflection equation which can be derived for this condition is

$$y = \frac{m}{24EI} x^2 (l - x)^2,$$

(7.46)

hence

$$\omega^2 = 504 \frac{EI}{ml^4}$$

(7.47)

$$f = \frac{\omega}{2\pi} = \frac{3.57}{l^2} \sqrt{\left(\frac{EI}{m}\right)} \, (Hz).$$

(7.48)

7.3.4 Concentrated load – general case

Neglecting the effect of the mass of the bar,

$$\text{strain energy} = \sum M \frac{y}{2},$$

(7.49)

$$\text{kinetic energy} = \sum \frac{M}{2}(\omega y)^2$$

(7.50)

where M = applied mass,

   y = deflection of the applied mass.

Equating 7.49 and 7.50

$$\frac{\omega^2}{2} \sum M(y)^2 = \frac{1}{2} \sum M(y)$$

$$\omega^2 = \frac{\sum W(y)}{\sum W(y)^2} \qquad (7.51)$$

7.3.5 Dunkerley's formula

An empirical formula devised by Dunkerley [6] to calculate the transverse frequency of a complexly loaded beam as in Figure 7.8 consists of splitting the beam into a number of simple sections for which the fundamental frequency can be easily calculated. These are known as 'partial' frequencies. The sum of the squares of the reciprocals of these partial frequencies is equal to the square of the reciprocal of the fundamental frequency for the whole system, thus

$$\left(\frac{1}{f}\right)^2 = \left(\frac{1}{f_1}\right)^2 + \left(\frac{1}{f_2}\right)^2 + \left(\frac{1}{f_3}\right)^2 + \dots \text{etc} \qquad (7.52)$$

where f = frequency of whole system,

   f_1 = partial frequency, as for case (a) Figure 7.9,

   f_2 = partial frequency, as for case (b) Figure 7.9,

   f_3 = partial frequency, as for case (c) Figure 7.9.

In practice, the partial frequencies appear in the signature spectrum for a complex loaded system.

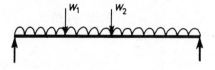

Figure 7.8 *Beam with complex loadings*

7.3.6 Loaded rod – axial vibrations

A rod which carries a mass M either at its free end as in Figure 7.10 and which is so large that the mass of the rod itself can be ignored produces a vibration which can be calculated on the bases of the general equation

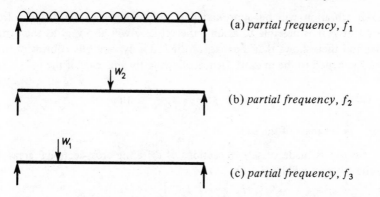

(a) *partial frequency,* f_1

(b) *partial frequency,* f_2

(c) *partial frequency,* f_3

Figure 7.9 *Partial frequencies of the beam with complex loadings (Figure 7.8)*

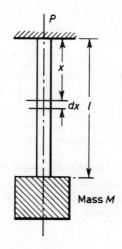

Figure 7.10 *Influence of mass of support*

$$f = \frac{1}{2\pi} \sqrt{\left(\frac{\text{stiffness}}{\text{mass}}\right)}$$

$$= \frac{1}{2\pi} \sqrt{\left(\frac{AE}{lM}\right)} \tag{7.53}$$

where A = cross-sectional area,

        E = Youngs modulus,

        l = length of rod,

        M = mass.

If the mass of the rod is small but needs to be allowed for then since the amplitude of axial vibration will be approximately proportional to the distance from the fixation at P it is possible to assume the velocity will also vary as the distance x; integration then shows that the mass of the rod is dynamically equivalent to $\frac{1}{3}$ mass of the rod added to the mass M. Hence, allowing for the mass of the rod

$$f = \frac{1}{2\pi} \sqrt{\left(\frac{AE}{l(M + \frac{1}{3}m)}\right)} \tag{7.54}$$

where m = mass of the rod.

If the rod is made of several sections of differing cross-sectional areas A_1, A_2, then the stiffness will be given by

$$s\left[\left(\frac{l_1}{A_1 E}\right) + \left(\frac{l_2}{A_2 E}\right)\right] \ldots \text{etc} = 1$$

i.e.

$$s = \frac{1}{\sum\left(\frac{l}{AE}\right)} \tag{7.55}$$

hence

$$f = \frac{1}{2\pi} \sqrt{\frac{1}{\sum\left(\frac{l}{AE}\right) \cdot M}} \tag{7.56}$$

7.4 More precise evaluations – overtones

Reliance on the concepts of simple harmonic motion simply provides a good approximation to the fundamental frequency. For many purposes this may be adequate although overtones have a significance in monitoring and therefore more precise evaluations may be required. For a detailed treatment see [7, 8].

7.4.1 Precise evaluation – transverse vibration of a beam

Consider the free undamped vibration of beam by relating the forces and moments for the equilibrium of an element as shown in Figure 7.11.

$$\text{Shear force on element, } F = -\partial M/\partial x$$

$$\text{Transverse force on element} = (\partial F/\partial x)\delta x$$

$$= -(\partial^2 M/\partial x^2)\delta x \tag{7.57}$$

$$\text{Restoring moment due to bending} = E(Ak^2)\partial^2 y/\partial x^2 \tag{7.58}$$

Substituting in 7.57,

$$\text{Restoring force on element} = -E(Ak^2)\partial^4 y/\partial x^4 \cdot \delta x \tag{7.59}$$

$$\text{Inertia force on element} = (\rho A \delta x)\partial^2 y/\partial t^2 \tag{7.60}$$

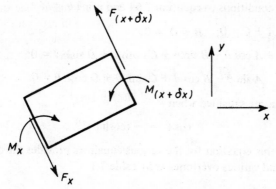

Figure 7.11 *Force and moment distribution on a vibrating element*

For equilibrium, equation equations 7.59 and 7.60

$$(\rho A \delta x)\, \partial^2 y/\partial t^2 = -E(Ak^2)\partial^4 y/\partial x^4 \cdot \delta x$$

hence

$$\partial^4 y/\partial x^4 + \frac{\rho}{Ek^2}\partial^2 y/\partial t^2 = 0$$

or

$$\partial^4 y/\partial x^4 + \frac{1}{C^2 k^2}\, \partial^2 y/\partial t^2 = 0. \tag{7.61}$$

Assuming that y varies with time such that

$$y = y_1 \sin(\omega t) \tag{7.62}$$

i.e. $dy/dt = y_1 \omega \cos(\omega t); \quad d^2 y/dt^2 = -y_1 \omega^2 \sin(\omega t) = -\omega^2 y$

substituting in equation 7.61

$$d^4 y/dx^4 - (\omega^2/c^2 k^2)y = 0. \tag{7.63}$$

The solution to this, put into a form containing four arbitrary constants A, B. C, D, is

$$y = A \cos(bx) + B \sin(bx) + C \cosh(bx) + D \sinh(bx) \tag{7.64}$$

which represents the general vibration contour equation.

7.4.2 Cantilever beam

The four arbitrary constants in equation 7.64 can be evaluated by reference to the end conditions,

when $x = 0 \quad y = 0 \quad dy/dx = 0$

$\quad x = 1 \quad M = 0 \quad$ (i.e. $d^2 y/dx^2 = 0$)

$\quad\quad\quad\quad F = 0 \quad\quad$ (i.e. $dM/dx = 0; d^3 y/dx^3 = 0$)

applying these conditions to equation 7.64 and using $\theta = bl$

when $x = 0$ $A + C = 0$; $B + D = 0$;

when $x = 1$ $-A \cos \theta - B \sin \theta + C \cosh \theta + D \sinh \theta = 0$;

$$A \sin \theta - B \cos \theta + C \sinh \theta + D \cosh \theta = 0.$$

These equations are satisified when

$$\cos \theta = -(\cosh \theta)^{-1}. \tag{7.65}$$

Evaluation of this equation for the various equalities provides the condition for the fundamental and various overtones as in Table 7.1.

Table 7.1

Frequency mode	rads	deg.
Fundamental (f)	1.87510406	107.4355487
2 × (f)	4.69409113	268.9516103
3 × (f)	7.85475743	450.0444498
4 × (f)	10.99554073	629.9980772
5 × (f)	14.13716839	810.0000830
6 × (f)	17.27875953	989.9999962

Thus the overtones are not in phase but have angular displacements.

By using the appropriate values of θ and further summation of the individual overtone forms to produce a cumulative profile identical to the original static contour it is possible to evaluate the significance and form of the individual vibration forms of the various contours.

7.4.3 Overtone frequencies

The value of θ is significant in containing terms relevant to the frequency of each mode, thus

$$\theta = bl = (\omega/ck)^{1/2}l$$

$$\text{frequency} = (\theta)^2 ck/2\pi l^2$$

$$= (\theta)^2 E^{1/2}k/2\pi l^2 \rho^{1/2} \tag{7.60}$$

where A = cross-sectional area of beam, assumed constant,

k = second moment of area of cross-section about neutral plane, assumed constant,

E = Young's Modulus (Modulus of Elasticity) of material,

ρ = density of material,

c = $(E/\rho)^{1/2}$,

ω = natural periodicity = 2π/natural periodic time,

t = time,

$b = (\omega/ck)^{1/2}$,

f = frequency = $\omega/2\pi$,

$\theta = bl$,

l = length of beam.

Hence, provided the values of E, ρ, k do not change for any given length l

$$\text{frequency} \propto (\theta)^2. \tag{7.67}$$

Accordingly, values of $(\theta)^2$ and the consequential ratio of mode to fundamental frequency have been calculated as in Table 7.2.

Table 7.2

Mode	$(\theta)^2$	Frequency ratio to fundamental
(f)	3.510615235	1.00
$2 \times (f)$	22.03449153	6.266893075
$3 \times (f)$	61.97214280	17.62567495
$4 \times (f)$	120.9019159	34.38606144
$5 \times (f)$	199.8595300	56.84260726
$6 \times (f)$	298.5555308	84.91301258

A correction for the effect of shear deflection was made by Timoshenko [9] which is of significance with high overtones.

7.5 Torsional oscillation of flywheel-bearing shafts

Torsional oscillation of shafts carrying concentrated flywheel loads can be evaluated by a number of methods according to their complexity, the simplest method being based on the location of nodal points, others being based on the equations of motion of the loads, tabulation techniques for shafts with multiple loads, Dunkerley's method, and various graphical systems [10–12].

The torsional rigidity (τ) of a plain shaft operating under stress reversals below the elastic limit is given by

$$\tau = \frac{GJ}{l} \tag{7.68}$$

For a stepped shaft of lengths l_1, l_2 and polar moments J_1, J_2 the torsional rigidity is given by

$$\frac{1}{\tau} = \frac{1}{G}\left(\frac{l_1}{J_1} + \frac{l_2}{J_2}\right) \tag{7.69}$$

where G = modulus of rigidity,

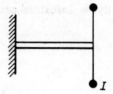

Figure 7.12 *Single flywheel, fixed end*

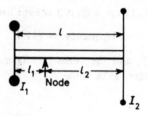

Figure 7.13 *Free shaft with two flywheels*

J = polar moment of area = $\pi d^4/32$ or $\pi r^4/2$ for a circular shaft

d = diameter, r = radius of cross-section,

l = length of shaft.

7.5.1 Single flywheel – one fixed shaft end (Figure 7.12)

The fixed end will locate the node position and thus:

$$f = \frac{1}{2\pi} \sqrt{\left(\frac{\tau}{I}\right)}$$

(7.70)

where I = inertial of flywheel

7.5.2 Two flywheels – shaft free (Figure 7.13)

This will have a single node at some position along the shaft, hence:

$$f = \frac{1}{2\pi} \sqrt{\left[\tau\left(\frac{1}{I_1} + \frac{1}{I_2}\right)\right]}$$

(7.71)

where I_1 and I_2 = polar inertia of each flywheel

7.5.3 Three flywheels – shaft free (Figure 7.14)

There are two possible modes of vibration (i) a two-node vibration in which the end loads I_1 and I_3 are always turning in the same direction as each other and load

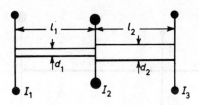

Figure 7.14 *Free shaft with three flywheels*

I_2 in the opposite direction, (ii) a single-node vibration in which the end load nearest the node turns in one direction while I_2 and the other load turns in the other direction.

The values of the frequencies f_1, f_2 are derived from the roots to the equation for f:

$$(f^2 - f_a^2)(f^2 - f_b^2) = (f_c \, f_d)^2 \qquad (7.72)$$

where f = system frequencies f_1, f_2,

$$f_a = \frac{1}{2\pi} \sqrt{\left[\tau_1 \left(\frac{1}{I_1} + \frac{1}{I_2}\right)\right]} \, , \qquad (7.73)$$

$$f_b = \frac{1}{2\pi} \sqrt{\left[\tau_2 \left(\frac{1}{I_2} - \frac{1}{I_3}\right)\right]} \, , \qquad (7.74)$$

$$f_c = \frac{1}{2\pi} \sqrt{\left(\frac{\tau_1}{I_1}\right)} \, . \qquad (7.75)$$

$$f_d = \frac{1}{2\pi} \sqrt{\left(\frac{\tau_2}{I_2}\right)} \, , \qquad (7.76)$$

$$\tau_1 = \frac{GI_1}{l_1}, \qquad (7.77)$$

$$\tau_2 = \frac{GI_2}{l_2}. \qquad (7.78)$$

7.5.4 Multiple flywheel loads – shaft free

A tabular method based on the equation of motion for each of the loads taken in turn can be used to determine the natural frequency of a multiple-load system.

7.5.5 Automobile transmission systems (torsional oscillations)

For a typical automobile system where the major masses are the engine crankshaft with its rotating and reciprocating parts, flywheel and road wheels an equivalent 3-mass system can be introduced as shown in Figure 7.15.

The crank masses are replaced by I_1, flywheel and attachments I_2, roadwheels adjusted to crankspeed, I_3. Torsional rigidities are τ_1 for the crankshaft, τ_2 the

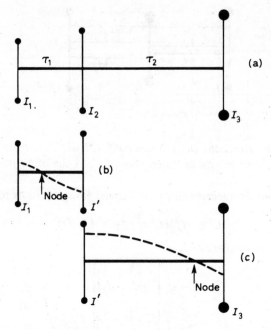

Figure 7.15 *Equivalent mass system – automobile transmission*

combined torsion rigidity adjusted to crankshaft speed of the propellor and axle shafts. Some interconnection exists between road-wheel frequencies, chassis stiffness, tyre pressure and flexibility, suspension stiffness and wheel/road adhesion which has an effect on the transmission shaft but negligible influence on the crankshaft system.

Typical values calculated by Ker Wilson [13] gave 1-node fundamental frequencies: crankshaft frequency 270 Hz (16 200rpm); propellor shaft frequency 59 Hz (358 rpm). The amplitude is small and of little practical significance for the higher frequency modes of vibration.

7.5.6 Direct-coupled generating sets (torsional oscillations)

Pistons, connecting rods and cranks can be replaced by equivalent masses depending upon the firing cycle so that a direct coupled generating set may be represented by the arrangement shown in Figure 7.16. If there is a bearing between flywheel and generator, three nodes will be located as shown with the 2-node positions close to the intermediate bearing as in case (a). If there is no intermediate bearing, flywheel and armature may be taken as an equivalent mass with one of the nodes of the 2-node deflection at middle position as in case (b).

Representative values for the one-node frequencies of diesel-generating sets based on case (b) are given in Table 7.4.

For engines with the same stroke/bore and ratio and of similar design with the

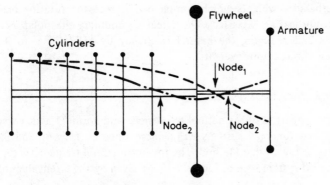

Case (a) bearing between flywheel and generator

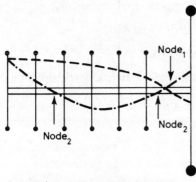

Case (b) no intermediate bearing

Figure 7.16 *6-cylinder direct-coupled generator arrangement.*

Table 7.4

Cylinder bore	10 in (254 mm)	20 in (508 mm)	30 in (762 mm)
stroke	15 in (381 mm)	30 in (762 mm)	45 in (1143 mm)
No cylinders	Frequency Hz (revs/min)		
3	70 (4200)	35 (2100)	23.3 (1400)
4	60 (3600)	30 (1800)	20 (1200)
5	53.3 (3200)	26.7 (1600)	18.3 (1100)
6	48.3 (2900)	25 (1500)	16.7 (1000)
7	45 (2700)	23.3 (1400)	15 (900)
8	43.3 (2600)	21.7 (1300)	14.1 (850)

same number of cylinders the natural frequencies are inversely proportional to cylinder bore if the engines run under equivalent conditions, namely piston speeds equal, indicator cards identical. Likewise natural frequencies are inversely proportional to the square root of the number of cylinders.

Counterbalance weights which completely balance the rotating mass and one-half the reciprocating mass have the effect of doubling the polar inertia of each crank mass and reducing the natural frequency by $(2)^{-1/2}$, i.e. to 0.707 of the frequency without counterweights.

7.5.7 Branched torsional systems

The general properties of the natural frequencies and normal elastic curves of multi-rotor branched torsional systems was demonstrated by Dawson and Sidwell [14] using a Matrix Eigenvalue Method. The torsional natural frequencies ω_1 of a multi-rotor branched system such as Figure 7.17 are given by the eigenvalues of

$$aAX = \omega^2 BX \qquad (7.79)$$

where X = column matrix of the angular displacement, θ_1,

$\quad\quad\quad B$ = a positive diagonal matrix of inertias,

$\quad\quad\quad A$ = a symmetric stiffness matrix.

The Jacobi rotation method simplifies the evaluation by transforming the foregoing equation into

$$(C)(Y) = \omega^2 \qquad (7.80)$$

where $(C) = (1/\sqrt{B})A(1/\sqrt{B})$

$\quad\quad\quad (Y) = (\sqrt{B})X.$

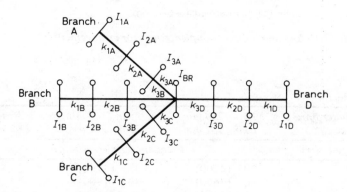

Figure 7.17 *General branch point system*

A previous study at the Polytechnic of Central London [15] proved that for such systems the following relationship applies between the eigenvalues of the total system given by $AX = \lambda'BX$ and those of the subsidiary systems given by the equations $A_i x = \lambda B_i x$, namely that if the values λ_i are arranged in ascending order $\lambda_1 \leqslant \lambda_2 \leqslant \lambda_3 \ldots \leqslant \lambda_n$, taking into account any multiplicity of eigenvalues, then the n eigenvalues $\lambda'_0 \ldots \lambda'_{n-1}$ of the system satisfy the relationship

$$\lambda_0' < \lambda_1 \leqslant \lambda_1' \leqslant \lambda_2 \leqslant \lambda_2' \leqslant \ldots \leqslant \lambda_{n-1} \leqslant \lambda_{n-1}' \qquad (7.81)$$

where λ_{n-1}' has an upper limit given by $\lambda_{n-1}' < \lambda_{n-1} + a/b$.

Examples of the use of this technique using equation 7.79, with eigenvalues and eigenvectors of the synthesized systems evaluated by computer techniques included the single branch point system without any identical branch frequencies (an untuned system) shown in Figure 7.18 produced the eigenvector information summarized in Table 7.5 in terms of the nodes in each arm for the particular nodal frequency of the system.

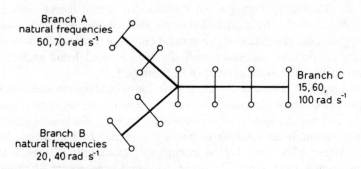

Figure 7.18 *No common branch frequencies*

Table 7.5 *System I calculated results*

λ_i number	System natural frequency rad/s	Number of nodes		
		50, 70 rad/s branch	20, 40 rad/s branch	15, 60, 100 rad/s branch
0	0	0	0	0
1	17.19	0	0	1
2	24.08	0	1	1
3	41.14	0	2	1
4	59.93	1	2	1
5	64.71	1	2	2
6	99.996	2	2	2
7	199.4	2	2	3

The procedure established the ranges within which the system natural frequencies fell when applied to more complex systems.

7.5.8 Effects of variable inertia on torsional vibrations

Torsional vibrations in the running gear of reciprocating engine systems involves variations in inertia torques which are normally neglected. Carnegie and Pasricha [16] showed that while the equation of motion taking the variable inertia effect into account is non-linear by assuming small displacements the equation could be linearized to predict important characteristics of the motion such as the 'Coriolis' effect arising from the movement of a mass in the vibrating system.

Such an equation, when solved by numerical methods using a digital computer, predicts the regions of instability and the manner in which the amplitude and frequency vary with the speed of rotation of the engine. The responses of the system show a modulation of amplitude and frequency at definite rotational speeds. The occurrence of such a modulation in amplitude and frequency is established by use of the process given by Wentzel, Kramers, Brillouin and Jeffreys generally known as the WKBJ approximation [17].

It was indicated by Carnegie and Pasricha that several recent cases of marine crankshaft failures had been attributed to the phenomenon of secondary resonance. This suggested the possibility of an nth order critical of small equilibrium amplitude occurring at, or near, resonance with the service speed being excited by large resultant engine excitations of order $(n-2)$ and $(n+2)$.

Draminsky [18, 19] stated that, in practice, secondary resonance in torsional vibration occurs only for resonance with the lower-order secondary component, $(n-2)$, and not the higher order component, $(n+2)$. The multi-cylinder engine with large second order variation in inertia is explained by the use of a non-linear theory. Archer [20] cited typical examples of crankshaft failures in large ten-cylinder engines from service in which measured stress values of certain orders were found to be much greater than those calculated by normal methods.

7.5.9 Crankshafts, torsional rigidity

Calculations of the torsional rigidity of crankshafts are usually based on empirical formula which modify the basic theoretical equations derived from an analysis of the combined effect of the elements, as follows:

Element	Torsional rigidity
Crankshaft journal	$\tau_1 = \dfrac{\pi}{32} G(D_1^4 - d_1^4) \cdot L_1$
Crankpin	$\tau_2 = \dfrac{\pi}{32} G(D_2^4 - d_2^4) \cdot L_2$
Crankweb	$\tau_3 = \dfrac{b \cdot t^3 JE}{12} X$

where b = width of crankweb,

   t = thickness of crankweb,

   E = Youngs modulus,

   X = crank throw ($\frac{1}{2}$ stroke).

The use of equivalent lengths L_e in terms of the length between bearings (L) are used in empirical formulae such as the Carter equation [21] and the Constant formulae [22], typical values of L_e/L range from 0.95 to 2.5 of which the distribution of flexibility is approximately as follows:

Element	% total flexibility
Journal (1)	45
Crankpin (1)	40
Crankwebs (2)	15
	100

7.5.10 Crankshaft elements – mass polar moment of inertia

Complex shapes such as crankshafts must be broken into their individual elements and the polar inertia summed, the 'transposition of axes' equation $I = I_x + Mx^2$ may be needed where I_0 = moment about a parallel axis, x = distance of the parallel axis of rotation, M = mass of the element.

Element	Mass	(Radius of gyration about axis of rotation)^2	Moment of Polar inertia
Crankshaft–solid	$\frac{\pi}{4}D_1^2 L_1 \rho$	$\frac{D_1^2}{8}$	$\frac{\pi}{32}D_1^3 L_1 \rho$
Journal – hollow	$\frac{\pi}{4}(D_1^2 - d_1^2)L_1\rho$	$\frac{D_1^2 + d_1^2}{8}$	$\frac{\pi}{32}(D_1^4 - d_1^4)L_1\rho$
Crankpin – solid	$\frac{\pi}{4}D_2^2 L_2\rho$	$\frac{D_2^2}{8} + X^2$	$\frac{\pi}{4}D_2^2 L_2^2\rho\left(\frac{D_2^2}{8} + X^2\right)$
– hollow	$\frac{\pi}{4}(D_2^2 - d_2^2)L_2\rho$	$\left(\frac{D_2^2 + d_2^2}{8}\right) + X^2$	$\frac{\pi}{4}(D_2^2 - d_2^2)L_2^2\rho\left[\left(\frac{D_2^2 + d_2^2}{8}\right) + X^2\right]$

where
D_1 = diameter of crankshaft journal,

L_1 = length of crankshaft journal,

ρ = density of material,

d_1 = bore of hollow crankshaft journal,

D_2 = diameter of crankpin,

L_2 = length of crankpin,

X = crank throw ($\frac{1}{2}$ – stroke),

d_2 = bore of crankpin.

The polar inertia of the crankweb can be evaluated by graphical integration. Considering the element at radius R shown in Figure 7.19 having an arc of α degrees with a web width Z.

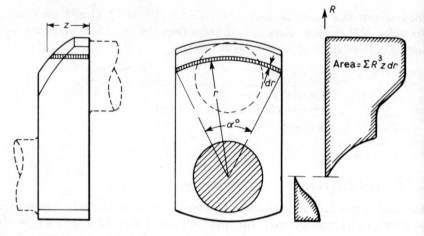

Figure 7.19 *Crankshaft polar inertia calculation*

mass of element $= 2\pi R \cdot \dfrac{\alpha}{360} \cdot Z(\delta R)\rho,$

radius of gyration of element $= R,$

polar moment of inertia of element, $\mathrm{d}I = mk^2 = 2\pi R^3 \cdot \dfrac{\alpha}{360} \cdot Z(\delta R)\rho$

Hence for the whole crankweb the total polar moment is the sum of that of the elements.

i.e.   polar moment of whole crankweb $= \dfrac{2\pi\rho}{360} \sum R^3 \alpha Z(\delta R)$

Component	% total polar moment of inertia
Crankpin (1)	28.5
Journal (1)	6.5
Crankweb (2)	65.0
	100.0

7.6 Belt drives

Power transmission by means of pulleys and belts provides a unique example of frequency calculations, somewhat analagous techniques can be used with chain and sprocket calculations.

7.6.1 Belt drives – transverse vibrations

A disturbance to a belt may produce a resonant flapping motion as shown in

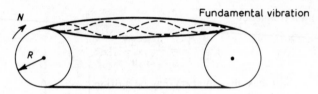

Figure 7.20 *Resonant flapping of belt drive*

Figure 7.20 the velocity of propagation of the disturbance along the belt depending on tension and belt density according to

$$u^2 = \frac{P}{\rho A} \tag{7.82}$$

where u = velocity of propagation of the disturbance,

p = tension is belt,

ρ = density of belt,

A = cross-sectional area of belt.

Thus for a stationary belt, time for a disturbance to travel along belt length L_s and back, $t_0 = 2L_s/u$

i.e. frequency of transverse vibration of a stationary belt, $f_0 = \dfrac{1}{t_0} = \dfrac{u}{2L_s}$. (7.83)

The time for a disturbance to travel from X to Y along the belt will be

(a) in direction of motion $\qquad t_1 = \dfrac{L_s}{u + v}$ (7.84)

(b) against direction of motion $\qquad t_2 = \dfrac{L_s}{u - v}$ (7.85)

where t_1, t_2 = time,

L_s = belt length,

v = linear velocity of belt = $\dfrac{2\pi R}{N}$.

Hence, time for a complete disturbance cycle,

$$t = t_1 + t_2$$

$$= L_s \left(\frac{1}{u + v} + \frac{1}{u - v} \right)$$

$$= \frac{2 L_s u}{u^2 - v^2} \tag{7.86}$$

and fundamental frequency,

$$f = \frac{1}{t} = \frac{u^2 - v^2}{2L_s u} \quad \text{Hz.} \tag{7.87}$$

To allow for m overtones, the equation may be written

$$f = m \frac{u^2 - v^2}{2L_s u} \quad \text{Hz.} \tag{7.88}$$

from equation 7.83

$$= m \frac{u}{2L_s} \left(1 - \frac{v^2}{u^2}\right) \tag{7.89}$$

$$= 2mf_0 \left(1 - \frac{v^2}{u^2}\right). \tag{7.89}$$

With high speed belts, the supplementation of belt tension arising from centrifugal tension should be included in calculations for u.

7.7 Whirling of marine line shafting

Whirl action in the shafting of marine vessels, aggravated by hull flexibility, produces sympathetic vibrations at several frequencies. Propeller blade orders are responsible for the creation of excessive vibrations in the tailshaft of marine systems, such vibrations being described as 'tailshaft whirling', 'transverse vibrations', 'lateral vibrations' and 'blade-order whirl'.

A propeller is not perfectly symmetrical no matter how much attention has been paid to its design and manufacture. Sea-water is not homogeneous because of the wake arising from the flow over and under the hull. As a result, each propeller blade experiences periodic forces of complex wave form. Harmonic analysis of this wave-form shows that some wake components act on the blades to produce a bending moment about the propeller i.e. the thrust forces on the propeller as a whole are unbalanced. Such bending moments produce the excitation for lateral vibration (or whirl) of the shafting.

For a propeller with n blades these bending moments emanate from the $(kn \pm 1)$ wake components, k being any integer. The most significant components are those where $k = 1$. It can be shown that the $(n - 1)$ component gives rise to a bending moment, of frequency $n \times$ propeller speed, relative to the propeller, which rotates in the same direction as the propeller. The $(n + 1)$ component causes a bending moment of the same frequency but rotating in the opposite sense to the propeller. The two forms of lateral vibration which ensue are termed forward whirl and counter, reverse or backward whirl respectively [23].

Thus propellers with an odd number of blades, taking their bending moment variation from the even order components, give greater excitation than propellers with an even number of blades. The reverse applies for torque and thrust variations as shown by Archer [20].

Volcy [24] reported two different types of propeller shaft vibration: lateral, i.e. either vertical or horizontal vibration; and precession (positive or negative).

7.7.1 Whirling calculations – bending moment/deflexion diagrams method

The effects of high orders of propeller excitation were first studied by Panagopulos [25] but current calculations are derived from Jasper [26]. Using computers, it is possible to consider the whole length of a shaft and the higher critical speeds.

Toms and Martyn [23] stated forthrightly that one fallacious notion that should be disposed of is the effect of gravity. This originated when the Rayleigh energy method used the gravity form of the bent shaft as a first approximation to the fundamental whirling shape.

In fact, whirling is a form of bending vibration occurring about the bent form of the shaft, in its bearings, whilst rotating in the steady state. The mathematical model deals with the shaft loaded at any point by the mass multiplied by the shaft deflexion at that point.

From elementary beam theory, the normal differential equation of the deflexion curve for a shaft, of variable section, whirling with its central axis in a bowed shape is:

$$\frac{d^2}{dx^2} EI \frac{d^2y}{dx^2} = u\omega^2 y \tag{7.90}$$

where x = distance along the shaft,

y = deflexion of shaft,

E = modulus of elasticity,

I = diametral moment of inertia of shaft cross-section,

u = mass per unit length,

ω = periodicity of vibration rad/s.

The slope of the deflexion curve depends also on shear which will be important if the cross-sectional dimensions are not small in comparison with the length. This may be the case for vibrations at higher frequencies when the shaft is subdivided by modal sections into comparatively short portions.

The variable angle of rotation of the cross-sections is equal to the slope of the deflexion curve and gives rise to a moment of the inertia forces termed the 'rotatory inertia'. The inclusion of these effects for the shaft leads to a complicated differential equation.

Fortunately, for the whirling of line shafting the scantlings and frequencies do not warrant their inclusion, hence from the foregoing equation it is reasonable to use

$$EI \frac{d^2y}{dx^2} = M \tag{7.91}$$

where M = bending moment.

Hence equation 7.90 may be rewritten as

$$\frac{d^2M}{dx^2} = u\omega^2 y. \tag{7.92}$$

Equation 7.90 is a fourth order differential equation and there are four boundary conditions to be satisified. Any speed satisfying this equation and also the four boundary conditions is a natural frequency of the system. A solution can be found by constructing the bending moment and deflexion diagrams using a step-by-step integration process in conjunction with equations 7.91 and 7.92.

In general, these diagrams can be constructed to satisfy three of the boundary conditions. By plotting the discrepancy in the fourth boundary condition as a function of the assumed speed and noting where this becomes zero, the natural frequency can be determined. In plotting such a curve it is necessary to select some arbitrary boundary condition and hold it constant.

7.7.2 Whirling calculations – the effect of supports

Ideally, all marine bearings should be downward loaded and some calculations such as those made by Panagopulos [25] assumed there was no support at the forward sterntube since due to the short span between the two stern tube bearings it is often difficult to obtain a downward loading with a straight alignment. The effect of bearing loading on the modal curve is shown in Figure 7.21.

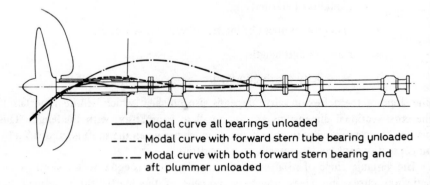

——— Modal curve all bearings unloaded
– – – – Modal curve with forward stern tube bearing unloaded
—.— Modal curve with both forward stern bearing and
          aft plummer unloaded

Figure 7.21 *Conventional shaft arrangement for aft-engined oil tanker depicting modal curves for lateral vibration*

Examination of the wear conditions of lignum vitae bearings of a length four times the shaft diameter indicate an initial point of support approximately one diameter from the aft-end. Since the effective point of support moves forward as the shaft 'beds in' it was considered by Toms and Martyn [23] preferable to calculate for the support at a quarter to a third of the length from the aft end according to whether the estimated critical speed occurs below or above the service speed. For the shorter white metal aft sterntube bearings the effective point of support should be considered to be at a third to a half of the length from the aft end. Other bearings are generally about one shaft diameter in length and it is sufficient to take the point of support at the mid-point.

The natural frequency of the line shafting is also dependent on the flexibility of the support bearings. Except where the length between bearings may result in local vibrations for a particular span, the relevance of flexibility decreases the further forward the bearing. In general, it is sufficient to consider only the aftermost three bearings within the ship as flexible. In any case, if the natural frequency of the bearing support is at least $3n \times$ service speed, where n is the number of blades, it can be considered rigid.

Structural flexibility may be determined by 'finite element' methods in order to assess the extent to which the surrounding structure contributes to the general flexibility. Since the dynamic movements may differ from those calculated by statical considerations the employment of such a complicated procedure is not always justified. In general, for a single screw vessel, even for the lightest structure, the bearing stiffness will exceed $50\,000\,\mathrm{kg\,mm^{-1}}$ ($1250\,\mathrm{ton\,in^{-2}}$, $49.3\,\mathrm{kN\,mm^{-1}}$).

When very flexible bearings are incorporated into the system the modal shape based on straight line may not indicate the modes with which many engineers are familiar. As shown by Figure 7.22, the base line has to be altered to pass through the equilibrium positions of the loaded bearings.

7.7.3 Whirling frequency calculations – Prohl method

A numerical method analogous to the Holzer method for torsional vibrations may be used for calculating whirling frequencies. Four integrations are involved instead of the two for torsional vibrations and additional complications arise in dealing with the boundary conditions. Such a method, described by Prohl [27] was illustrated by Toms and Martyn [23].

In this method the actual system is transformed into an idealized form consisting of a series of discs connected by sections of elastic but massless shaft. The mass of the discs and their spacing is chosen so as to approximate the mass distribution of the actual system. The bending flexibility of the connecting sections is taken to correspond to the actual flexibility of the system.

7.8 Gear excitation

Imperfections in individual gears create frequency impulses depending on the number of teeth on a wheel, the wheel speed and the wheel combination. For even load distribution the number of teeth should correspond to a prime number with the minimum possibility of repeated tooth-to-tooth contact repetition.

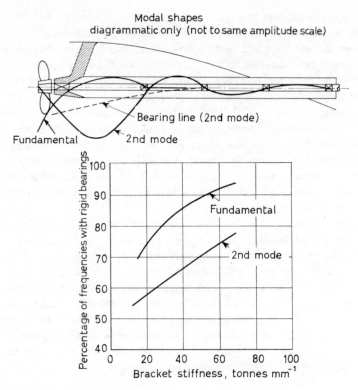

Figure 7.22 *Effect of 'A' bracket stiffness in a twin-screw vessel*

7.8.1 Specific gear assembly excitations

The frequency of specific gear excitation frequencies are as follows:

(1) 2-Wheel gear train, (Figure 7.23).

Tooth meshing frequency $= t_1 N_1/60\,\text{Hz}$

High spot, wheel 1 $= N_1/60\,\text{Hz}$

High spot, wheel 2 $= \dfrac{t_1}{t_2} N_1/60 = N_2/60\,\text{Hz}$

(2) 3-Wheel gear train, (Figure 7.24).

Tooth meshing frequency $= t_1 N_1 = t_2 N_2 = t_3 N_3$

High spot, wheel 1 $= N_1$

High spot, wheel 2 $= N_2$

High spot, wheel 3 $= N_3$

(3) Planetary gear, (Figure 7.25).

(i) Fixed cage $\dfrac{N_2}{N_1} = -\dfrac{t_2}{t_1} : \dfrac{N_3}{N_1} = -\dfrac{t_1}{t_3}$

Tooth meshing frequency $= t_1 N_1 = t_2 N_2 = t_3 N_3$

High spot on sun $= n N_1$ (where n = number of planets)

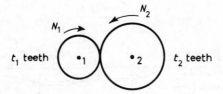

Figure 7.23 *Simple gear pair*

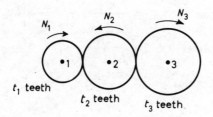

Figure 7.24 *Three-wheel gear train*

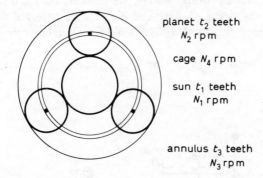

Figure 7.25 *Planetary gear*

High spot on planet $\quad = 2\dfrac{t_1}{t_2}N_1 = 2N_2$

High spot on annulus $\quad = n\dfrac{t_1}{t_3}N_1 = nN_3$

(ii) Fixed sun $\left(\dfrac{N_2}{N_3} = \dfrac{t_3}{t_3 - t_1} : \dfrac{N_4}{N_3} = \dfrac{t_3}{t_1 + t_3}\right)$

Tooth meshing frequency $= \dfrac{t_1 t_3}{t_1 + t_3} \cdot N_3 = t_1 N_4$

High spot on sun $\qquad = \dfrac{n t_3}{t_1 + t_3} \cdot N_3 = n N_4$

High spot on planet $\quad = \dfrac{2t_1t_3}{t_2(t_1 + t_3)} \cdot N_3 \; = \; 2\dfrac{t_1}{t_2}.N_4$

High spot on annulus $\quad = \dfrac{nt_1}{t_1 + t_3} \cdot N_3 \; = \; n\dfrac{t_1}{t_3} \cdot N_4$

(iii) Fixed annulus $\left(\dfrac{N_2}{N_1} = \dfrac{t_1}{t_1 - t_3} : \; \dfrac{N_4}{N_1} = \dfrac{t_1}{t_1 + t_3} \right)$

Tooth meshing frequency $= \dfrac{t_1t_3}{t_1 + t_3} \cdot N_1 \; = \; t_3.N_4$

High spot on sun $\quad = \dfrac{nt_3}{t_1 + t_3} \cdot N_1 \; = \; n\dfrac{t_3}{t_1}N_4$

High spot on planet $\quad = \dfrac{2t_1t_3}{t_2(t_1 + t_3)} N_1 \; = \; 2\dfrac{t_3}{t_2}N_4$

High spot on annulus $\quad = \dfrac{nt_3}{t_1 + t_3}.N_1 \; = \; N_4$

7.8.2 Geared system— neglible gear inertia

Provided there is no backlash in the gearing, the teeth are rigid and do not distort under load and that the inertia of shafts and gears can be neglected the geared system shown in Figure 7.26 can be reduced to an equivalent single system shaft Figure 7.27.

The length l_2 is replaced by a length of dynamically equivalent torsional

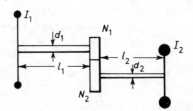

Figure 7.26 *Gear system*

Figure 7.27 *Equivalent single shaft*

stiffness, thus if a torque T is applied at J_1 the gear ratio $r = N_2/N_1$ produces the need for a corresponding torque $r.T$ to be applied to the rotor. The angle θ_0 with which J_1 twists relative to J_2 is given by

$$\theta_0 = \theta_1 + r\theta_2,$$

where $\quad \theta_1 = $ angle of twist of shaft (1) $= \dfrac{Tl_1}{GJ_1}$,

$\qquad \theta_2 = $ angle of twist of shaft (2) $= \dfrac{rTl_2}{GJ_2}$,

$$\therefore \quad \theta_0 = \frac{Tl_1}{GJ_1} + \frac{r^2 Tl_2}{GJ_2}$$

$$= \frac{T}{GJ_1}\left(l_1 + r^2 l_2 \frac{J_1}{J_2}\right), \tag{7.93}$$

where $\quad J = \dfrac{\pi d^4}{32}$,

$\qquad \dfrac{J_1}{J_2} = \left(\dfrac{d_1}{d_2}\right)^4$,

$$\theta_0 = \frac{T}{GJ_1}\left[l_1 + r^2 l_2 \left(\frac{d_1}{d_2}\right)^4\right] = \frac{T}{GJ_1}l_e$$

$$\text{Equivalent length } l_e = r^2 l_2 \left(\frac{d_1}{d_2}\right)^4 \tag{7.94}$$

An equivalent mass moment of inertia $J_{2'}$ must be introduced in addition to the equivalent shaft length. This equivalent inertia $J_{2'}$ provides a kinetic energy equivalent to that which the actual load would apply. It can be shown that

$$J_{2'} = \frac{J_2}{(r)^2} \tag{7.95}$$

A geared system with negligible shaft and gear wheel inertia can accordingly be reduced to this equivalent single system shaft with a single node.

7.8.3 Geared system – gear inertia included

Gear wheel inertia introduces a third equivalent inertia to an 'equivalent single system' as shown in Figures 7.28 and 7.29.

$$\text{Equivalent gear inertia} = J_{g1} + \frac{J_{g2}}{(r)^2} \tag{7.96}$$

where $\quad r = $ speed ratio $=$ teeth ratio N_1/N_2.

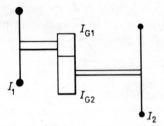

Figure 7.28 *Gear system with gears of high inertia*

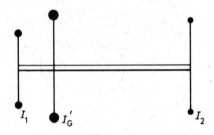

Figure 7.29 *Equivalent to gear system with high gear inertia*

This 3-rotor system can then be analysed for node positions and frequencies in the normal manner.

7.8.5 Backlash in gears

The effect of backlash following a vibratory torque amplitude in excess of the mean transmitted torque was studied by Tuplin [28] on the basis that during the period of tooth separation there would be zero torque as shown in Figure 7.30 for the break-away half-cycle of the back-lash motion involving a reversal of torque and twist; re-engagement provides the other half-cycle of the motion under similar conditions.

Teeth separate at B and re-engage after time t_2 at position C. The time t_2 is given by

$$t_2 = \frac{2\delta}{\alpha\omega} \tag{7.97}$$

where $\delta = \frac{1}{2}$ angular backlash, rad,

$\alpha = \theta - \delta$, rad,

θ = maximum amplitude of twist due to vibratory torque, rad,

ω = relative periodicity of natural vibration without backlash, rad^{-1}.

From C to D the motion is subject to the system mass and elasticity for a time $t_1 = T_0/4$ where $T_0 = 2\pi/\omega$, thus the time for a complete cycle is given by

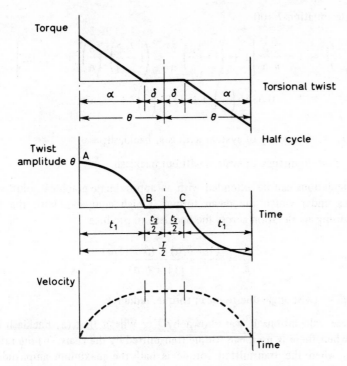

Figure 7.30 *Half-cycle showing the effect of back-lash between gear teeth*

$$T = 2(2t_1 + t_2)$$

$$= 2\left(2\frac{2\pi}{4\omega} + \frac{2\delta}{\alpha\omega}\right)$$

$$= \frac{2(\pi\alpha + 2\delta)}{\alpha\omega} \tag{7.98}$$

Taking backlash into account, the phase velocity of natural vibration ω_1 is given by

$$\omega_1 = \frac{2\pi}{T}$$

from equation 7.98

$$= \frac{\pi\alpha\omega}{\pi\alpha + 2\delta}$$

i.e.

$$\frac{\omega_1}{\omega} = \frac{\pi\alpha}{\pi\alpha + 2\delta} \tag{7.99}$$

The ratio ω_1/ω is equal to the ratio of frequencies f_1/f, this from equation 7.99

$$\frac{f_1}{f} = \frac{1}{1 + (2\delta/\pi\alpha)} = \left(1 + \frac{2\delta}{\pi\alpha}\right)^{-1}. \tag{7.100}$$

Expanding equation 7.100

$$\frac{f_1}{f} = 1 - \frac{2}{\pi}\frac{\delta}{\alpha}\left[1 + \left(1 - \frac{2}{\pi}\right)\frac{\delta}{\alpha} + \left(1 - \frac{2}{\pi}\right)^2\left(\frac{\delta}{\alpha}\right)^2 + \ldots \text{etc.}\right]$$

$$= 1 - 0.6366\left(\frac{\delta}{\alpha}\right) + 0.2313\left(\frac{\delta}{\alpha}\right)^2 + \ldots \text{etc.} \qquad (7.101)$$

where f_1 = frequency of system with gear backlash,

f = frequency of system without backlash.

These calculations can be extended with an approximate graphical solution for the conditions under which the mean torque is high compared with the vibratory torque causing acceleration across the backlash to produce

$$\frac{f_1}{f} = \frac{(1 + \delta/\alpha) + (\beta/d)(\delta/\alpha)}{(1 + \delta/\alpha)} \qquad (7.102)$$

where β = twist angle due to mean torque, radians.

From these calculations it was shown by Ker Wilson that (a) backlash effect is greatest when there is no mean torque transmitted by the gears, (b) the ratio f_1/f is minimum when the transmitted torque is half the maximum amplitude of the vibratory torque, (c) with mean torque equal to or greater than the maximum amplitude of the vibratory torque there is no gear separation.

Considered farther in terms of condition monitoring, the development of increasing backlash results in the creation of increasing side-band frequencies about the fundamental frequency.

7.9 Rolling element bearings

Most rolling element frequency calculations of interest to normal monitoring analysis arise from shock forces. These are mainly the result of surface defects which produce linear frequency relationships discussed below.

7.9.1 Characteristic rotational frequencies – rolling element bearings

From a kinematic analysis of the defect-impacting of the action of bearings [29] the following equations may be found for the rotational or 'pass' frequencies

Let d = diameter of rolling element,

D = bearing pitch diameter,

β = contact angle, deg.,

n = number of rolling elements.

$$\text{Outer race frequency } f_{or} = \frac{n}{2}\frac{N}{60}\left(1 - \frac{d}{D}\cos\beta\right) \qquad (7.103)$$

$$\text{Inner race frequency } f_{ir} = \frac{n}{2}\frac{N}{60}\left(1 + \frac{d}{D}\cos\beta\right) \qquad (7.104)$$

$$\text{Roller element frequency } f_b = \frac{D}{d}\frac{N}{60}\left[1 - \left(\frac{d}{D}\right)^2\cos\beta\right] \qquad (7.105)$$

$$\text{Cage frequency } f_c = \frac{N}{120}\left(1 - \frac{d}{D}\cos\beta\right). \qquad (7.106)$$

Typical values for a 2-inch, single row angular contact bearing are given in the following table ($d = 0.3125$ in, $D = 1.540$ in, $\beta = 15$ deg., $n = 12$).

Speed revs/min	Frequencies, Hz			
N	f_{or}	f_{ir}	f_b	f_c
1750	140.7	209.3	138.2	11.7

7.9.2 Significant frequencies -- rolling element bearings

Discrete frequency	Significance
Shaft rotational frequency, $$f_1 = \frac{N}{60} \qquad (7.107)$$	Appears at the slightest imbalance usually a minor influence.
Rolling element train frequency $$f_2 = \frac{(R-\Omega)}{120R}N \qquad (7.108)$$	Caused by an irregularity (rough spot or indentation) of a rolling element or the cage.
Rolling element spin frequency $$f_{2A} = \left(\frac{R+\Omega}{\Omega}\right)\left(\frac{R-\Omega}{R}\right)\frac{N}{60} \qquad (7.109)$$	
Rolling element defect $$f_3 = \left(\frac{R+\Omega}{\Omega}\right)\left(\frac{R-\Omega}{R}\right)\frac{N}{30} \qquad (7.110)$$	Irregularity strikes the inner and outer races alternately.
Inner race irregularity on the inner raceway $$f_4 = (f_1 - f_2)n = \left(\frac{R+\Omega}{120R}\right)Nn \qquad (7.111)$$	
Outer race irregularity on the outer raceway $$f_5 = f_2 \cdot n = \frac{(R-\Omega)}{120R}Nn \qquad (7.112)$$	Likely to arise if there is a variation in stiffness around the bearing housing.

where R = pitch circle radius,

$\quad\quad\quad\Omega$ = ball/roller radius,

$\quad\quad\quad N$ = shaft speed, revs/min,

$\quad\quad\quad n$ = number of balls/rollers.

7.9.3 Elastic resonance of elements – structure resonance

Structural resonance may be excited by the periodic impacting of defective race elements at the rotational or 'pass' frequencies. These induced frequencies cause a 'ringing' which is characterised by an exponentially decaying high-frequency oscillation.

The rolling element resonance is one of the typical resonances which may be excited as the rolling element itself pulsates as a rigid body due to the impact. From an analysis by Love [31] the frequency of a ball may be calculated by:

$$\text{Ball resonant frequency,}\ f_{br}\ =\ \frac{0.848}{2r}\frac{E}{2\rho} \tag{7.113}$$

where E = modulus of elasticity,

$\quad\quad\quad\rho$ = density of rolling element.

Typically, for steel balls 0.3125 inch diameter, f_{br} = 387.5 kHz.

Races exhibit resonant characteristics for which the free resonant frequency equation is, according to Stokey [32] given by:

$$\text{Race resonance frequency,}\ f_{rr}\ =\ \frac{k(k^2-1)}{2\pi\sqrt{(k^2+1)}}\frac{1}{a^2}\sqrt{\left(\frac{EI}{m}\right)} \tag{7.114}$$

where k = order of the resonance,

$\quad\quad\quad a$ = radius to the neutral axis,

$\quad\quad\quad I$ = second moment of area of cross-section,

$\quad\quad\quad m$ = mass of race per linear inch.

Typically, for a 2-in single row angular contact bearing, d = 0.3125 in, D = 1.540 in, the following race resonance frequencies apply:

	Inner race				Outer race		
k (order)	2	3	4	5	6	2	3	4
f_{rr} (kHz)	3.94	11.14	21.36	34.54	50.66	9.74	27.56	52.83

These frequencies relate to the resonance of ball elements or races in a free-isolation In operation, mounted in a structure the exact frequency changes slightly from the free state.

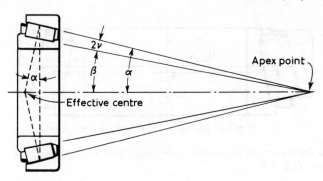

Figure 7.31 *Geometry of a tapered roller bearing*

7.9.4 Tapered roller bearings – discrete frequencies

Considering the bearing geometry shown in Figure 7.31, the angular speed of the various parts is as follows [32]:

$$\omega_1 = -\left(\frac{\sin\alpha}{\sin\alpha + \sin\beta}\right) \cdot \omega_4 \qquad (7.115)$$

$$\omega_2 = \left(\frac{\sin\beta}{\sin\alpha + \sin\beta}\right) \cdot \omega_4 \qquad (7.116)$$

$$\omega_3 = -\left(\frac{\sin\beta}{\sin\nu}\right) \cdot \left(\frac{\sin\alpha}{\sin\alpha + \sin\beta}\right) \cdot \omega_4 \qquad (7.117)$$

where ω_1 = relative angular speeds of cage to cone,

   ω_2 = relative angular speeds of cage to cup,

   ω_3 = relative angular speeds of roller to cage,

   ω_4 = relative angular speeds of cone to cup,

   α = $\frac{1}{2}$ included cup angle,

   β = $\frac{1}{2}$ included cone angle,

   ν = $\frac{1}{2}$ included roller angle.

The negative sign indicates anticlockwise rotation in relation to positive rotation of the cone and shaft.

The foregoing motion equations can be related to the dimensions of the bearing shown in Figure 7.32 as follows:

$$\omega_1 = -\left(\frac{D_{\text{cup}}}{D_{\text{cup}} + D_{\text{cone}}}\right)\omega_4 \qquad (7.118)$$

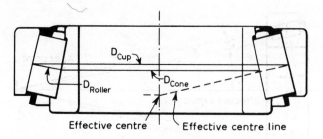

Figure 7.32 *Diameters of a tapered roller bearing (The Timken Roller Bearing Co, Canton, Ohio, U.S.A.)*

$$\omega_2 = \left(\frac{D_{\text{cone}}}{D_{\text{cup}} + D_{\text{cone}}}\right)\omega_4 \tag{7.119}$$

$$\omega_3 = -\left(\frac{D_{\text{cone}}}{D_{\text{roller}}}\right)\left(\frac{D_{\text{cup}}}{D_{\text{cup}} + D_{\text{cone}}}\right)\omega_4 \tag{7.120}$$

A discontinuity on a roller produces a periodic frequency, f_1, given by

$$f_1 = \frac{1}{30}\omega_3. \tag{7.121}$$

A discontinuity on the cup produces a periodic frequency, f_2, equal to the number of times the roller passes over the discontinuity each second, i.e.

$$f_2 = \frac{a}{60}\omega_2 \tag{7.122}$$

where a = number of rollers in the bearing.

Similarly a discontinuity on the cone produces a periodic frequency, f_3, given by

$$f_3 = \frac{a}{60}\omega_1. \tag{7.123}$$

These equations assume that all rollers are continually in contact, as would occur with a thrust load; with combined or radial loads continuous roller contact is less likely to be assured [34].

7.10 Blade vibration

The analysis of turbine blade vibrations presents a difficult theoretical problem involving factors such as pre-twist, asymmetry of cross-section, disc speed, stagger angle, non-uniform distribution of the mass, rotary inertia, etc. Solution of the differential equations of the motion of a uniform cantilever is relatively simple, that for cantilever turbine blade is complex, one of the earliest solutions being that by Carnegie [35] using an energy method.

7.10.1 Computerized blade-frequency/mode analysis program

A computer programme available from Professor W. Carnegie of the University of Surrey, Guildford uses the Extended Holzer numerical process derived by Carnegie [36] to determine the natural frequencies and corresponding mode shapes of lateral vibration of a cantilever. The method [37] determines the coupled bending-bending vibration of a blade mounted on the periphery of a rotating disc.

A typical analysis divides the turbine blade into a number of cantilever elements each of length dZ such that each element is regarded as a weightless cantilever with its mass concentrated at the root and equal to half the sum of the masses of the two adjacent elements. From the resulting force and moment equations the relevant frequencies were calculated.

Using this analysis the natural frequencies and some typical mode shapes were determined for a cantilever blade having the following data:

Breadth, B =   0.5 in,

Thickness, D =   0.25 in,

Length, L =   10.0 in,

Density, ρ =   0.283 lb in^{-3},

Young's modulus E = 30×10^6 lb in^{-2}.

The blade was divided into 30 elements.

The problem was analysed for seven values of angle of pre-twist viz, $\alpha = 0$, 15, 30, 45, 60, 75 and 90 degrees and the results obtained for coupling bending-bending vibration frequencies are plotted in Figure 7.33 for the first six modes of vibration. It can be seen that modes 1, 3 and 5 show an increase in frequency with increase of pre-twist angle, while modes 2, 4 and 6 show a decrease in frequency with increase of pre-twist angle. Two typical mode shapes for frequencies of $f = 84.38$ and $f = 1564$ Hz with 15 and 45 degrees of pre-twist respectively are plotted in Figures 7.34(a)(b).

The uncoupled natural frequencies for the first three modes of vibration in the $y-z$ plane were determined for seven values of rotational speed, viz, $\omega = 0$, 50, 100, 150, 200, 250, and 300 Hz and the results plotted in Figure 7.35. The curves show an increase in frequency of all three modes of vibration with increase in rotational speed. A typical mode plot at a frequency $f = 743$ Hz corresponding to a rotational speed $\omega = 200$ Hz is shown in Figure 7.34(c).

7.11 Cam mechanism vibration

A comprehensive study with relevant computer programmes was made by Koster [38] in his thesis to the Technical University, Eindhoven, October 1973. From the cam-operated transfer mechanism, Figure 7.36, Koster produced the dynamic model (Figure 7.37) showing sub-systems which was analysed in terms of varying degrees of freedom. Typical computer program flow diagrams are shown in Figures 7.38 and 7.39.

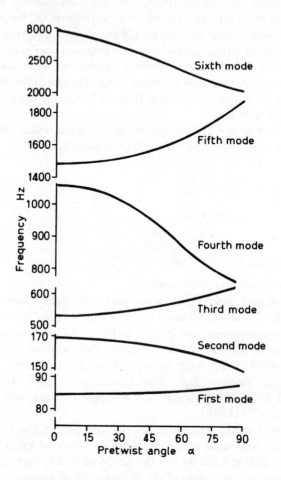

Figure 7.33 *Frequency against pre-twist angle. A numerical procedure for the determination of the frequencies and mode shapes of lateral vibration of blades allowing for the effects of pre-twist and rotation*

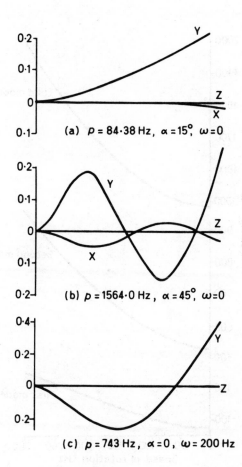

Figure 7.34 *Typical vibration mode shapes*

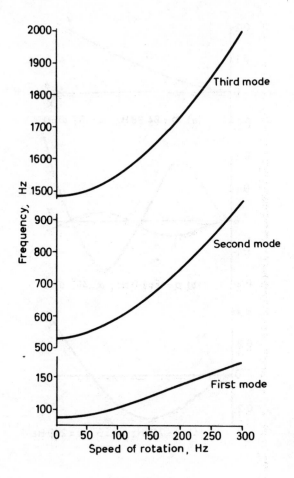

Figure 7.35 *Frequency versus speed of rotation of the disc*

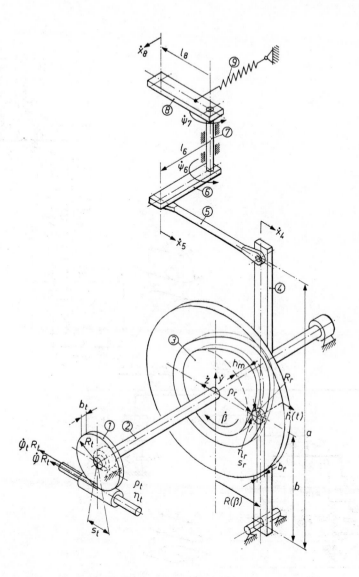

Figure 7.36 *Cam-operated transfer mechanism*

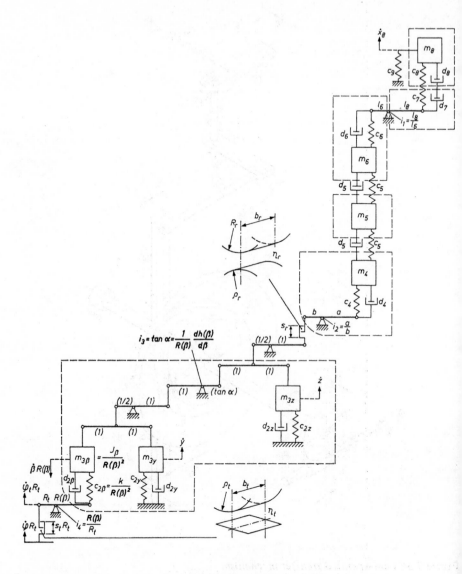

Figure 7.37 *Dynamic model of the cam-operated transfer mechanism*

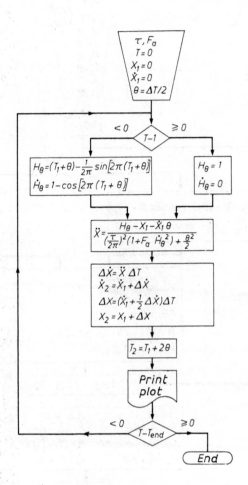

Figure 7.38 *Flow diagram of the CAMSHAFT-1 program*

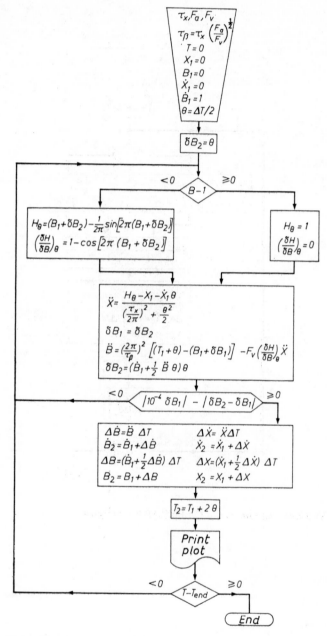

Figure 7.39 *Flow diagram of the CAMSHAFT-2 program*

References

1 McClachlan, N.W. (1956), *Theory of Vibrations,* Dover Publications Inc.
2 Rankine, W.J. (1869), 'On the centrifugal force of rotating shafts', *The Engineer,* **27**, 9 April 249.
3 Jeffcott, H.H. (1919), 'The lateral vibration of loaded shafts in the neighbourhood of a whirling speed; the effect of want of balance', *Phil. Mag,* **37**, March.
4 Morley, A. (1938), *Strength of Materials,* Logmans, Green & Co.
5 Morley, A. (1909), 'The calculation of the transverse vibration of beams', *Engineering,* 30 July, 13 August.
6 Dunkerley, W. (1894), *Phil. Trans. Roy. Soc.* (A) **185**.
7 Collacott, R.A. (1976), 'Monitoring to determine the dynamics of fatigue', *ASTM, Journal of Testing & Evaluation,* May.
8 Collacott, R.A. (1974), 'Sonic fault diagnosis', British Steel Corporation, Research Fellowship Report.
9 Timoshenko, S.O. (1921), 'On the correction for shear of the differential equation for transverse vibrations of prismatic bars', *Phil. Mag.* 41, 744–749.
10 Bevan, T. (1950), *The Theory of Machines,* Longmans, Green & Co Ltd.
11 Bishop, R.E.D. (1960), *The Mechanics of Vibration,* Cambridge University Press.
12 Myklestad, N.O. (1956), *Fundamentals of Vibration Analysis',* MgGraw Hill.
13 Ker Wilson, W. (1956) *Practical Solution of Torsional Vibration Problems,* Vol. 1, Chapman & Hall Ltd.
14 Dawson, B. and Sidwell, J.M. (1974), 'The vibrational properties of branched torsional systems having one or more branch points' *Trans. I. Mar. E,* **86**, Series A Part 12.
15 Dawson, B., Goodchild, C.J. and Sidwell, J.M. (1974), 'Natural frequency bands of branched vibrating systems', Polytechnic of Central London, Dynamics Group, Research Report No. 2
16 Carnegie, W. and Pasricha, M.S. (1972), 'An examination of the effects of variable inertia on the torsional vibrations of marine systems', *Trans. I. Mar. E,* **84**, Part 6, 160–167.
17 Cunningham, W.J. (1958), *Introduction to Non-Linear Analysis',* McGraw-Hill Book Co Ltd.
18 Draminsky, P. (1961), 'Secondary resonance and subharmonics in reciprocating engine shafts', Acta polytechn. Scand., **10**,
19 Draminsky, P. (1965), 'An introduction to secondary resonance', The Marine Engineer and Architect, Jan.
20 Archer, S. (1964), 'Some factors influencing the life of marine crankshafts', *Trans. I. Mar. E.* 76, 73–134.
21 Carter, B.C. (1928), 'An empirical formula for crankshaft stiffness', *Engineering,* 13th July.
22 Constant, H. (1929), 'On the stiffness of crankshafts', *Engineering,* 1st November.

23 Toms, A.E. and Martyn, D.K. (1972), 'Whirling of line shafting', *Trans. I. Mar E.*, **84** 176–191.

24 Volcy, G.C. Discussion on [23].

25 Panagopulos, E. (1950), 'Design stage calculations of torsional, axial and lateral vibrations of marine shafting', *Trans. A.S.M.E.*, **58**.

26 Jasper, N.H. (1954), 'A design approach to the problem of critical whirling speeds of shaft-disc systems', David Taylor Model Basin Report 890, also International Shipbuilding Progress **3** (1956).

27 Prohl, M.A. (1945), 'A general method for calculating critical speeds of rotors', *J. App. Mech.*, **67**,.

28 Tuplin, W.A. (1943), 'Torque reversals in gear drives', *Engineering*, 24th December, 503.

29 Balderston, H.L. (1969), 'Detection of incipient failure in bearings', *Materials Evaluation*, June, 123.

30 Carmody, T. (1972), 'The measurement of vibrations as a diagnostic tool', *Trans. I. Mar. E.*, **84**, no. 1.

31 Love, A.E.H. *A Treatise on the Mathematical Theory of Elasticity*, The University Press, Cambridge, Mass. 286.

32 Stokey, W.F. (1961), 'Vibration of systems having distributed mass and elasticity', *Shock and Vibration Handbook* Vol. 1, Section 7, McGraw-Hill, New York, 36.

33 Allen, R.K. (1945), *Rolling Bearings* Sir Isaac Pitman & Sons Ltd, London.

34 Ballas, T. (1969), 'Periodic noise in bearings', *Trans S.A.E.* Paper 690756, Cleveland, Ohio October.

35 Carnegie, W., (1967), 'The application of the variational method to derive the equations of motion of vibrating cantilever blading under rotation', *Bull. Mech. Eng. Ed.* **6**, no. 29.

36 Carnegie, W. (1967), 'Solution of the equations of motion for the flexural vibration of cantilever blades under rotation by the Holzer method', *Bull. Mech. Eng. Ed.*, **6**, No. 225.

37 Rao, J.S. and Carnegie, W. (1973), 'A numerical procedure for the determination of the frequencies and mode shapes of lateral vibrations of blades allowing for the effects of pre-twist and rotation', *Int. J. Mech. Eng. Ed.*, **1**, no. 1, November.

38 Koster, M.P. (1973), 'Vibrations of cam mechanisms and their consequences on design', N.V. Phillips Gloeilampenfabrieken, Eindhoven, Holland Research Reports Suppl. 6.

8 Contaminant analysis

8.1 Introduction

Liquid-washed systems transport the products of deterioration and these products may themselves interact with the carrier liquid to cause its further deterioration. By means of qualitative and quantitative monitoring both of the products of deterioration and of the carrier liquid itself it is possible to assess the extent and source of the deterioration within the system itself. Gas-transported systems may also be monitored by the analysis of their contaminants.

Lubricating oil contamination analysis involving spectrometric oil analysis procedures (SOAP) is already extensively and successfully employed. The extension of contamination analysis to the monitoring of other fluids is also well-advanced.

8.2 Contaminants in used lubricating oils

Oil used for the lubrication of internal combustion engines is likely to pick up some concentrations of the following elements, (the extent of the concentration present being related to the rate of engine component deterioration):

Aluminium	Iron
Antimony	Lead
Boron	Silicon
Chromate	Silver
Copper	Tin

To consider the sources of these contaminants in general terms:

(1) Aluminium is likely to be present when aluminium pistons are used and it indicates wear of the piston;

(2) Antimony helps to indicate whether a rise in copper content is due to wear in the crankshaft bearings or in the camshaft bearings;

(3) Borates are used as coolant anti-corrosion inhibitors and oil samples containing boron indicate coolant leakage into the oil. Special care has to be taken during emission spectrometry with the quality of the graphite used for the electrode because boron is often the impurity present in the greatest concentration in graphite, and this can lead to erroneous results;

(4) Chromates are also sometimes used to inhibit the engine cooling water and suppress corrosion, accordingly the presence of chromate in the lubricating oil indicates that a leakage is occuring from the water cooling side, possibly due to faulty gaskets or cracks in castings although this effect may be masked if chromium-plated piston rings or cylinder liners are used;

(5) Copper and lead concentrations occur in engines fitted with copper–lead bearings, increased concentration suggests incipient failure of one or more bearings.

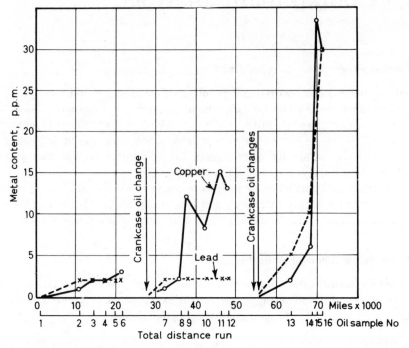

Figure 8.1 *Big-end failure based on contaminant trend*

Figure 8.1 shows a big-end bearing failure development and the associated oil contamination. With bearings made of white metal, a simultaneous rise in lead and tin would occur as a result of wear. Copper and tin increases could be caused by high wear of bronze small-end bushes. Copper-lead bearings are often finished with a lead and/or indium electro-plated flashing about 0.0002 in thick which helps the bedding-in. It may endure for several hundred hours producing a high lead concentration in proportion to that for copper. This might cause for alarm if the reason was not known.

(6) Iron concentration usually increases as a consequence of cylinder liners, piston rings or pistons wear (if they are made from ferrous materials). Piston rings which have stuck in their grooves with consequent blow-by of combustion gases and burning of the oil film (leading to scuffing, piston seizure or breakage) are characteristic consequences of iron concentration increase.

(7) Iron and silicon together in large concentrations suggest excessive liner and ring wear rate caused by dust in the intake air due to inefficient or choked air filters. Air filters fitted relatively low in the body of a motor vehicle may pass both silica and iron particles (the latter originating from brake shoe dust and contributing to the higher iron concentration, both directly, and by further increasing the abrasive wear of cylinder liner and piston rings).

(8) Silicon is mainly present as a result of silica dust aspirated with the induction

air and passing the piston rings into the sump. Some silicon concentration may arise from the presence of silicon to suppress foam in the oil, the level of silicon in the fresh oil must be allowed for when interpreting results.

(9) Silver is used to plate bearings and as a soldering material and its presence in an engine lubricating oil indicates deterioration from such sources.

8.2.1 Evaluation of contaminant sources – I.C. engines

Techniques of contaminant analysis are advanced and available. To apply a procedure with satisfaction an inventory must be prepared of the elements which go to make up any machine. A list of common contaminants and their possible origins is given in the Table 8.1.

Table 8.1 *Common contaminants and their possible sources*

Contaminant	Origin
Aluminium	Pistons (aircraft engines), bearings, air-borne dust.
Boron	Coolant leak (where borax inhibitor used)
Chromium	Platings (e.g., cylinder liners, piston rings), coolant (where chromate inhibitor used).
Copper	Bearings, bushings, sleeves, washers, etc.; piping (corrosion).
Iron	Piston rings, ball and roller bearings, gearing.
Lead	Bearings, bushings.
Magnesium	Aircraft engine components (e.g., abraded gearbox housing).
Nickel	Platings.
Silicon	Air-borne dust, silicone lubricant.
Silver	Platings (aircraft engines) bearings, piping (silver-soldered joints).
Sodium	Coolant leak, e.g., salt water, chromate inhibitor, or borax inhibitor in glycol anti-freeze.
Tin	Journal bearings.
Zinc	Brass components, neoprene seals, (measurable if not masked by zinc-containing additive in oil).

8.2.2 Permissible contaminant concentrations

The limit which can be set on the concentration level of a particular contaminant does not lend itself to simple rational specification. The decision to take a machine out of service as a consequence of a particularly high concentration level must be based on the source diagnosis, the concentration trend and the relevance of the deterioration source to the overall system reliability.

The concentration levels of metallic contaminants vary widely depending on their origin; that is, whether they are major or minor alloying constituents, the type of equipment, its operating characteristics, and the nature of incipient defects. In extreme cases, copper and iron concentrations of several hundred parts per million (p.p.m.) – by weight – have been observed while the machinery is still operable. In other instances, a concentration of less than 50 p.p.m. may reflect imminent failure.

A useful guide can be provided from the limits in Table 8.2 set up for Caterpilla engines.

Table 8.2 *General Wear Limits (Caterpillar Engine)*

System	Contaminant		Limit allowed	
Engine:	Iron	Fe	50 p.p.m.	(Direct injection engin 100–150 p.p.m.
	Aluminium	Al	10 p.p.m.	
	Silica	Si	15 p.p.m.	
	Copper	Cu	10 p.p.m.	
	Chromium	Cr	5 p.p.m.	
	Lead	Pb	Variable	Caused by main bearin coating, high values at run-in and prior to fail- ure.
	Sodium	Na	Variable	Sodium is always present in the oil. The rise in sodium indicate a water or anti-freeze leak.
Transmission:	Iron	Fe	50–200 p.p.m.	
	Aluminium	Al	10 p.p.m.	
	Silica	Si	20–50 p.p.m	
	Copper	Cu	100–500 p.p.m.	
	Magnesium	Mg	Variable	An indication of brake lining wear on track- type tractors
Final Drive:	Iron	Fe	100–200 p.p.m.	
	Silica	Si	20–50 p.p.m.	
	Copper	Cu	50 p.p.m	
Differential:	Iron	Fe	40–500 p.p.m.	
	Silica	Si	20–50 p.p.m	
	Copper	Cu	50 p.p.m	
Hydraulics:	Silica	Si	10–15 p.p.m (max)	

Apart from these general limits it is sometimes possible to express the extent t which contamination should be allowed in terms, of high, medium, low and norma concentration, expressed as a multiple of normal readings, these may be:

Al, Cu, Fe, Si:	High	2 to 3 times normal
	Medium	1.5 to 2 times normal
	Low	1.25 to 1.5 times norma
Cr:	High	above 5 times normal
	Medium	3 to 5 times normal
	Low	2 to 3 times normal

The only positive condemning limit is that for silica. When values reach medium levels, an engine is progressing towards failure. Deviation from the normal conditions may dictate corrective action before high values are reached, thus, copper particles due to engine thrust washer wear can reach 100 to 200 p.p.m., and the engine will operate satisfactorily; on the other hand, if aluminium values indicate failure of the oil pump bushings on an engine with aluminium bronze bearings, even a small increase is a danger signal.

8.3 Carrier fluid degradation

Quite separately from the presence of transported contaminant, the carrier fluid may itself be degraded due to plant or engine deterioration. Monitoring of this degradation may in some instances provide the means of diagnosing the condition of the whole system

During the course of time, crankcase lubricating oils display changes in properties which are directly related to the condition of the engine; these may be summarized as in the Table 8.3.

Table 8.3 *Pollutants and their effect on lubricating oil*

Lubricant property	Pollutants influencing changes in properties
Acidity	Oxidation products; sulphurous products
Alkalinity	Possible additives
Ash	Base mineral constituents
Flash point	Fuel dilution
Insolubles	Carbonaceous products; dust wear products; corrosion products; additive degeneration products
Specific gravity	All
Viscosity	Fuel dilution, water oxidation products
Viscosity index	Different oil mixes

Acidity in an oil in service frequently arises from the presence of weak organic acids. There is no consistant relationship between these acids and the corrosion of bearings such as to warrant the monitoring of this property with a view to assessing plant condition, although tests by Perrier [1] using the STM Control Kit at the French Fleet Fuel and Lubricant Testing Station (SECLF) indicated increased metal-in-oil content when the acidity exceeded a pH of 4.0 and strip-down after 300 hours following the development of this acidity revealed extensive corrosive wear of bearings and deterioration of tappets, rocker arms, shafts and piston rods.

Depletion of additives from lubricating oil may be assessed from the sulphonated ash content. In service, a proportion of the additive is removed by adsorption on the insoluble matter which is being maintained in suspension; some of the additive decomposes and becomes insoluble in the oil. By comparing the sulphated ash content with that of the new oil, the fraction of the additive which has remained soluble, is indicated, but not the active additive content, since some of the degradation products may still be soluble.

Fuel oil dilution causes a decrease in viscosity. Oxidation and the presence of insolubles causes the viscosity to increase. Flash point tends to increase due to the removal of the light fractions from the oil; this increase is particularly noticeable with low viscosity oil. Fuel dilution causes a decrease in flash point.

Viscosity will change throughout the various parts of an engine according to the local temperature of the parts being washed. Thus the Redwood No. 1 seconds viscosity for various temperature were given by Bairstow [2] below.

Table 8.4 *Effect of local temperature on lubricating oil viscosity*

Temperature °F.	S.A.E. No.				
	10W	20	30	40	10W/30
50	812	1712	3250	5280	1185
70	406	770	1380	2150	650
140	77	113	165	234	129
210	42	48	56	69	56
210(cS)	6	8.5	11	15	11

8.3.1 Insoluble matter in oils

The principal insoluble matter which passes into the engine lubricating oil is:
(1) Carbon: from incomplete combustion of the fuel,
(2) Oxidation products: from the oil and fuel,
(3) Wear products: from moving parts,
(4) Air-borne dust residues: which pass through the air filters,
(5) Degradation products: from the lubricating oil additives.
Analysis of the 'insoluble matter' may be derived in two separate ways and give two particular values:

(1) Insolubles in some appropriate diluent such as I.P. Spirit, pentane or *n*-heptane, which does not have any appreciable solvent action on any of the products mentioned. This value is sometimes described as 'total insolubles,' it is important for assessing the ability of the detergent oil to maintain the total insoluble matter in a dispersed form.

(2) Insolubles in benzene or chloroform which do not include the oxidation products soluble in these diluents.

The difference between the two values often referred to as 'asphaltenes,' represents the quantity of oxidation products present and in some measure of general oil degradation. A rapid increase can have serious implications, particularly if this is not accompanied by a similar increase in total insolubles. Under these conditions it shows that the oil is not only overloaded with insoluble matter, but has also lost the power to maintain in dispersion that which has already accumulated.

8.3.2 Water contamination

Lubricating oil may involve a foam or suspension of insoluble water due to the condensation of products of combustion which blow past piston rings or by leakage from the cooling system through defective gaskets or cracks in the cylinder head block. Cooling system leaks are more usual. Water in crankcase oil promotes the formation of sludge and may interfere with the dispersive properties of detergent additives.

8.4 Contaminant monitoring techniques (wear processes)

Spectrometric oil analysis procedure (SOAP) and magnetic plug inspection systems are currently in extensive use as means of condition monitoring. Ferrograph and X-ray spectrometry are becoming of increasing interest while for special selective monitoring a range of alternative techniques are available.

A brief description of various contaminant monitoring techniques is given in alphabetic order in the Table 8.5.

Table 8.5 *Various contaminant monitoring techniques*

Technique	Brief description
Electron spectroscopy	Samples of wear have been examined by the 'Electron Spectroscopy for Chemical Analysis' (ESCA) technique. The spectra of Fe_3O_4 dominates, probably because ESCA is a 'surface' technique and restricts the analysis to a surface layer of about 30 A thickness. The application of ESCA has been fruitful in the investigation of material transfer between pairs of bearings of dissimilar materials.
Electron spin resonance spectroscopy	Used for the investigation of the chemical nature of ferromagnetic products in oil debris and presence of organic free radicals.
Ferrography	Separates particles from lubricating oils according to their particle size and provides a measure of particle size distribution. The total density of particles and the ratio of large to small particles may indicate the type and extent of wear.
Induced radioactivity	Components which have been previously 'seeded' or irradiated with thermal neutrons in a reactor have their debris examined by radiochemical monitoring.
Magnetic plug	Suitable magnetized plugs fitted into the oil-washed parts of a lubricating oil system collect ferrous and other debris. The quantity of debris collected, its shape and colour can provide a useful guide to deterioration activity.
Mass spectrometry	Quantitative analysis of milligram amounts of iron and iron oxides by means of a high vacuum apparatus coupled to a small mass spectrometer. Chemical reactions on the wear debris produces gaseous products which relate to the source material.

Table 8.5 (continued)

Technique	Brief description
Mossbauer spectroscopy	To identify the different chemical forms of iron in wear debris.
Particle size distribution	Particles collected from the fluid either by removal from a filter or separation from a specimen are counted in size groups and their statistical distribution used as a measure of the suspension density.
Scanning electron microscopy	This technique is used to determine the shape of debris. The wear particles formed during the progression of a wear process are believed to have characteristic shapes, for example spherical particles are formed during rolling contact fatigue [3,4].
Spark source mass spectrography	This produces an elemental analysis of solid materials and in favourable cases has a sensitivity of 0.001 p.p.m. and can provide gross analysis of very small samples.
Spectrometric oil analysis programme	This uses the build-up of metallic elements as measured by an emission or atomic absorption spectrometry, it indicates the location and degree of wear. This technique is already established in industry, aviation and marine transportation.
X-ray fluorescence	A useful, although somewhat expensive instrument which can be used to analyse for the same metallic elements that can be analysed by atomic absorption spectrometry with the exception of lithium and magnesium. In addition, sulphur and chlorine (often present in EP and antiwear additives) as well as iodine and bromine can be analysed. X-ray fluorescence is not as accurate in the parts-per-million range.

8.5 Oil degradation analysis

In addition to the physical effects of carrier fluid degradation described in Section 8.3 it is possible for iron particles to dissolve, crystalline organic products to evolve and anti-oxidant additives in oil to be oxidized by the catalytic action of the wear debris.

These complex secondary and tertiary consequences of carrier fluid degradation involve a complex laboratory investigation. The techniques which are most applicable to service monitoring are (a) thin layer chromatography, (b) STM control kits [3].

Some of the oil analysis techniques in use are given in Table 8.6.

8.5.1 Straight and additive used oil degration analysis methods

Products which pollute or degrade straight base-stock oil are insoluble and thus straight oils can be analysed by adding a solvent to a specimen, centrifuging to remove clear layer from the top, evaporating the solvent and weighing the insolubles which remain.

Additive oils contain detergent-dispersant anti-agglomeration compounds and thus by their very nature the insoluble products cannot be centrifuged. Based on ASTM Method D893, a coagulant may be added and the specimen subjected to high-speed centrifuging. Problems associated with this technique in relation to engine lubricants were discussed by Metzger [6] – the coagulent makes the specimen a 'foreign product' and is included in the weighed insolubles.

A calibrated-membrane filtration proposed by Frassa, Siegfriedt and Houston [7] is a satisfactory method for additive oils. Membranes of cellulose esters with pores of uniform size and regular distribution separate insolubles according to their size between 0.1 and 8 μm. Although this method indicates particle size and the anti-flocculant, dispersive properties of an oil it involves delicate techniques which limit its use.

Table 8.6 *Oil degradation analytical techniques*

Technique	Application
Electron spin resonance spectroscopy	To indicate the presence of long-level free radicals in samples of used oil containing wear product debris.
Gas chromatography	A sensitive technique of electron capture for the detection of certain derivative products in oil or for detecting very low levels of added tracer elements.
Gradient elution chromatography	An alternative technique to gas chromatography which also separates component and products in oils.
Photoluminescence	Fluorescence and phosphorescence spectra obtained by this technique provide a particular sensitive method for assessing anti-oxidants and oxidation products.
STM kit	Used to monitor for oil dilution, acidity and pollution.
Thin layer chromatography	Separates anti-oxidants and their products for identification by calorimetric or spectrometric methods.
U.V. and visible absorptionmetry	Concentrations of certain anti-oxidants and their oxidation products can be determined by direct absorptiometry at specified wavelengths or by means of colorimetric reaction.

Standard infra-red spectrometry evaluates the molecular structure of an oil, and produces a spectrum of wavelengths which 'fingerprints' the oil. Differential spectrometry uses two beams, one of which passes through a new oil and the other through a used oil so that the two absorption spectra can be simultaneously plotted in order that significant differences can be identified.

8.5.2 Spectral bands of typical oil pollutants

The wavelengths of typical spectral emissions for various oil pollutants used as a basis of degradation 'finger printing' were given by Callat [8] (Table 8.7).

Table 8.7 *Typical used oil pollutant finger-printing spectra*

Wave length (μm)	No. waves per cm	Contaminant
2.9	3400	Water
5.85	1750	Oxidation products
6.1	1630	Nitrogen-containing products
7.9–11.6	1280– 860	Organic nitrates
9.3– 9.7	1030–1070	Ethylene glycol
12.4–12.8	810– 780	Hydrocarbons

8.6 Abrasive particles in lubricating oil

Any foreign solid substance found in the lubricant can cause abrasion, where the degree of which depends upon the hardness and size of the particles. Hard particles, such as sand from improperly cleaned castings or road dust which has entered through inadequate sealing, can cause severe wear. Gearmarking compounds containing titanium dioxide are particularly abrasive.

Abrasive particles will wear or lap the bearing surfaces. This wear causes changes in bearing adjustment which can in turn affect other parts of the system in which the bearing is used. Although used lubricants are usually analysed to determine the type and amount of debris present, it is often helpful to examine the debris under a medium-power microscope, 100-X magnification is usually adequate.

Particles which can sometimes be identified by microscope include sand, coal dust, paper or rag fibres, seal wear debris and many more. Lubricants from damaged bearings often contain wear debris caused by different contaminants. Molybdenum disulphide, sometimes added to lubricants, appears as metallic particles under high magnification and may be mistaken for dirt or bearing wear particles. The *Particle Atlas* [9] aids in microscopic identification for the more experienced microscopist. Automatic particle counters make a quick count of the size and distribution of the particles in oil.

A simple way to tell if road dirt or sand is present is to determine what percentage of the ash is insoluble in acid. Boiling the ash in six normal hydrochloric acid will usually dissolve everything in the ash except silicon dioxide, the major constituent of sand and many soils. The solution is then passed through ashless filter paper. The residue remaining after ashing the paper and contents is called an 'acid insoluble.' A preferred and more reliable method would be instrumental elemental analysis.

8.7 Abrasive particles in bearings

Post mortem examination of bearings may at times be needed to relate contamination monitoring to failed components. Particles in lead overlay bearings can be extracted with an acetic acid/hydrogen peroxide mixture; those in aluminium–tin bearings with caustic soda. Both mixtures can loosen particles in tin–lead white metals. Embedded ferrous particles can be quickly identified by 'iron printing'. Unglazed paper, previously soaked in a 5 wt% solution of potassium ferricyanide to which a few drops of hydrochloric acid and wetting agent have been added, is placed in contact with the degreased bearing surface for about 30 seconds. The paper is then removed, and blue spots indicate the presence of ferrous particles. Similar techniques are available for the identification of particles of most common non-ferrous metals [10]. Fine particles, smaller than normal bearing clearances, can circulate with the oil and erode the bearing surface. Hard particles erode deep, well-defined channels whereas soft particles give rise to more general erosion, particularly on soft electrodeposited overlays. Erosion by fine particles is most prevalent on high-speed bearings, and may be associated with cavitation-erosion damage.

Lead can be leached out of copper–lead on lead–bronze bearings by the weak organic acids that may accumulate in used oils. The presence of lead in the surface layers of the bearing and hence the absence of corrosion can be verified by lead printing. In this test unglazed paper is immersed in a weak (3 wt%) sodium hydroxide solution and after the surplus liquid has been removed the paper is placed in contact with the degreased bearing surface. After about 30 seconds the paper is removed and 'developed' in a 3 wt% solution of freshly prepared sodium sulphide. A dark-brown colour indicates the presence of lead.

8.8 Abrasive particles in hydraulic systems

Contaminating particles enter a hydraulic system as a result of:

    (a) in-built particles,

    (b) particles which enter the system,

    (c) contaminating particles generated by the system,

    (d) particles or contaminants arising from reactions within the system, such as corrosion or cavitation products. [11]

Olsen [12] concluded that in aircraft systems using synthetic hydraulic fluids an electro-chemical corrosion of metal removal occurred. This was reputedly caused by the electrical current generated by the streaming potential resulting from the searing of oil through a valve.

Water presence in hydraulic oil has a marked effect on the wear produced by contamination products according to a study of Swain [13] confirmed by studies in the U.S.S.R. [14] showing that the damaging effects of cavitation are substantially increased if abrasive contamination is introduced into the cavitational field.

The bodies which contaminate a hydraulic system exist as a silt which contains solid particles of various chemical compositions in an infinite range of shapes, sizes and degrees of hardness. Particle sizes for most hydraulic systems are considered to range from 5 to $100 \, \mu m$ although in some critical systems particles down to $1 \, \mu m$ may be considered ($1 \, \mu m = 0.00004$ in approximately). Since particle shapes vary, dimensions are usually considered as the horizontal dimension when viewed under a microscope, Figure 8.2(a). If the longest dimension exceeds $100 \, \mu m$ and the length/breadth ratio exceeds 10, the body is called a fibre, Figure 8.2(b).

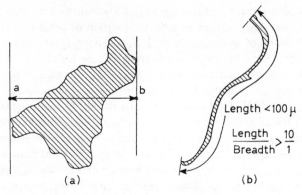

Figure 8.2 *Particle measurement (a) maximum dimensions; (b) fibre measurement maximum dimensions*

Considering the characteristics of ultrafine silt particles, these are only from 1 to 5 μm in size (1 μm = 10^{-6} m) and invisible to the human eye (resolving power of the unaided eye is approximately 4 μm; the diameter of the average human hair = 75 μm; the diameter of red corpuscle of human blood = 7.5 μm).

The particles gradually wear away components and a silt particle actually generates more contamination as it works its way into the moving clearances and acts as a lapping compound to grind away at surfaces – a phenomenon often termed the 'chain' reaction of wear. A system can function satisfactorily if the wear-causing silt particles can pass through the clearances between the actual moving parts of each component. According to research by Fitch [12] components are sensitive to specific sizes and concentration of contamination. Clearances in a typical hydraulic system are mostly ultra fine ranging down to 0.5 μm.

A typical contamination level record quoted by Hutchings [15] quoted a 3 month field test of a vehicle in which the 25 μm nominal traditional OEM filter was retrofitted with a 3 μm absolute filter element. In 20 minutes it reduced the total count exceeding 5 μm per 100 ml from over 2 million to 20 000. During the subsequent 300 hours contamination level was further reduced to less than 2 500 particles/100 ml a new OEM 25 μm nominal filter was then installed in place of the 3 μm absolute filter and after a further 100 hours the count increased to 800 000 particles/100 ml.

In a hydraulic system the distribution of particle sizes settles down to an equilibrium contamination level such that the number of particles of a size being taken into suspension by the operating liquid is approximately balanced by the number of particles being removed by the contamination control devices such as filters. Each system produces its own characteristic particle distribution pattern.

Particle size measurements of one substance obtained by a variety of methods is almost certain to differ because of observational errors, differences in sample preparation, fundamental disagreements concerning the parameter actually measured and the method of presentation [17]. In a recent study, Busky [18] of the Research

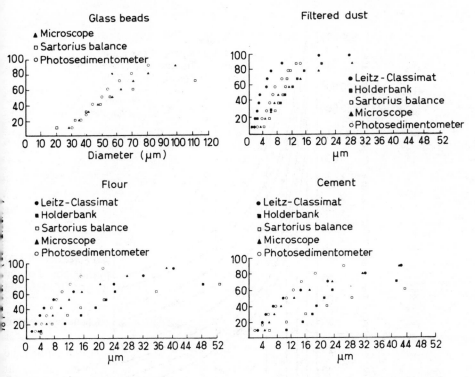

Figure 8.3 *Cumulative distribution curves for four materials – glass beads, filtered dust, flour and cement – each determined by up to five different, standard particle-size measuring techniques*

Institute for Technical Chemistry at the Hungarian Academy of Sciences, Vezprém considered the critical factors for glass beads, filtered dust, flour and cement using measuring techniques involving a common Leitz-Classimat automatic microscope, a Holderbank air separator, a Sartorius sedimentation balance and a modified photo-sedimentometer[19, 20] and produced the cumulative distribution curves shown in Figure 8.3.

The procedure adopted to monitor the particle distribution in hydraulic systems in the Royal Navy and reported by Kitch [21] requires a 100 millilitre sample of fluid to be extracted from the system. This is filtered through a membrane filter and the contamination on the membrane either visually compared on a microscope with a Master Slide which has been contaminated to represent each R.N. Contamination Class or alternatively counted under a microscope.

Some classification systems have been devised such as ASTM/SAE or NAS 1638, but these were not entirely satisfactory for the Ministry of Defence (N) who accordingly took hundreds of samples of hydraulic fluid from different systems over a period of years and analysed particle distribution patterns from which the R.N. range of contamination classes was prepared Figure 8.4. Each class is defined in

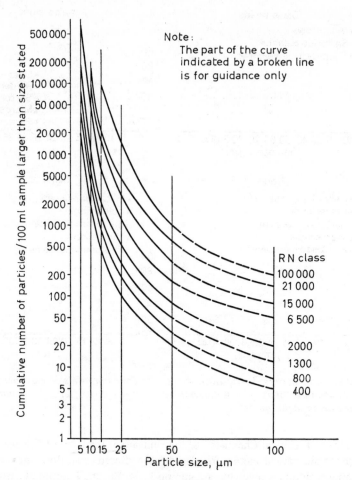

Figure 8.4 *Contamination classes approved by the Royal Navy*

terms of the maximum number of particles larger than the specified particle sizes in a 100 millilitre sample.

Optical sizing can be achieved for particles in the 1.0 to 250 μm range by using a double image technique [22]. The instrument has two basic parts in addition to the microscope itself, namely:

(1) A control unit which provides (a) power to the various devices and circuits, (b) preset resistors for setting up the limits of the desired size groups, (c) a manually operated switch to select a particular size group, (d) eleven counters, one for each of ten channels and one totalizing counter, (e) a micrometer control for varying the shear between the two images, (f) a meter for measuring the current to the vibrator and, after calibration, the size in microns of the particle under examination, (g) controls such as mains on/off etc.

(2) A vibration unit which creates two images whose shear is dependent on the magnitude of the current through the vibrator.

In the condition of edge-to-edge shear the meter current is a measure of one diameter of the particle under observation. Other diameters of irregularly shaped particles may be measured by rotating the stage of the microscope and adjusting the amount of shear, as necessary.

8.9 Dissolved gas fault monitoring

Gases produced as a consequence of faults have been monitored as a basis of failure prevention in electrical systems. The fault gases may be monitored under conditions of gas-in-liquid, gas-in-solid and gas-in-gas.

8.9.1 Gas-in-liquid failure monitoring

Dissolved gas analysis was first applied by Waddington and Allan of G.E.C. Power Engineering Ltd., Trafford Park, Manchester, [23] early in 1961 when a series of faults developed in certain small distribution transformers. These compact oil/paper units were not fitted with Buchholz relays so that the evidence of arcing or overheating could be obtained only by extracting the dissolved gases from a sample of the oil by vacuum treatment. With these early analyses infrared spectroscopy only was used to check for acetylene and carbon monoxide in the oil; the absence of these was deemed sufficient to show that neither serious arcing nor insulation breakdown had occurred.

The first application of dissolved gas analysis to hot spot problems occurred with transformers containing core/earth faults; such faults can produce localized very high temperatures which slowly break down the oil and produce ethylene gas.

Gas-in-oil analysis can be used to pin-point the position of a gas-evolving fault. A typical example of this occurred with a 25 kV transformer which had produced a considerable volume of gas sufficient to activate the Buchholz. In this design of transformer the oil for the three-phase selectors, though having a common supply, is in tanks separate from each other and from the main oil tank. Analysis of the dissolved gas in the oil from the four tanks gave the results detailed in Table 8.8.

Table 8.8 *Analysis of Dissolved Gas in a 275 kV Transformer*

	H_2	CO	CH_4	C_2H_6	C_2H_4	C_2H_2	%CO_2
Main tank (p.p.m.)	26	4000	240	60	500	–	0.4
Phase A tank (p.p.m.)	36	2300	475	110	520	–	0.3
Phase B tank (p.p.m.)	31	4580	300	100	250	–	0.37
Phase C tank (p.p.m.)	665	5900	9150	1600	20 240	50	0.33

From this analysis it was evident that the fault was metal-to-metal arcing causing a severe hot spot and was located in the phase C tank. The CO present was not limited to the fault area and was attributed to normal thermal ageing of the oil rather than insulation breakdown. A simple inspection showed this diagnosis to be correct, the arcing being due to the burnt contacts in the tap changer.

Data from several current transformers and several voltage transformers were compared with those from transformers known to be discharging under over-voltage testing in the factory. It was found that only the methane component of the partial discharge gas could be relied on for diagnostic work. Hydrogen could be formed without discharge, a correlation existing between moisture content of the cellulose and hydrogen evolution under stress only. Carbon monoxide was always present as a product of normal ageing of cellulose/oil.

Experience shows that a discharge level of 10 pC can be detected after a few hours' discharge by degassing approximately a pint sample from a current transformer and by degassing 1 or 2 gallons from a power transformer.

Partial degassing can be achieved by simply displacing the dissolved gases from the oil sample by bubbling air or nitrogen through the sample or by a piston type expansion chamber, while a portable combustiometer or katharometer can be used to analyse the gas evolved.

8.9.2 Gas-in-solid failure monitoring

Gases which arise from electric discharge or thermal breakdown in appreciable thickness of solid insulation can take a very long time to diffuse out of the material.

In certain respects, gases dissolved in insulating solids may be compared with gases dissolved in metals, but only thermoplastic materials such as polystyrene, polythene and Perspex can be readily degassed by a vacuum fusion technique similar to that used for metals. The majority of insulating materials are thermoset compounds and a special technique is needed to remove dissolved gases for analysis.

The insulation is first cut into thick sections ($\simeq 25$ mm) and these are then reduced to wafers or chips about 3 mm thick, by rapid delamination (in the case of laminates) or by hammer shattering the case of brittle plastic or porcelain, and immediately immersed in degassed transformer oil. The gas then diffuses into the liquid over a period of several hours, and is subsequently extracted by vacuum technique and analysed by gas chromatography. Such methods have been used by Waddington and Allan [23], Brown Boveri [24] and the Westinghouse Electric Co. of America [25]. Typical gas analyses involving oil and various solid insulants are given in Table 8.9.

Carbon monoxide (CO) concentrations are particularly characteristic of the type of insulation involved; acetylene (C_2H_2) predominates when oil alone is involved; methane (CH_4) is high in the case of arcing epoxy glass cloth.

8.9.3 Gas-in-gas failure monitoring

Odours from fumes or decomposing insulation can be detected to sensitivities of 1 part in 10^9 by using long path infrared spectroscopy technique. Sulphur

Table 8.9 *Arc gas composition by volume (%)*

	H_2	CO	CO_2	CH_4	C_2H_6	C_2H_4	C_2H_3	O_2	N_2
Oil only	60	0.1	0.1	3.3	0.05	2.1	25	2.4	6.3
Oil/kraft paper	52	14.0	0.2	3.8	0.05	8.0	12	3.0	6.7
Oil/pressboard laminate	48	27.0	0.4	5.0	–	5.0	6	2.0	6.2
Oil/alkyd paint	55	20.0	0.2	4.0	–	5.0	8	2.4	7.0
Oil/polyurethane enamel	60	1.0	0.1	9.0	–	11.0	10	2.0	6.0
Oil/p.v.a. enamel	61	5.0	0.1	6.0	–	14.0	5	2.5	6.5
Oil/epoxy glass cloth	57	2.0	0.1	14.0	–	10.0	8	2.5	6.5
Oil/isophthalate cotton tape	55	11.0	4.0	8.0	–	8.0	5	–	–

hexafluoride (SF_6), nitrogen, air, and hydrogen are typical of gaseous systems into which fault gases may evolve either from breakdown of the gas itself (as with SF_6) or from solid/liquid insulation breakdown within the system. Adequate samples for highly sensitive analysis are difficult to obtain. In a large system (e.g. a nuclear reactor or turbine generator) any fault gases may be rapidly diluted.

Gas chromatography is a convenient method of analysing traces of organic and permanent gases. Gases are selectively absorbed by being passed through long columns of finely divided absorbent material. The separate gases are subsequently desorbed by the passage of an inert carrier gas through the column; light gases appear first from the column and more complex gases last. During the 1960s it became possible to 'condense' samples of the gas from a large system by 'freezing' the adsorbed gases directly from the system on to convenient short lengths of chromatographic column known as pre-columns.

Equilibrium conditions can be achieved for typical fault gases, with a sensitivity which has allowed us to detect one part in 10^9 by volume. This technique has been applied with some success to nuclear power systems and turbine generators, and is well suited to the monitoring of SF_6 and nitrogen sealed systems.

It is necessary, when applying this technique, to choose a diagnostic gas which is not detectable in any quantities in the original insulating gas.

References

1 Perrier, P. (1973), 'The French Navy Control Kit', *Rapid Methods for the Analysis of Used Oils* Scientific Publications (GB) Ltd., Broseley, Shrops.

2 Bairstow, S. (1961), 'Control of quality of crankcase lubricating oils of locomotive diesel engines in service', *Proc. I. Locomitive Engineers* (I. Mech. E) **17**, January.

3 Endo, K. and Kotani, S. (1973), 'Observations of steel structures under lubricated wear', *Wear*, **26**, 239–251.

4 Loy, B. and McCallum, R. (1973), 'Mode of formation of spherical particles in rolling contact fatigue', *Wear*, **24**, 219–228.

5 Collacott, R.A. (1946), 'Acidity and alkalinity of liquids', *Mechanical World*, January 4.

6 Metzger, G. (1963), 'The use of high speed centrifugation to demonstrate the lack of a simple relationship between engine cleanliness and suspended crank-case insolubles', S.A.E. Preprint 776B, October.

7 Frassa, K.A., Siegfriedt, R.K. and Houston, C.A. (1965), 'Modern analytical techniques to establish realistic crankcase drains', S.A.E. Preprint 951D, January.

8 Callat, R. (1973), 'Rapid laboratory methods; differential infrared and calibra-ted-membrane filtration', *Rapid Methods for the Analysis of Used Oils*, Scientific Publications (GB) Ltd., Broseley, Shrops.

9 Crone, W.C. (1967), *The Particle Atlas*. Ann. Arbor Science Publishers Inc.

10 Hunter, M.S., Churchill, J.R. and Mear, R.B. (1942), 'Electrographic methods of surface analysis', *Metal Progress*, **42**, 1070.

11 Farris, J.A. (1970), 'The control of silt and abrasive wear in hydraulic systems', Field Service Report No. 47, September, Pall Corporation, Glen Cove, NY 11542.

12 Olsen, J.H. (1969), 'Flow-induced damage to servo valves using phosphate ester hydraulic fluids', Vickers, Aerospace Fluid Power Conference, (ref: Pall Corporation, Glen Cove, NY 11542).

13 Swain, J.C. (1970), 'Some effects of dirt and water contaminants on vane pump life', National Conference on Fluid Power, October, (ref: Pall Corporation, Glen Cove, NY 11542).

14 Kozyrev, S.P. (1962), 'Cavitational wear of materials in a hydro-abrasive stream', A.S.M.E. booklet, *Friction and Wear in Machinery*, **17** (translated from the Russian).

15 Filtch, E.C. Jr. (1969), 'Representation of contamination levels and tolerances', *A.S.L.E. Transactions*, **12**, No. 3 July Paper 69AM1B-1.

16 Hutchings, C.S. (1974), 'Contamination control in machine tool hydraulics systems', *Machine Tool Review*, **62**, No. 359, May.

17 Filtch, E.C. Jr. (1963), 'Classification of methods for determining particle size', *Analyst* **88**, 156.

18 Bucsky, G. (1975), 'Accounting for the differences between particle sizing techniques', *Process Engineering*, July, 52–54.

19 Cadle, R.D. (1965), 'Particle size, theory and applications', Reinhold.

20 Allen, T. (1968), 'Particle Size Measurement', Chapman & Hall Ltd.

21 Kitch, D.L. (1974), 'Hydraulic systems cleanliness – design, manufacture, operation and maintenance aspects', *Trans. I. Mar. E.* **86**, 156–162.

22 Anon (1974), 'Particle size micrometer and analyser type 526', Fleming Instruments Limited, Stevenage, Herts. SG1 2DE.

23 Waddington, F.B. and Allan, D.J. (1969), 'Transformer detection by dissolved gas analysis', *Electrical Review*, 23 May 3–7.

24 Howe, V.H., Massey, L. and Wilson, A.C.M. (1956), 'The identity and significance of gases collected in Buchholz protectors' Metropolitan-Vickers Gazette, 27 May 139–148.

25 Waddington, F.B. (1974), 'Fault protection in power systems by tracer-gas analysis', *GEC J. Sci. Tech.*, **41** Nos. 2/3 89–93.

26 Barrett, G.M. (1961), 'Spectrographic analysis of crankcase lubricating oils as a guide to preventive maintenance of locomotive diesel engines', *Proc. I. Loco. Eng.*, 17th January.

9 SOAP and other contaminant monitoring techniques

9.1 Introduction

Continuous contaminant analysis is realistically and regularly employed by some major industries. At this date, the techniques are operated on laboratory-centered procedures but methods are being developed to produce miniaturized equipment which can be fitted to a machine or 'plugged-in' and used as a normal operational control.

9.2 Spectrometric oil analysis procedure

Four independent basic operations are involved in the practice of a spectrometric oil analysis procedure as follows:

(1) sampling,
(2) spectrometric analysis,
(3) diagnosis — data interpretation,
(4) validation of the diagnosis.

Responsibility for one or more of these operations must be given to a clearly delegated individual and confusion avoided at all stages by accurate and complete documentation.

It is usual for the equipment user to draw the sample and despatch it with the pertinent history to the laboratory. The analyst usually obtains the spectrometric results and assists in the diagnosis/data interpretation. In the event of equipment being dismantled on the basis of a SOAP diagnosis a validation report should be obtained for proper assessment of the validity of the previous procedure.

9.2.1 Scheduled oil sampling

Routines and techniques necessary to obtain genuinely representative oil samples for spectrometric analysis were described by Klug [1]. There are three accepted methods of obtaining a sample:

(1) sampling valve;
(2) vacuum sampling;
(3) drain stream.

Samples should be taken when the oil is hot and well-mixed with possible contaminants after a running period.

A sampling valve fitted in an oil pile-line provides a means of obtaining a sample between oil changes without shutting down the machine or wasting a large quantity of oil. The valve should be readily accessible and located away from moving parts. U.S. railroad diesels use a miniaturized plug-shut-off valve with 'O' rings to prevent frictional metallic contact with the valve stems.

Vacuum sampling involves the use of a manually operated suction 'gun' (pump) to draw off oil samples from a crankcase and various other machine compartments. Drain stream sampling is the procedure of taking samples when a sump is being drained during an oil change. For purposes of oil contamination analysis neither the first nor the final drainings are representative, the oil drained at about half-way is preferred.

A tight control on sampling is necessary. Many plant engineers may be inclined to omit sampling after the first one or two and a diary monitor check should be introduced by supervisory engineers to ensure that machines are regularly sampled. The maximum benefit from contamination analysis is only obtained when sampling is done on a continuous basis.

9.2.2 Sampling frequency

The frequency with which sampling is carried out depends upon trend changes shown by the spectrometric analysis together with such operational factors as:

(1) Function — i.e. importance of the machinery;

(2) Age — time since overhaul;

(3) Operating schedule, loading characteristics;

(4) Safety considerations;

(5) Rapidity of failure from defect initiation;

(6) New equipment with possible infant mortality requires frequent sampling, this also provides an adequate basis for trends studies.

With motor vehicle engines, samples should be taken before any 'topping up' of lubricating oil, at 250 hour periods, before and after scheduled oil changes. For aero-engines the frequency is 10 ± 2 hours for engines and transmissions in jet air-craft, 25 ± 5 hours for reciprocating engines.

9.2.3 Sampling precautions

Essential precautions to be observed in sampling are as follows:

(1) The sample container must be absolutely clean, showing no visible trace of dirt, water, or other matter; it should be discarded after use;

(2) Extreme care is necessary during sample withdrawal to ensure that no foreign contaminants are introduced; if the sample is taken by gravity flow, the first few millilitres should be discarded before filling the sample bottle.

(3) Sampling must be done during operation or shortly after shut down of the machinery while the lubricant is at normal operating temperature, and before particulate settling can occur.

(4) A complete identification must accompany the sample, including: date of sampling, hours since overhaul, hours since change, and amount of oil added (since

last sample). Also any other pertinent information relating to the operation of the machine (e.g. minor repairs) should be reported.

9.2.4 Spectrometric analysis

Two methods of spectrometric analysis are available and used:
   (1) Emission spectrometry (electric spark)(ESS);
   (2) Atomic absorption spectrometry (flame)(AAS).
   Both the emission and absorption methods are based on related underlying principles of atomic physics involving complementary energy changes within atoms which give rise to the emission or absorption of light.

9.2.5 Principles of spectrometry

In the conventional concept, an atom consists basically of a central nucleus about which one or more electrons revolve in fixed orbits. The precise number of electrons differentiates the various elements; for example, nickel has 28 electrons, and copper 29. Energy changes within an atom may involve either the nucleus or the electrons.

In the present context, only energy changes accompanying outer electron transitions are of interest. The light absorbed or emitted as a result of these transitions corresponds to the ultraviolet and visible regions of the energy spectrum with wave-lengths ranging from 2000 to 8000 Angstrom (Å) where 1 Angstrom = 10^{-8} cm. A spectrometer is designed to detect and measure radiation within these wavelengths.

Typical wave-lengths are given in Table 9.1.

Table 9.1 *Spectral wavelengths used in atomic absorption determination of metals in lubricating oils*

Element and Chemical Symbol	Wavelength (Å)	
Copper (Cu)	3247	(3274)[*]
Iron (Fe)	3270	
Chromium (Cr)	3579	
Nickel (Ni)	3415	
Lead (Pb)	2833	
Tin (Sn)	2354	
Sodium (Na)	5890	(3302)[*]
Aluminium (Al)	3092	
Silicon (Si)	2516	
Magnesium (Mg)	2852	
Silver (Ag)	3281	

[*] Alternative lines for determining high concentrations.

9.2.6 Emission spectrometry (electric spark)

A direct high-voltage (15 000 V) excitation, of the metallic impurities in the oil sample causes them to emit their characteristic radiations which can be spectrally analysed.

The spectrometer consists essentially of a narrow slit, a grating or prism to separate the component wavelength of the radiation after it has passed through the slit, and a photoelectric system to detect and measure the spectral radiation (Figure 9.2.) By using separate photodetectors for each element and a typewriter readout, measurements can be made in rapid sequence, and a complete analysis is almost immediately available.

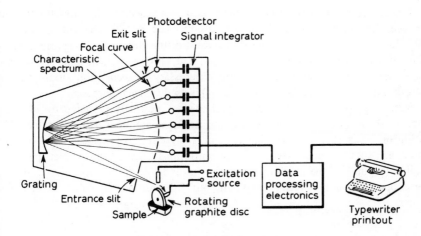

Figure 9.1 *Method of analysis by emission spectrometry*

9.2.7 Atomic absorption spectrometry (flame)

Five basic instrumental components are essential to conduct atomic absorption analysis, as follows:

(1) An energy source which emits light characteristic of the element to be determined – in practice, a hollow-cathode discharge lamp;

(2) An energy source to transport and atomize the metallic elements in the sample – usually a flame, e.g. an air–acetylene gas combination;

(3) Wavelength selector, that is, a monochromator, to isolate an appropriate wavelength of light for measurement; consists of a grating or prism and associated lens and slit systems;

(4) An energy detector which converts the light energy to an electrical signal – normally a photomultiplier tube and associated amplifier circuitry;

(5) A readout device which quantitatively measures the magnitude of the electrical signal, for example, a meter, chart recorder, digitizer, or printer.

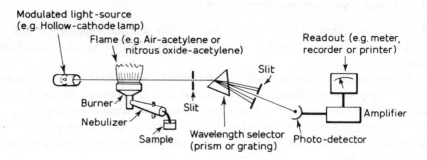

Figure 9.2 *Schematic illustration of the atomic absorption principle*

The arrangement of these components is shown schematically in Figure 9.2. An air—acetylene flame is normally used to atomize the metallic elements in the sample. This type of flame is satisfactory for most of the elements of diagnostic significance in oil analysis, including copper, iron, lead, chromium, nickel and sodium but for some elements an even hotter flame is necessary to accomplish atomisation. A nitrous oxide—acetylene flame ($N_2O-C_2H_2$) is used for such purposes.

Standards for an oil analysis programme are normally prepared by blending appropriate weights of oil-soluble organo-metallic compounds with a suitable lubricating oil — usually the type that is being analysed — to provide a series of standards, e.g. six in number, with metal concentrations ranging from 1 or less parts per million (p.p.m.) to the maximum content that is likely to be encountered, which, for copper or iron, may reach 500 p.p.m., but, for most metals, seldom exceeds 100 p.p.m.

9.2.8 Comparison of spectrometric methods

The high cost of emission spectrometry equipment has been the main obstacle to its widespread use. Despite competition from the atomic absorption method EES remains a preferred method where its higher capital cost can be justified EES has great speed (e.g. 40 samples/hour for 20 element analyses) but atomic absorption spectrometry is claimed to be more reproducible at low concentrations.

9.2.9 A typical SOAP organization

U.S. railroads introduced SOAP as the basis for condition monitoring programme during 1942 and it is now used extensively throughout the world particularly in connection with aircraft operation and some marine and other applications. In a report published by the U.S. Bureau of the Navy, Pensacola, Florida 32508 [2] Bond, Superintendent of the SOAP laboratory stated that their programme, officially known as the BuWeps Astronautical Oil Analysis Programme., monitored (in 1964) approximately 3500 aircraft engines and transmissions. The laboratory then had a permanent staff of five people, including a supervising chemist, two analytical chemists, a physical science technician and a card punch-operator. The laboratory operated on an 8 hour shift, five days a week and analysed more than 200 samples per day.

The BuWeps Programme started with one chemist, using an emission spectrograph, and monitored 20 engines for two years before any tangible results were obtained; then two defective engines were detected. Operations are now carried out with a spectrometer built to their own specifications which translates its analyses from analog to digital information, and automatically feeds the information in p.p.m. to an IBM870 document writing system consisting of a card punch and a slave typewriter.

Analysis is carried out for a range of 1 to 500 p.p.m., and the metallic elements that might be indicative of internal engine wear: aluminium, iron, chromium, silver, copper, tin, magnesium, lead and nickel. Some of these metals, such as aluminium, iron, silver and copper, are very important as indicators; the others have, so far, been of little or no importance. Also, in reciprocating engines, they analyse for silicon as an indicator of atmospheric sand contamination, but the value of silicon analysis has been nullified in jet engines because of the uncontrolled use of silicone additives by some manufacturers of jet engine oils.

Analytical results are delivered from the spectrometer in p.p.m. directly to IBM card punch and typing equipment. This avoids loss of time and the possibility of error involved both in the manual translation of analog to digital information and in the manual transcription of the data to permanent records. The data is also available in a form that can be used for statisitical purposes by automatic data processing equipment. The basic labroatory record is an 8 × 8 inch card, on which all the sample and analytical data history for one engine are recorded.

Three kinds of information are shown on this card:

(1) The basic data, including the model and serial numbers of the engine and the operating activity identification;

(2) The data for each individual sample, including any laboratory request or recommendation preceding the sample, the reason for submission of the sample, the sampling date, the in-route time for transmission of the sample to the laboratory, the operating hours since overhaul and since oil change for the engine, and any corrective maintenance performed on the engine since the previous sample;

(3) The sample analysis.

Basic data do not change from sample to sample; it is therefore typed at the top of each laboratory record card, and punched into an IBM master card which is kept at all times with the laboratory record card. The other data is entered on the record card and on an IBM analytical data card at the time of sample analysis.

9.2.10 SOAP establishments

In the U.S.A. there are over 100 military SOAP laboratories (including some on aircraft carriers) of which the principal are Naval Air Station, Pensacola, Florida; Fort Rucker; McDill Airforce Base.

In the U.K. the main SOAP laboratories are:

Naval Aircraft Materials Laboratory
Fleetlands
Hampshire.

MQAD Laboratory
Harefield
Uxbridge
Middlesex.

Westland Helicopters Ltd.
Yeovil
Somerset.

British Airways
Maintenance Division
Heathrow
Middlesex.

The concept of SOAP was adopted by Canadian National Railways in 1951 and a laboratory established for the RCAF at the Air Maintenance Development Unit, Trenton, Canada and at the Naval Dockyard Laboratory (Atlantic), Halifax. Other organisations are reported in Australia, New Zealand, Holland and France.

It was reported by E.R. Holtiveg, Chief of the Materials Branch of the Aircraft Division, Air Material Command, Royal Danish Air Force that exchange arrangements had been organised to ensure that analysts themselves continued to work accurately. For this purpose an arrangement exists with the United States Air Force to the effect the the RDAF (AMC) takes part in the periodic exchange of so-called SOAP Correlation Samples. This is issued monthly from the U.S.A.F. Aerospace Fuels Laboratory, McDill Air Force Base to all USAF SOAP laboratories within continental U.S. (CONUS) and overseas as well. The analytical results obtained are circulated among all participating laboratories, thus enabling each laboratory to keep a constant check on its own 'professional standard'.

A similar exchange programme, though on a more individual basis, has been established with the Royal Norwegian Air Force, the Belgian Air Force (BAF) and the Royal Swedish Air Force (RSAF) who have also introduced the SOAP programme.

From the USAF SOAP wearmetal guidelines (formerly called 'threshold limits') — which should not be exceeded — have been obtained for the various types of components. These figures are based on a statistic evaluation of all USAF–SOAP data and are currently kept up to date. The Air Material Command, Royal Danish Air Force has also prepared a statistical evaluation of the Danish SOAP data in order to establish 'normal' wear metal levels for the various types of components. These pay due regard to the operational and maintenance conditions characteristic of the Danish Air Force which need not necessarily be similar to those of the USAF.

9.2.11 SOAP diagnosis and validation experience

According to Davies [3] operators of military aircraft including helicopters, and most of the civil airlines have experimented with this technique with varying results. The problem is not so much the retrieval and analysis of samples, but in the expression of results in terms of rate of wear metal increase in the system.

During Concorde powerplant flight and bench development samples were taken regularly at 15 hour intervals, yet at the beginning failures occurred without warning for which this technique was appropriate. Various analytical treatments were tried to achieve better correlation and with a feeling that at least they provide a data bank which would be useful to the airlines. Eventually a fairly simple procedure was developed which is now giving excellent correlation with defects in the oil-wetted areas.

Davies suggested that it is a mistake to over-sophisticate the procedure for correcting results; apparently advanced methods have failed to detect trends which have been picked up by the less sophisticated Bristol Engine Division system. The metals currently measured are Fe, Cr, Cu, Ni, Ag, Al, Mg, Ti and Si. The most important is certainly iron, but to Rolls Royce (1971) Ltd. titanium and magnesium can be extremely useful if the installations include casings of these materials; tungsten is potentially of help in detecting bearing metal failure.

9.2.12 Wear rate evaluation

Calculations used by Rolls Royce (1971) Ltd. to establish wear rates are based on the total metal in the system and take account of normal oil loss, with its quota of metal, and the dilution effect of oil added to top up the system. The following is a typical calculation:

> The Olympus engine has a total capacity (oil tank plus system) of 25 litres (5.5 gal). Therefore 1 p.p.m. of Fe indicates a content of 25 mg. As an example of analysis, if the second sample 15 hours later gave 2 p.p.m. and there had been a total top-up of ten litres (2.2 gal) between sample a correction is made as follows: actual metal content at time of sampling is 50 mg. (i.e. 2 × 25), to this has to be added the used ten litres with an assumed 2 p.p.m. content, equivalent to 20 mg of metal, giving a total of 70 mg. From this is subtracted the original 25 mg which was based on the previous reading of 1 p.p.m. giving an increase of 45 mg metal content in 15 hours. The wear rate is therefore 3 mg h^{-1}. This is then compared with successive samples.

A typical plot of the iron (Fe) wear rate trend in an Olympus engine installed in the Concorde prototype aircraft is shown in Figure 9.3. The rapid increase at 220 hours was found to arise from heavy fretting wear on the port gear-box driven bevel spline location on the layshaft.

9.2.13 Case Studies (Marine)

Failure predictions in marine equipment resulting from an interpretation of SOAP

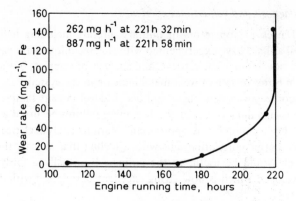

Figure 9.3 *Spectrometric analysis – Olympus engine 59332 in Concorde prototype aircraft*

information and the resultant damage findings were given by Waggoner [4] of which the following is a very small selection from a large collection:

Case 1. Diesel generator

	Cu	Fe	Cr
Normal SOAP p.p.m.	10	10	2
Abnormal SOAP,	40	55	10

Consequence: Restricted to emergency use only. Subsequent inspection revealed need to renew piston rings, cylinder liners, bearings.

Case 2. Low pressure air compressor

	Cu	Fe	Pb
Normal SOAP, p.p.m.	10–75	2–25	5
Abnormal SOAP,	210	10	11

Consequence: Excessive wrist pin war revealed by inspection.

Case 3. High pressure air compressor

	Cu	Fe	Ni	Cr
Normal SOAP, p.p.m.	20	7	2	2
Abnormal SOAP,	59	150	3	13

Consequence: Unit failed due to piston ring collapse before recommended inspection could be made

Case 4. Fuel oil pump

	Cu	Fe
Normal SOAP, p.p.m.	15	15
Abnormal SOAP,	24	40

Consequence: Cracked lower bearing cage and badly damaged upper bearing on gear box shaft revealed by inspection.

Case 5. Turbo-driven forced lubrication pump (top bearing)

	Cu	Pb
Normal SOAP, p.p.m.	10	15
Abnormal SOAP	80	30

Consequence: Inspection recommended; excessive bearing wear discovered.

Case 6. Stern plane tilting machinery in submarine

	Cu	Fe
Normal SOAP, p.p.m.	5–20	10–35
Abnormal SOAP	30	57

Consequence: Examination showed detachment of two ball bearing assemblies from main shaft, and extensive grooving of bronze gland spacer at end of shaft.

Case 7. Main thrust bearing

	Cu	Fe	Pb
Normal SOAP, p.p.m.	2–25	10–50	5
Abnormal SOAP	35	135	37

Consequence: Inspection revealed seizure of oil seal ring on propulsion shaft.

Case 8. H.P. Air compressor

	Na
Normal SOAP, p.p.m.	25
Abnormal SOAP	300

Consequence: Salt water leak discovered in cooler.

Note: Numerous failures of salt water cooling systems resulting in leakage into the lubrication system have been detected by SOAP.

9.2.14 Case Studies (Aircraft)

Aeroengines experience in the use of SOAP as quoted by Bond [2] included the following case histories-

Case 9. Engine   0-435-6A SerNo L1337-31

        Hours since overhaul   518

        Aircraft model   TH-13N

        Abnormal oil analysis   Very high iron and aluminium

        Defect found   The oil scavenge pump and oil pressure pump impellers were spalled; the oil pump shaft was excessively worn; four of the intake valve faces were cupped.

Case 10. Engine   R985-14B SerNo 104276

Hours since overhaul   1410

Aircraft model   TC-45J

Abnormal oil analysis   High aluminium, iron and copper

Defectfound   The impeller shaft splines were excessively worn; the front and rear bearing journals were moderately worn.

Case 11. Engine   J75P19W

Serial   610597

Aircraft model   F-105D

Operting hours since new   440

Abnormal oil analyses:
Before failure (2 hours)   Very high iron; high copper

After failure   Extremely high iron and nickel; very high copper: high chromium

Action taken: None. The laboratory warning was received too late. The engine failed in flight and the airplane crashed three days before the sample was received in the laboratory.

Comments: This aircraft (cost $2,388,583.00) was lost, but it could have been saved. Thus, this case history points very dramatically both to the potential cash value of the program and to the urgent necessity for keeping the sample reaction time (the time interval between sample taking and the receipt of laboratory advice by the operator) at the absolute minimum. Although the sample lead time before failure was only 2 hours engine operating time, it was 4 days calendar time. The sample was taken on Thursday, the airplane crashed on the following Monday, the sample was received and analysed by the laboratory on the following Thursday, and the operator was warned by telephone the same day. This airplane probably would have been saved if the sample reaction time had been 3 days (normal for the West Coast from this laboratory), but this was not possible because the laboratory does not operate on Saturdays and Sundays. The airplane might still have been saved if the sample had been received, analyzed and reported on Monday a 4-day sample reaction time), but this would depend on the actual hour of the crash. As it was, the aircraft was lost by a margin of 3 days.

9.3 Magnetic chip detectors

Chip detector monitoring and SOAP are complementary in that the two techniques of contaminant monitoring are looking for rather dissimilar failure modes. Spectrometric oil analysis is aimed at fine metal particles in suspension such as arises from spinning bearings, fretted splines, etc. Chip detectors will pick up metal flakes such as arise from fatigue break up. The capture of particles, the bulk of which are ferrous, by the use of a magnet provides a simple and effective method for monitoring the contaimination of liquids and assessing the condition of a system.

One such detector comprises a body which is permanently mounted in the lubrication system and a magnetic probe which can be inserted so that the magnet is exposed to the circulating lubricant; a self-closing valve is contained within the body so that the probe can be readily removed for examination without loss of lubricant. Use of this detector involves removal of the magnetic probe at regular intervals, typically every 25 machine-hours and the insertion of a fresh probe while the original is retained for the assessment of the particles adhering to it. Typical normal wear particles are shown adhering to the end of the plug before removal for assessment in Figure 9.4.

Figure 9.4 *Typical magnet plug debris collection —normal gear wear*

The detector must be placed in a position in the lubrication system to give the maximum possible capture — the outside of a bend in a pipe is a suitable position as the centrifugal force will carry the particles on to the magnet. A strong magnet will attract particles in an oil stream but with a straight pipe assistance can be given by providing an enlargement on plenum chamber at the position of the detector. A suggested installation is shown in Figure 9.5.

9.3.1 Typical detectors

A typical magnetic ship detection unit is shown in Figure 9.6. 'A' is a probe and body assembled where 'B' is the self-closing valve, 'C' the magnet, 'D' the probe seals, 'E' is the body, 'F' is the magnet, 'G' is the magnetic probe.

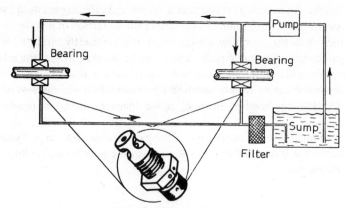

Figure 9.5 *Magnetic plug installation positions*

Figure 9.6 *Magnetic plug components*

9.3.2 Detector Filter

The Tedeco oil monitor filter is most effective in the detection of non-ferrous metal debris such as bronze, white metal, aluminium and its alloys or certain high quality stainless steel. The oil monitor filter consists of a cone-shaped filter placed in the return line in order that all the scavenger oil passes into the open end of the filter and out through the sides: a magnetic chip detector is fitted at the bottom.

9.3.3 Magnetic debris evaluation

When evaluating debris, experience suggests that during the running-in and normal operation phases of a machine's life the debris is of a fine nature, short and of irregular cross-section mixed with some metallic dust. Considerable debris is collected during the running-in stage but the quantity diminishes considerably when normal conditions are established.

During the earlier stages of running-in a new engine or machine a proportion of large particles will appear which will be mainly swarf and other foreign matter left behind during assembly. When the quantity of particles begins to increase at the same time as the particle size there is an indication of impending failure.

Low power magnification ($\times$ 10 to $\times$ 40) can aid in the assessment of such debris. With rolling element bearings and gears, as the original surface breaks up the debris are flake-like, bright and relatively smooth on one side and of a matt-like texture on the other; flakes from rolling element bearings have a finer, greyer structure than those from gears.

9.3.3 Identification of magnetic debris

A review of some of the debris which can be identified from magnetic plugs is given in Table 9.2 which was prepared [5] with the assistance of British Airways, European Division (B.E.A.), Heathrow who have very considerable experience of this technique.

Table 9.2 *Magnetic plug debris characteristics*

Source	Debris characteristics
Ball bearings (Figures 9.7 and 9.9)	(1) Rounded, 'rose petal', radially split shape.
	(2) Highly polished surface texture with faint criss-cross lines and 'pop' marks.
	(3) Fine granular, light grey, scintillating appearance.
	Ball debris
	(a) Initially (especially on lightly loaded ball races) the flakes are roughly rounded in shape with radial splits and markings produced by point to point contact of balls and tracks. Fine criss-cross surface fatigue lines sometimes appear on the surface of the balls.
	(b) Under $\times$ 10 to $\times$ 20 magnification the particles have tiny 'pop' marks on the surface due to the fine granular structure of the metal having polished high spots which produces a scintillating effect. Easily recognized as high grade steel. Flakes tend to have 'body' which is thicker in the centre. One side has usually a highly polished surface while the other side is of even grey granular texture.
	(c) With heavy initial loading the particles are darker but scintillate when move relative to the light source.
	(d) Subsequent underlay materials is produced which is darker, more irregularly shaped and with a coarser structure.
	Track debris
	(a) Surface break-up debris has a high polish on one side often with faint criss-cross scratch lines as on ball material; roughly rounded in shape it has similar characteristics to roller bearing track material.

Table 9.2. (Continued)

Source	Debris characterisitcs
Roller Bearings (Figure 9.9)	*Roller debris* (1) Generally curled and rectangular of length = 2 to 3 × width. (2) High-polished surface texture (3) Fine granular, light grey, scintillating appearance. (4) Because of the rolling action one side often has a series of parallel lines running across the width of the particles. (5) Underlay material is long and has a splintered appearance, it is darker than surface debris. *Track debris* (1) Irregular rectangular shape. (2) Highly polished surface texture with scratches running lengthwise. (3) Fine granular, light grey, scintilating appearance. (4) Since the surface is essentially flat with a rolling contact scratch marks run along the track. (5) The outer edges of both track and rollers tend to break up first, appearing first as generally rectangular flakes and as further deterioration develops they become more irregular 'chunks'. (6) The inner track deteriorates first, then the rollers, finally the outer track.
Ball and Roller bearings	*Spinning and skidding debris* (1) Generally granular in shape. (2) Debris appears as balck dust. *Cage Debris* (1) Appears as large, thin, petal-like flakes. (2) Has a polished surface texture. (3) Copper-coloured appearnace. (4) The initial debris appears as a fine bronze dust followed by large petal-shaped flakes, copper coloured. The flakes do not signify a serious failure condition until such time as steel particles appear either separately or embedded in the flakes, or as thicker chunky bronze particles.
Needle roller bearings	(1) Sharp needle-like shape similar to splinters. (2) Rough surface texture. (3) Dark grey scintillating appearance.
White-metal bearings	(1) Flat or spherical general shape. (2) Smooth surface texture. (3) Appearance resembles solder spatter or silver. (4) Under normal wear conditions debris rarely appears in the scavenge oil as the localized heat melts and spreads the material into minute cavities in the bearing surface.

Table 9.2 (Continued)

Source	Debris characteristics
	(5) When bearing failure commences, fine hair cracks appear in random directions, producing a generally crazed effect on the surface of the bearing. Local oil pressure exerted on the bearing and often in the region of 2–3000 lb in^{-2} forces the oil into the hairline cracks and eventually loosens the particles which when subjected to heat, break up and become flattened. This debris is often deposited either on the opposite side of the bearing, or in the oil return line. Because of their molten condition when entering the oil stream, they frequently form tiny balls resembling solder.
Aluminium/20% tin bearings	(1) Irregular shape. (2) Smooth surface texture with fine parallel lines. (3) Solder-like appearance, silver with dark lines. (4) These bearings have a good fatigue resistance, and in general have to be in a fairly advanced state of failure before particles actually break away and enter the scavenge oil stream.
Gears (Figure 9.10)	*Scuffing debris* (1) Irregular shape. (2) Lustrous surface texture with many tiny indentations. (3) Grey appearance resembling splash of solder. (4) Gear teeth pressure marks are occasionally visible accompanied by score lines due to the grinding-up of debris between the gears, alternatively only the scoring lines may be seen. *Normal wear Debris* (1) Tiny hairline strands of irregular cross-section, very short and mixed with metallic dust. (2) Rough surface texture. (3) Dark grey appearance. (4) The small hairline strands are usually bunched together giving the appearance of being thicker when on the magnetic probe. *Failure debris* (1) Irregular shape. (2) Surface texture is polished with score marks. (3) Coarse appearance, dull grey with bright spots. (4) These particles are produced as ground up irregular shapes, dark in appearance with bright high spots. Flakes sometimes appear showing the surface outlines of gear teeth. Usually more polished on the outer side with score marks in evidence and sometimes accompanied by heat discolouration. The material lacks lustre and is coarser than debris produced from bearings.

Table 9.2 (Continued)

Source	Debris characteristics
	(5) Due to the rolling contact characteristics of gearing, the point to point contact of the teeth produces on the tip, markings similar to those of ball bearings, and the sliding contact imparted to the flank of the teech gives rise to the parallel scratches typical of debris from rollers in bearings.
	(6) Underlay debris is very irregular, being long and splintered, this condition is accentuated by further grinding action of the gears. The debris collected on the magnetic probe, appears when viewed as separate particles to be strands of metal, shredded, long and thin with irregular outlines and can be likened to rough hairs.

9.3.4 British Airways metallic particle monitoring system

British Airways centralize metallic particle monitoring at the Detection Centre, Engineering Base, Heathrow, London. This was arranged to ensure uniformity of control and decision-making, also to provide that the engine history cards were always ready for inspection. According to Hunter [6], this arrangement was eminently suitable because British Airway's network of operation was such that the majority of our aircraft cycled through London at least once per day. Therefore the sample monitoring and transmission of these was easily controlled and of minimum delay. (Air Canada operate a decentralized system carrying their engine metallic particle history cards around in the holds of their aircraft).

Magnetic (turbomag) plugs are removed at prescribed monitoring periods, currently (1975):

Rolls Royce Spey engine plugs   50 hour intervals;
Trident main oil pressure filters   60 hour intervals;
BAC S1-11 main oil pressure filters 75 hour intervals.

(On the first showing of failure evidence sampling intervals are down graded to every 10 hours.) Removed plugs are identified with colour sleeves, denoting the specific location on the engine and placed, without cleaning, in allocated compartments in the container box. Where a filter sample is also due the filter cone, in a plastic envelope, is also placed in the container. The box is then identified with the aircraft registration, date, station and technical log hours, and dispatched, without delay, to the Early Failure Detection Centre.

The first activity at the Detection Centre is to recognize background debris by examining the plugs and filters visually — in the majority of occasions the plug catch and filter deposits will not be significant to the naked eye, but if 'loaded' or there is any doubt, immediate action is initiated to change the engine or carry out an extended ground run, with clean plugs and filters, and then re-examine the

Figure 9.7 *Ball bearing debris particles (× 20 magnification)*

evidence in conjunction with the past monitored history of the particular engine in liaison with and the advice of the Early Failure Detection Centre.

The plugs are rinsed in a vat containing a cold cleaning solvent to remove all traces of oil, carbon and other soluble matter. Originally carbon tetrachloride was used but even with a funnelled vent system and extractor fan this cleaning agent was rather overpowering, so 'Inhibisol' or Atdrox 551 which is non-inflammable and about 30 times less toxic than carbon tetrachloride is now used. After rinsing, the plugs are carefully viewed under a binocular microscope ranging from 10 to 20

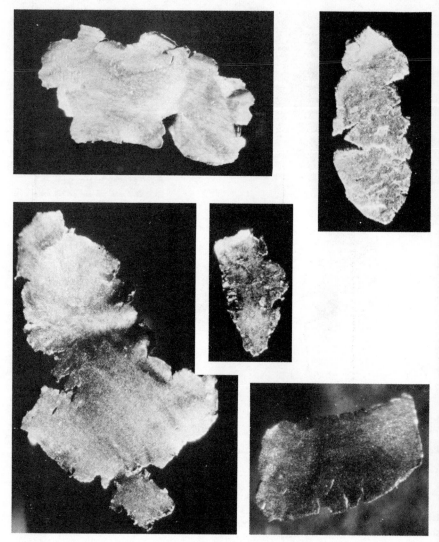

Figure 9.8 *Roller bearing debris particles (× 20 magnification)*

times magnification, with a light source illuminating the detector head of the plug to which the particles are magnetically attached.

A pair of non-magnetic tweezers are used to separate the particles for more detailed observation. After this examination the flakes are removed from the plug using a small piece of adhesive tape, which is then attached to the concerned engine's history card together with the interpretation details. A similar procedure is followed with filter deposits.

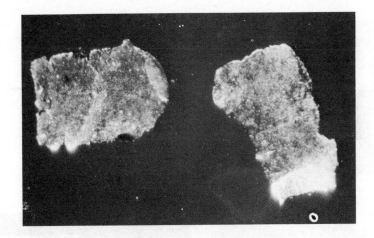

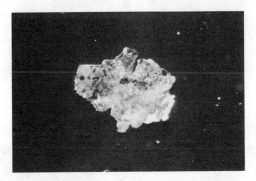

Figure 9.9 *Bearing track debris particles (× 20 magnification)*

When evidence of failure first appears in a metallic particle monitoring examination a red spot is endorsed against the particular sample. If the subsequent check shows development of failure evidence two red spots are added and immediate action is initiated to downgrade the monitoring periods to 10 hours. If the evidence from these additional samples confirms a positive failure trend growth, the appropriate authority is alerted to arrange an engine change within a recommended safe period.

The display of engine history cards, mounted on the walls of the Detection Centre, show at a glance those engines which are under 'alert' monitoring by reference to the red spots. The system cycle is only considered complete when the 'detected' engine has been strip-examined, a full report issued, usually with photographic presentation of the failure and follow-up action taken.

Figure 9.10 *Gear debris particles (× 20 magnification)*

9.4 'Ferrograph' particle precipitation

Wear particles taken from lubricating oil can be precipitated on a transparent substrate and arranged according to size by means of a technique incorporated in an instrument known as a Ferrograph. As a consequence large particles appear predominantly at one end of the substrate (slide) and smaller particles at the other; this provides a valuable basis for analysis.

Most of the ferrous particles in a used oil arise for wear inside the machine, other particles such as salt, dust, carbon paricles etc. arise from external sources. The Ferrograph was accordingly developed to collect those wear particles within the oil which exhibit magnetic properties. A magnet, designed to develop a field of ex-extremely high gradient near the poles forms the basis of this instrument. As shown in Figure 9.11 and described by Seifert and Westcott [7], a pump delivers the oil sample at a slow steady rate (e.g. $0.25 \, cm^3 \, min^{-1}$) on to a substrate on which the magnetic particles are deposited and on which they can be examined by optical densitometer or high performance microscope. The substrate is treated so that the oil runs down the centre of the slide and the particles will adhere to the surface on the removal of the oil. The slide is aligned with the distance between the slide and the magnet decreasing as shown. The slide is mounted at a slight angle to the hori-zontal. The net effect of the forces on the particles is to sort them out by size while depositing them out on the slide. Large particles are deposited first. Smaller par-ticles migrate the longest distance down the slide as shown in Figure 9.12.

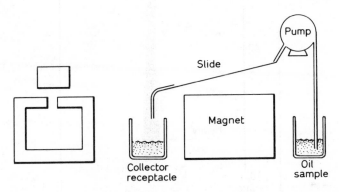

Figure 9.11 *Schematic diagram of the FERROGRAPH – the strength of the magnetic field increases to a maximum at the bottom of the slope*

9.4.1 'Ferrogram' particle deposit measurement

Diffusion of the optical density or alternatively, bichromatic microscopic examina-tion can be used to measure the density of the various deposits on a 'Ferrogram'.

With an optical densitometer light is passed through the slide from the Ferro-graph on which ferrous particles have precipitated and some falls onto the cathode of a photomultiplier which generates a voltage proportional to the logarithm of the light intensity. This voltage is measured by an integral digital voltmeter which registers zero when a material has been deposited and can record on two density ranges – 'low' to 0.400, 'high' to 4.00.

Metal particles in contaminated oil arise from abrasive wear, others from adhes-ive wear; other metallic products include those arising from corrosion. Ferrographic examination is more sensitive than emission spectrometry at the early stages of engine wear, when particles remain in the oils as stable colloids (Figure 9.13).

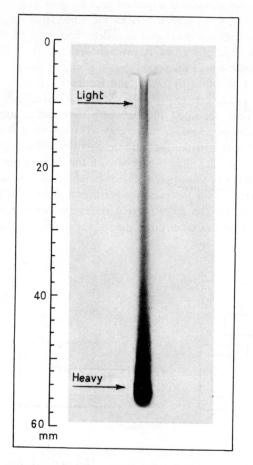

Figure 9.12 *Typical Ferrogram – the largest particles are arranged at one end and the smallest at the other. The band shown has the largest particles between 50 and 55 mm and the smallest near the 10 mm position*

9.4.2 Photomicrograms from 'Ferrograms'

A series of photomicrograms by Siefert and Westcott [7] was able to distinguish strings of particles resulting from abrasive wear and finally the spherical particles which precede failure. Very large particles observed in oils from a jet engine and spectrographically reported as having a 'large iron content' displayed longitudinal ridges as if the surface had been scuffed.

Other photomicrograms by these same authors [7] monitor the development of a main bearing failure in a jet engine and are accompanied by the figures for the time since overhaul (representing the total operating time of the oil) and the quantity of iron as indicated by a spectrograph.

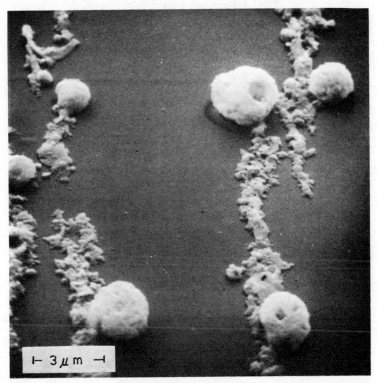

Figure 9.13 *Ferrograph of debris in oil*

	(A)	(B)	(C)	(D)
Time, hours	5	10	20	25
Iron content, p.p.m.	2	2	3	210

Machined particles (abrasive wear) appeared in all photographs. A very high number arose at (B) which were considered to imply incipient failure. At (C) several heavy chips showed signs of extreme cold working with shadow-like particles of oxide in the background. At (D) a solid deposit of particles several layers thick appeared with the introduction of spherical debris; it was thought likely that some spheres were formed by the melting of metal followed by subsequent cooling in the oil.

Spherical debris is optically unique in that they are so smooth as to be difficult to observe under a microscope unless the angle of illumination is sufficiently high; with small angles of illumination only the top of the sphere is illuminated and appears as a point of light rather than a round body.

9.4.3 Spherical debris and fatigue monitoring

By the use of the electron scanning to investigate fracture surfaces and lubricant

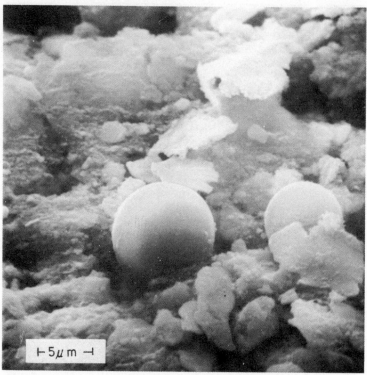

Figure 9.14 *Spherical debris in oil showing onset of fatigue*

debris, Scott and Mills [8] of the National Engineering Laboratory, East Kilbride, Glasgow, revealed the spherical debris is a characteristic feature associated with rolling contact fatigue. The detection of such debris was suggested as a diagnostic aid for the indication of distress in critical rolling mechanisms (Figure 9.14).

The examination of Ferrograms taken from a jet engine which eventually failed by rolling contact fatigue of the main roller bearings revealed, just prior to failure, the presence of lathe-like and spherical debris. Scott and Mills in another communication [9] reported that electron microscopic examination of Ferrograms showed the formation of spherical debris from tongues of metal during the deformation process. There is evidence to suggest that a large preponderance of spherical debris accumulates just prior to the failure of a ball bearing; that this material arises from the steel of the bearing itself has been confirmed by X-ray energy analysis in the scanning electron microscope.

9.5 STM Control Kit

Surveys completed by the French Navy in 1963 showed that the service time of oils could be greatly increased with some types of engine if a quality control test system were employed. The French Navy has adopted an STM Control Kit [15] which can be used aboard naval vessels whose missions take them far from bases with laboratory facilities; this kit enables the essential oil parameters — dilution, acidity, pollution — to be checked.

Dilution is determined by a falling-ball viscosity comparator. A reference oil is obtained by mixing new oil with a given precentage of diesel fuel. Acidity is measured by colour variations of an indicator placed in an aqueous solution, thus

pH	Colour
Under 4	Yellow
5 to 6	Green
Over 7	Purple

Pollution is measured by the used blotter spot test. The maximum insoluble content corresponds to a spot with an intensity of 4 on the Munsell scale using Durieux 122 paper.

9.6 Used oil blotter test

Early concepts of this test were made by Alphonse Schilling [11] in which a drop of oil is deposited on a filter paper. The oil initially spreads out mainly on the surface so that large pollutants particles will remain within a circular corona of small radius, this removes many organometallic and detergent-dispersent additives. Further dispersion involves an oil-penetration and filtration through the paper and that clearly defined circular zones corresponding to the size of particles transported by the filtering oil. A period of up to 24 hours is needed for the oil to 'blot' fully after which the results may be analysed photometrically. A typical blot is shown in Figure 9.15.

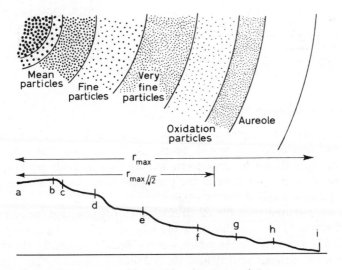

Figure 9.15 *Used-oil blotter test showing development of rings*

A photometric method of analysis described by Sibenaler [12] and used by the Mechanical Transport Laboratory, Royal Militray School, Brussels, essentially comprises a conventional smoke-density meter of the type used to monitor Diesel exhaust emissions. Measurements are made with blotting paper laid on a light box illuminated from below with the illumination sensed by a photocell. The paper position can be adjusted according to the deposit being analysed.

9.7 Thin-layer chromatography

This is a particular application of general chromatography based on the fact that the components of a mixture may have different affinities for a particular absorbent. Accordingly, a selective absorbent is applied as a thin coating 0.25 mm thick on an inert support. The plate is stood vertically in a dish containing solvent as shown in Figure 9.16. This solvent ascends by capillary action and entrains components of the sample originally spotted on the bottom of the plate. Selective or general colour developers may be used to develop the various components. Details of the experimental methods were given by Morot-Sir [13] of Societe Francais des Petroles B.P. included the identification and measurement of glycol from antifreeze formulations which might leak into the cylinder case from the cooling water.

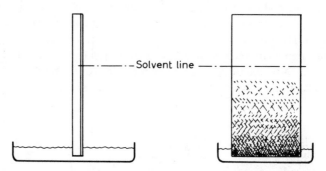

Figure 9.16 *Principle of chromatographic separation*

9.8 Capacitative oil debris monitor

The quantity and rate of generation of metallic debris in recirculating lubricating systems was developed and tested by the Applied Physics Laboratory of the Franklin Institute Research Laboratories and reported by Hollinger [14]. The device was based on the collection of the particulate debris in the annular space between the plates of a cylindrical capacitator.

Debris was removed from the oil stream by centrifugal action caused by tangential introduction of the oil flow into the monitor cell and by passing the flow through a 50-micrometre screen which was continually washed by the fluid flow. The particulate material settled into the measuring capacitor. The fluid which passed through the screen was allowed to flow through a reference capacitor of identical dimensions as the measuring capacitor. By using a bridge circuit with the measuring and the reference capacitors, the effect of oil properties such as temperature and water content could be cancelled, and only the collected metallic debris would affect the bridge output.

9.9 X-ray fluorescence detection of contamination (XRF)

A nucleonic metal sensor employing radioisotope-excited X-ray fluorescence technology has been developed for detection of both ferromagnetic and non-ferromagnetic particles in oils with a detection capability of 0—550 p.p.m. with an accuracy better than ± 6 p.p.m. over a temperature range from ambient to 400°F. The basic concept of this method has developed from the radioisotope gauge developed by Wright [15] which uses Krypton 85 (Kr^{85}).

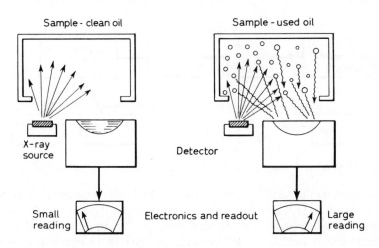

Figure 9.17 *Principles of X-ray fluorescence*

X-ray fluorescence (XRF) involves the exposure of a specimen to a source of radiation which causes it to emit characteristic X-rays. A typical arrangement is shown in Figure 9.17 which shows that successful detection depends on the correct combination of source and detector. Full particulars of a project involving the design, development and testing of an in-line debris sensor is given in a report by Bertin [16]. This was required to monitor recirculating lubricating oil systems in which the rate of build-up of the metallic debris is a function of the degradation of oil-wetted components with possible applications to continuous surveillance of helicopter and aircraft engines in flight.

9.10 X-ray photoelectron spectrometry

The two novel features of this technique of chemical analysis are (i), that it gives chemical information directly from the electronic core levels, and (ii), it can give detailed information about the surface of the sample (i.e. the top few atomic layers) in addition to the bulk material; it is applicable to solids, liquids and gases.

The principle of X-ray photoelectron spectrometry is the X-ray bombardment of a specimen causing the ejection of photoelectrons at various electron levels in the atoms according to the approximate expression:

$$E_b = E_x - E_k$$

where: E_b = binding energy of the electron ejected by the photoelectron process,

E_k = kinetic energy of the ejected photoelectron,

E_x = energy of the bambarding X-rays.

X-rays of known energy bombard the specimen and the energy of the ejected photoelectron is measured in the manner shown schematically in Figure 9.18.

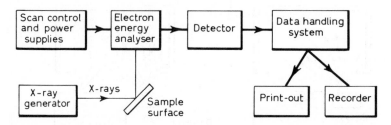

Figure 9.18 *Schematic diagram of the X-ray photoelectron spectrometer*

Early work on this technique was carried out by Louis de Borglie [17], Robinson [18], Kretschmar [19] and Siegbahn [20] who coined the acronym ESCA to denote electron spectroscopy for chemical analysis. The development of a modern ultra-high-vacuum model, the ES200 was described by Drummond, Errock and Watson [21]. This instrument deals with electrons of energies up to 4000 eV. A cross-section through the X-ray tube, sample chamber, lens system and hemispherical electron-energy analyser is shown in Figure 9.19 and electrostatic analyser is used because it is easier to shield from stray magnetic fields than is a magnetic analyser. The X-ray gun, sample chambers and analyser are independently pumped, each by a separate system consisting of a liquid-nitrogen cold trap and polyphenyl ether oil diffusion pump backed by a rotary pump.

9.11 Particle classification

Microscopic identification of particles may be assisted by a numerical classification system based on characteristic properties and aided by a particle atlas [22]. The basic classification characteristics are, transparency, colour (transmitted), colour (reflected), birefringence, refractive index (relative to medium) and shape.

The binary numerical system uses '0' to signify the presence and '1' to signify

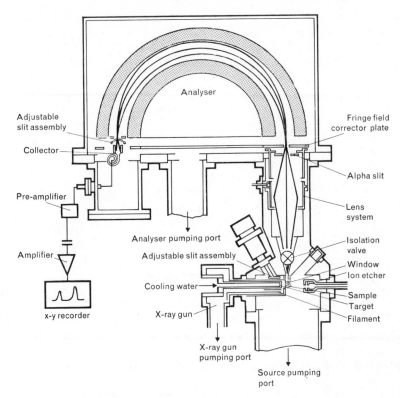

Figure 9.19 *Cross-section of ES200*

the absence of each of the six characteristics. Accordingly a six digit number establishes one particle or a group of particles.

The rules of this code are as follows:

(1) First digit (transparence) — the digit '1' means the particle is opaque

(2) Second digit (colour) — the digit '1' means the particle has some colour. The digit '0' means the particle is colourless (e.g. if transparent). Substances showing a colour due to impure atoms (e.g. sand containing iron), non-stoichiometry (e.g. silicon carbide, titanium dioxide) or some structural defect (diatoms) are classified as colourless. Some particles such as power plan flyash may be either coloured or colourless and should have both (1) and (0) classification.

(3) Third digit (anisotropy) — the digit '1' means anisotropic, strain birefringence is classified as isotropic.

(4) Fourth digit (refractive index) — the digit '1' is used for a refractive index greater than 1.662.

(5) Fifth digit (shape) — the digit '1' means that one dimension is one-quarter or less than either of the other two; shell-shaped (conchoidal) fractures such as produce flakes are classified as '0'.

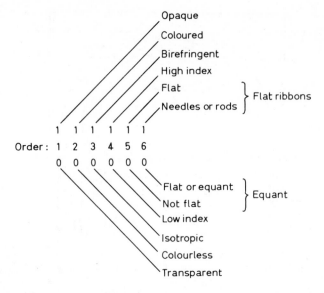

Figure 9.20 *Definition of a 6-digit particle classification code*

(6) Sixth digit (shape) — the digit '1' means that it is needle or rod shaped with one dimension four times (or more) the other two.

Shape is defined by the fifth and sixth digits in the following manner:

(1) (1) = an elongated, flattened rod, ribbon, balde or lathe;
(1) (0) = plate or tablet;
(0) (1) = needle or rod;
(0) (0) = cube.

These classifications are depicted in Figure 9.20.

9.11.1 Particle size determination

Optical microscopes may be used with dry debris which has been brushed onto a glass slide; fine debris of a size which cannot be resolved optically may be examined by electron microscope [23, 24]. Dry debris may be brushed onto a prepared carbon Formvar or similar film supported on a microscope grid. With heavy metals, shadowing *in vacuo* improves definition and enables a better interpretation to be made of particle size, shape and agglomeration. To obtain good dispersion, particles may be agitated with ether and a drop placed on the supporting film so that when the ether evaporates the debris is well dispersed over the film.

9.11.2 Television microscopy

Quantitative television microscopy reduces the tedium and time involved with the determination of particle size using microscopes. Quantitative television microscopy provides instantaneous readings for the number and size distribution of particles in a selected field of vision. The image is projected into a television camera and the output displayed on a monitor for viewing and also fed into an electronic detector which responds to the changes as the scanning spot passes over the particles in the field. Signals from these areas are fed into a computer which derives the number of particles into a series of size ranges.

Scott [25] described the use of quantitative television microscopy in which the computer output is attached to an electron microscope while using the Coulter principle [26, 27] which counts the number and size of particles in an electrically conductive liquid by forcing the suspension to flow through a small aperture with the simultaneous flow of electrical pulses each being proportional in magnitude to the volume of the particle causing the pulse. Particles ranging in size from 0.5–200 μm can be counted by the use of this technique.

9.11.3 Television with computer scan microscopy

Particles which result from combustion, particularly coal particles, oil droplets or soot particles, can be analysed in terms of shape and size by the use of a Quantimet analyser. Analysis is performed by using the Quantimet analyser to count particles of a given size, calculate the statistical distribution within a size range and measure various shape parameters, intercepts and areas. Particles to be examined are placed on a conventional microscope slide and can be viewed in the normal way. They can also be viewed through an electronic vidicon camera and displayed on a cathode ray screen. The electronic image can be modified, in particular it can be enlargened or one particular portion of the slide examined.

The analyser has an integral computer which can be programmed for slide scanning and therefore produce a considerable amount of information relating to any one slide. This information is delivered on paper tape through a teletype and subsequently processed on the Nova computer. The Nova consists of a 32 k core store, a single disc and a 16-channel multiplexer. Extensive use is made of this system [28] at the Department of Fuel and Combusion Science, Leeds University.

9.11.4 Particulate scattering monitor

An oil condition monitor recently developed in the U.S.A. by the Environment One Corporation [29] monitors the scattering of light by particulate matter. Light attenuation is also caused by chemical and thermal degradation of the oil. An airborne real-time system has been tested by the U.S. Navy.

An automatic particle counter manufactured by High Accuracy Products (U.S.A.) counts the pulses of reduced signals at a photocell as a light beam is interrupted by particles in oil flowing through a cell. The particles can be sorted into size distribution by analysis of the pulsed signal. The trends which are observed can be roughly correlated with incipient failure of mechanical units and the counts in the smaller ranges correlated well with results from spectrographic oil analysis.

A device manufactured by K. West (U.S.A.), monitors the incidence of metallic particles on a conducting wire mesh screen in a recirculating oil system. It is claimed that the relatively large mesh size of the screen is designed to minimize warnings from normal wear debris and to collect only particles which are significant in terms of mechanical degradation.

An interesting discovery made during combustion experiments at A.F.F.D.L. Wright-Patterson [30, 31] is the fact that an electrostatic probe can detect minute metallic debris in the gas stream. A failing engine, e.g. because of impact damage or bearing breakup, may have a 'spike' count rate of several orders of magnitude above that of a healthy engine.

References

 1 Klug, R.L. (1972) 'Scheduled oil sampling as a maintenance tool', S.A.E. Paper 720372, April, also: Caterpillar Tractor Co., Peoria, Illinois 61611, USA Report FEG 00023-01 (1973).
 2 Bond, B.B. (1965), 'Spectrometric oil analysis', Report O & R – P – 1, Naval Air Station, Pensacola, Florida 32508, October.
 3 Davies, A.E. (1972), 'Principles and practice of aircraft powerplant maintenance', *Trans. I. Mar. E.* **84**, 441–447.
 4 Waggoner, C.A. (1971), 'Spectrometric oil analysis – principles and practice' Materials Report 71A, Defence Research Establishment Pacific, Victoria B.C. Canada. March.
 5 Barugh, J.K. (1972), 'Equipment reliability through oil monitoring', Vactric Control Equipment Ltd., Morden, Surrey SM4 4LL.
 6 Hunter, R.C. (1975), 'Engine failure prediction techniques', *Aircraft Engineering*, March, 4–14.
 7 Seifert, W.W. and Westcott, V.C. (1972), 'A method for the study of wear particles in lubricating oil', *Wear*, **21**, 29–42.
 8 Scott, D. and Mills, G.H. (1973), 'Spherical debris – its occurrence, formation and significance in rolling contact fatigue', *Wear*, **24**, 235–242.
 9 Scott, D. and Mills, G.H. (1973), 'Spherical particles formed in rolling contact fatigue', *Nature*, **241**, 115–116.
10 Perrier, P. (1973), 'The french navy STM control kit', *Rapid Methods for the Analysis of Used Lubricating Oils*, Scientific Publications (GB) Ltd., Broseley Shrops.
11 Schilling, A. (1973), Motor oils and engine lubrication', Scientific Publications (GB) Ltd.
12 Sibernaler, E. (1973), 'The photometric analysis of blotter spot tests', *Rapid Methods for the Analysis of Used Lubricating Oils*, Scientific Publications (GB) Ltd.
13 Morot-Sir, F. (1973), 'Application of thin layer chromatography to the analysis of additives in lubricants', *Rapid Methods for the Analysis of Used Lubricating Oils*, Scientific Publications (GB) Ltd.

4 Hollinger, R.H. (1973), 'Advanced capacitative oil debris monitor', UASSMRDL Technical Report 73–53, Fort Eustis, Virginia 23604, August.

5 Wright, D. (1971), 'Nucleonic oil quantity indicating system to MIL-0-38338A', Nucleonic Data Systems, Santa Ana, California.

6 Bertin, M.C. (1972), 'A nucleonic sensor for detecting metal in recirculating lubricating oil systems', UASAMRDL Technical Report 72–38, September, Fort Eustis, Virginia.

7 De Broglie, L. (1921), 'Les phenomenes photo-electriques pour les rayons X et les spectres corpusculaires des elements', *J. Physique* **2**, 265.

8 Robinson, H. (1923), 'The secondary corpuscular rays produced by homogenous X-rays', *Proc. R. Soc.*, A104 455.

9 Kretschman, G.G. (1933), 'Determination of e/m by means of photoelectrons excited by X-rays', *Phys. Rev.*, **43**, 417.

20 Siegbahn, K.L. (1967), 'E.S.C.A. atomic molecular and solid-state structure by means of electron spectroscopy', Almquist & Wiksels, Uppsala.

21 Drummond, I.W., Errock, G.A. and Watson, J.M. (1974), 'X-ray photoelectron spectrometry – a tool for chemical analysis', *GEC J. Sci Tech.*, **41**, No. 2/3 94–103.

22 McCrone, W.C. (1967), *The Particle Atlas*, Ann Arbor Science Publishers Inc., P.O. Box 1425, Ann Arbor, Michigan 48106.

23 Scott, D. and Scott, H.M. (1957), 'The electron microscope in the study of wear', *Proc, Conf. Lub. and Wear* (I. Mech. E.) pp 609–612.

24 Bradley, D.E. (1965), 'Techniques for mounting, dispersing and disintegrating speciment', *Techniques for Electron Microscopy*, Blackwell Scientific Pubs. pp 63–81.

25 Scott, D. (1968), 'Examination of debris and lubricant contaminants', *Proc. I. Mech. E.*, **187** Pt. 3G, 83–86.

26 Berg, G.H. (1958), 'Electron size analysis of sub-sieve particles by flowing through a small liquid resistor', *Proc, Symp. Particle Size Measurement*, Sp. Tech. Publ. **234**, ASTM pp 245–248.

27 Kubitschek, H.E. (1960). 'Electronic measurement of particle size', *Research* **13** (4) 128–135.

28 ———— (1974), 'Particle size analyser uses NOVA', *Systems*, November, 9.

29 Ginty, A.G. (1964), 'Survey of crosshead bearing failures in slow speed marine oil engines', *Marine Engineering Review*, November, 41–43.

30 Couch, R.P., Rossback, D.R. and Burgess, R.W., Jr. (1972), 'Sensing incipient jet engine failure with electrostatic probes', *Symposium for Airbreathing Propulsion*, Monterey CA, 19–21 September. (AF Flight Dynamics Laboratory, Wright Patterson AFB, Ohio.)

31 Burgess, R.W., Jr (1972), 'An investigation of the detection of charged metal particles in a jet engine exhaust by a cylindrical electrostatic probe', Air Force Institute of Technology (AFIT-EN), Wright Patterson AFT, Ohio, AFIT Thesis, June, AD 745540.

10 Performance trend monitoring

10.1 Primary monitoring – performance

In technical classification, the monitoring of system performance is the primary method of relating deterioration to consequence. It is primary in classification but not necessarily simple in application.

The simplest form of such monitoring is the regular recording of relevant parameters in a 'log'. This may be a manually recorded log-book, the typed record or 'print-out' from machines or computers, or the regular trace of a graph from an instrument. Regular logging of important parameters and their derivatives compared with reference information provides the basis for performance trend monitoring.

10.1.1 Thermodynamic and fluid dynamic analysis

Elementary trends are demonstrated by such simple methods as relating the pressure-drop and through-flow in a heat exchanger. The build-up of deposits in the flow lines will be revealed by increases in the pressure-drop needed to maintain a certain flow rate or 'through-put'. Similarly, temperature differences on both inlet and outlet of both hot and cold sections can indicate the thickening of films and reduction in heat transfer coefficient.

To interpret the significance of performance trend monitoring it is important that the inter-relationship between pressure, temperature, velocity, specific volume etc. should be understood for all parts of a process. These can be theoretically evaluated in terms of performance such as specific fuel consumption rate, etc. so that the influence of performance trends may be well established by further in depth studies of the thermodynamics and fluid mechanics of systems with defective components.

10.1.2 Otto and Diesel cycles

These cycles which form the basis of spark-ignition and compression-ignition engines are extensively covered by many text books [1,3] provide, from an analysis of their indicator diagrams, information describing their performance (or lack of performance through deterioration).

Ideally, the efficiency of each of these cycles as measured by the ratio of the mechanical work output to the thermodynamic heat input (Figures 10.1 and 10.2) is given by:

$$\text{Otto cycle} = 1 - \frac{T_4}{T_3} \tag{10.1}$$

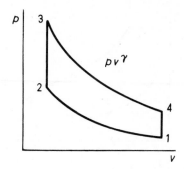

Figure 10.1 *Otto cycle*

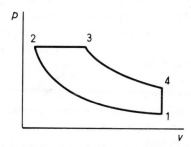

Figure 10.2 *Diesel cycle*

$$= 1 - (r)\frac{P_4}{P_3} \tag{10.2}$$

$$= 1 - \left(\frac{1}{r}\right)^{\gamma-1} \tag{10.3}$$

where: r = volumetric compression ratio,

γ = ratio of specific heats.

$$
\left.
\begin{aligned}
\text{Diesel cycle} &= 1 - \frac{(T_4 - T_1)}{\gamma(T_3 - T_2)} \\
&= 1 - \frac{1}{\gamma}\frac{T_1[(P_4/P_1)-1]}{T_2[(V_3/V_2)-1]} \\
&= 1 - \frac{1}{\gamma}\frac{[(P_4/P_1)-1]}{[(V_3/V_2)-1]}\left(\frac{1}{r}\right)^{\gamma-1}
\end{aligned}
\right\} \tag{10.4}
$$

Deterioration in either cycle leading to a reduction in the compression ratio (r) is

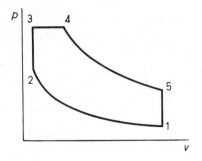

Figure 10.3 *Dual combustion cycle*

immediately followed by performance loss. Such deterioration can result from a lack of seal in the cylinder due to faulty piston rings, valve-seating or gaskets. A high value of the exhaust temperature T_4 or a low value of the combustion temperature T_3 both reduce efficiency; the monitoring of these parameters may indicate deterioration in combustion performance and assign a designated source of deterioration.

10.1.3 Dual combusion cycle

Similar considerations apply to the cycle in which combustion occurs both at constant volume and constant pressure. Figure 10.3, a cycle which is more characteristic of the modern diesel engine (indeed with some engines the constant pressure combustion is small). Thermodynamic analysis of this cycle give the following expressions for efficiency:

Dual combustion cycle

$$\eta = 1 - \left[\frac{(T_5 - T_1)}{(T_3 - T_2) + \gamma(T_4 + T_3)}\right] \tag{10.5}$$

$$= 1 - \left\{\frac{T_1\,[(P_5/P_1) - 1]}{(T_3 - T_2) + \gamma(T_4 - T_3)}\right\} \tag{10.6}$$

$$= 1 - \left[\frac{\beta(\alpha)^{\gamma-1}}{(\beta - 1) + \gamma\beta(\alpha - 1)}\right]\left(\frac{1}{r}\right)^{\gamma-1} \tag{10.7}$$

where: $\left.\begin{array}{l}\beta = P_3/P_2 \\ \alpha = V_4/V_3.\end{array}\right\}$ values related to combusion timing.

From these equations, a similar interpretation may be made of the effect of changing values of observed pressures. Temperatures may serve to indicate the extent of deterioration in an engine working on the dual combustion cycle as was described for the Otto and Diesel cycles.

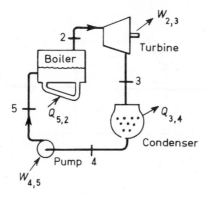

Figure 10.4 *Simple steam plant*

10.1.4 Vapour power (steam) cycles

A system comprising a boiler, turbine, condenser and feed-pump as shown in Figure 10.4 involves changes which are related to the thermal properties of steam and water at the salient parts of the cycle. These are inherently related to the pressure and temperature such that it is only possible to express the heat and work transfers in terms of the point conditions.

Boiler heat transfer $= Q_{5,2} =$ (steam enthalpy, 2) − (water enthalpy, 5)

Turbine work transfer $= W_{2,3} =$ (steam enthalpy, 2) − (steam enthalpy, 3)

Condenser heat transfer $= Q_{3,4} =$ (steam enthalpy, 3) − (water enthalpy, 4).

Meaningful interpretation of these terms can be best obtained by referring to the relevant temperature − entropy (T–s) diagram for the cycle. This is shown by the straight Rankine cycle (Figure 10.5) which is progressively modified by the introduction of superheating the steam at the boiler, Figure 10.6 and also by the further addition of reheat during expansion in the turbine, Figure 10.7. The areas of the T–s diagram give the heat transferred at the boiler and the work out of the system. The effect of the progressive complication of the cycle is to increase the efficiency of the system, any deterioration in any component process will have a related effect on the performance of the other parts of the system and thus the advantages of the complex design will be lost.

Thus, considering Figure 10.7, if the superheat from point 2 does not reach 2′ but some lower temperature, the temperature after first stage expansion will be lower and the steam will have condensed into droplets causing erosion of the low pressure blades in this first stage. Reheat will only reach a much lower temperature since the condensed droplets must be re-vapourized and energy used to raise the temperature will have been used up in this re-evaporation. The performance of the condenser may be reduced due to the lower temperatures at final discharge from the turbine. This is indicated on the T–s diagram (Figure 10.8).

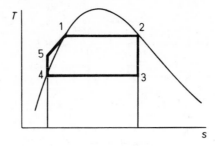

Figure 10.5 *Simple Rankine cycle*

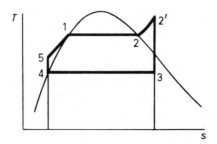

Figure 10.6 *Rankine cycle with superheat*

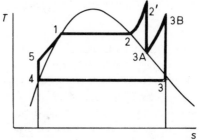

Figure 10.7 *Superheat and reheat*

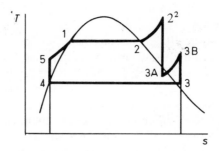

Figure 10.8 *Effect of inefficient superheat*

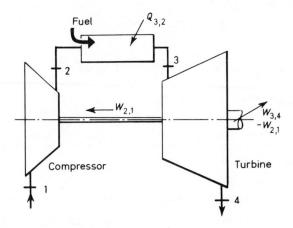

Figure 10.9 *Open cycle I.C. turbine*

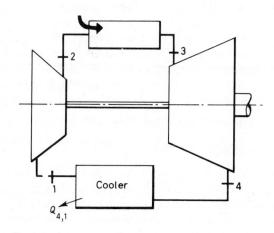

Figure 10.10 *Closed circuit I.C. turbine*

10.1.5 Gas-turbine plant

Cycles for gas turbine plant may be open as shown in Figure 10.9. which is typical of the aircraft propulsion system, or closed, as in Figure 10.10, which has been used for some commercial systems and is thermodynamically the Joule cycle.

$$\text{Net work transfer from turbine} = W_{3,4} - W_{2,1}$$

$$= C_p(T_3 - T_4) - (T_2 - T_1) \qquad (10.8)$$

$$\text{Heat transfer to turbine} = Q_{3,2}$$

$$= C_p(T_3 - T_2) \qquad (10.9)$$

$$\text{Cycle efficiency} = \frac{W_{3,4} - W_{2,1}}{Q_{3,2}}$$

$$= 1 - \left(\frac{T_1}{T_3}\right)(r)^{\gamma - 1/\gamma} \tag{10.10}$$

where: r = pressure compression ratio,

γ = ratio of specific heats (assuming air as the working substance).

Again the value of the compression ratio is a dominant performance factor together with the combustion-related factor, temperature T_3.

10.2 Primary and secondary performance parameters

Basic parameters may be taken to be: length; mass; time; electric current; temperature and luminous intensity. The primary parameters which are derived from these basic parameters are: force; pressure; work; energy; power; electric charge; potential difference; electric resistance; electric capacitance; inductance and heat transfer rate.

Further parameters, which may be regarded as secondary parameters derived from further inter-relationship between the quantities are: torque; impulse; rate of flow; rate of fuel consumption; specific fuel consumption rate; etc. Such secondary parameter quantities, obtained by the ratio of selected basic and primary quantities and the interpretation of performance trends analysis provide a link-up between the multiply-dependent parameters of a system.

10.2.1 Mean effective pressure

This may be expressed either in terms of the brake power or the indicated power and thus may be either a measure of the overall performance, the cycle performance or, by the ratio b.m.e.p/i.m.e.p. provide a measure of the mechanical deficiencies of an engine.

The S.I. unit for pressure is $N\,m^{-2}$ where $1\,N\,m^{-2} = 145 \times 10^{-6}\,lbf\,in^{-2} = 0.102 \times 10^{-6}\,kg\,mm^{-2}$ although the 'bar' (b) is also used where $1\,b = 10^5\,N\,m^{-2} = 14.5\,lbf\,in^{-2} = 0.0102\,kg\,mm^{-2}$.

10.2.2 Indicator diagram

Produced by one of the various mechanical instruments, or by means of a pressure transducer and cathode ray oscilloscope (using a 'Polaroid' camera to produce a permanent record), the indicator diagram provides information relating to the pressure-displacement variations. The shape of this diagram changes according to variations of timing, delays in valve closing or ignition and can provide a basis for interpretation if a systematic record of indicator diagrams is maintained. As with most power plant records, provision must be made to compare records under similar conditions of power, speed and ambient conditions.

10.2.3 Specific fuel consumption rate

This is a direct measure of the overall efficiency of a power system, the parameter being derived from:

specific power consumption rate = (fuel consumption rate)/(power output)

$$(10.11)$$

The unit for this parameter is $Kg(kW\text{-}h)^{-1}$ which may be obtained from Imperial Unit figures by the conversion

$$1 \text{ kg}(kW\text{-}h)^{-1} = 1.6440 \text{ lb}(hp\text{-}h)^{-1}.$$

By embracing the effect of consuming fuel and relating this to the final output power, this parameter includes all operational factors including the chemico-thermal properties (calorific value) of the fuel as well as the individual performance characteristics of combustion and the heat/work transfers within the machine. Monitoring this parameter against mean effective pressure provides an immediate reference criteria to detect deterioration particularly in combustion efficiency.

10.2.4 Specific steam consumption rate

This is similar to 10.2.3. and applied to a steam plant provides a monitoring parameter for all components other than the boiler. The magnitude of this parameter is derived from:

specific steam consumption rate = (steam generation rate)/(power output)

$$(10.12)$$

The unit for this parameter is thus $kg(kW\text{-}h)^{-1}$ as before.

10.2.5 Exhaust (flue) gas analysis

Chemical reactions during the combustion of hydrocarbon fuel in air produce exhaust gases of CO_2, H_2O, O_2, N_2 etc. which can be monitored by various automatic analysers. These usually give volumetric quantities which can be related to the mass content of carbon in the fuel to give the air/fuel ratio by mass as follows:

$$\text{air/fuel ratio (by mass)} = \frac{NC}{33(CO_2 + CO)} \qquad (10.13)$$

where: N = volumetric analysis of introgen,

CO_2 = volumetric analysis of carbon dioxide,

CO = volumetric analysis of carbon monoxide,

C = mass analysis of carbon in fuel.

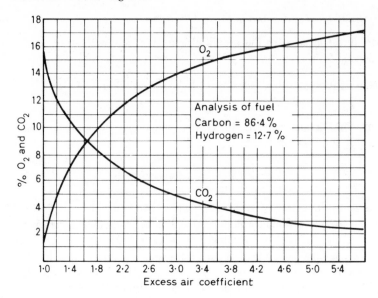

Figure 10.11 *Exhaust gas analysis – excess air coefficient*

Typical curves relating the oxygen (O_2) and carbon dioxide (CO_2) contents to the excess air coefficient are given in Figure 10.11, the excess air being the ratio

$$\text{excess air coefficient} = \frac{\text{actual air/fuel ratio}}{\text{stoichiometric air/fuel ratio}}. \tag{10.14}$$

The stoichiometric air/fuel ratio, itself, is the ideal which is theoretically needed for the complete combustion of the fuel and given by

$$\text{stoichiometric air/fuel ratio} = 34.56 \left[\frac{C}{3} + \left(H - \frac{O}{8} \right) \right] \tag{10.15}$$

where: C = mass analysis of carbon in the fuel,
        H = mass analysis of hydrogen in the fuel,
        O = mass analysis of oxygen in the fuel.

As applied to an engine, deviations in CO_2 concentration may indicate combustion deterioration requiring corrective action. The curves given in Figure 10.12. are characteristic of the relationship which can exist between exhaust gas concentrations and other engine performance data; the combustion factor being a slight modification of the Willan line, given by

$$\text{combustion factor} = (s)(\text{b.m.e.p})/(\text{b.h.p}) \tag{10.16}$$

where: s = fuel consumption rate,

b.m.e.p. = brake mean effective pressure,

b.h.p = brake horse-power.

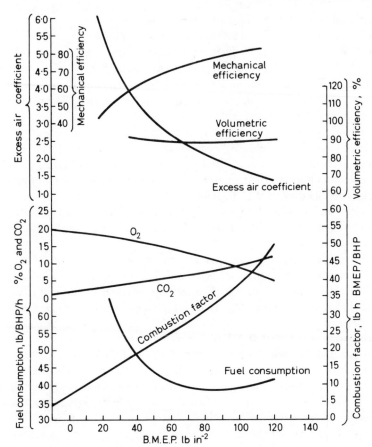

Figure 10.12 *Typical engine characteristic curves*

(Ideally, the plot of combustion factor versus b.m.e.p. should be a straight line with deviations only resulting from deterioration in air-induction 'breathing', combustion defects and similar air/fuel relationships).

10.2.6 Lubricating oil consumption rate

Lubricating oil which adheres to the cyclinder walls of engines is burnt and thus some lubricating oil consumption must be anticipated and is part of the overall fuel consumption. Abnormal lubricating oil consumption indicates a large quantity of oil is passing the piston rings and indicates that such rings are failing to provide adequate gas seals.

Oil consumption rate is generally independent of oil viscosity until a critical value is reached when it begins to increase rapidly. As a general rule with engines in average condition:

SAE 30, 40 and 50 = average oil consumption rate,
SAE 20W, 20　　　 = 1.2 × average oil consumption rate,
SAE 10W　　　　　= 1.5 × average oil consumption rate,
SAE 5W　　　　　 = 2.0 × average oil consumption rate.

Fuel dilution of the oil produces a reduction of viscosity and therefore an increase in oil consumption rate so that abnormal consumption may be related to both faulty combustion control and faulty piston rings.

A report on operational experience with medium speed marine diesl engines by Short [4] included the following observations on four ships with Pielstick PC2V medium speed diesel engines made by S.E.M.T., Societe d'Etudes de Machines Thermiques which were experiencing excessive lubricating oil consumption rates (in some cases double the S.E.M.T. figures).

After eliminating excessive losses of lubricating oil through maloperation of the lubricating oil separator and filter automatic cleaning, it was concluded that high consumption was due to:

(1) Excessive leakage at gudgeon pin cap. Examination of gudgeon pin caps showed that several of them could be turned. The 'O' ring seal which seals the gudgeon pin from the piston was found to take up a permanent set and it was found that two different size (cross-section) 'O' rings had been provided as spares.

(2) Ineffective scraping of oil by scraper rings. Although no abnormal piston ring or groove wear was reported, S.E.M.T. recommended the replacement of all rings with a new set and the blocking of the top row of drain holes in the piston.

(3) The constant speed engine. Lubricating oil consumption rate is largely a function of engine rev/min, so that with this constant speed engine the reduction in consumption with power will not be obtained, that would be expected with a variable speed engine where power and rev/min decrease together.

Aircraft engine performance monitoring by Alitalia includes an analysis of the oil consumption rate [5]. On the arrival of an aircraft the maintenance personnel immediately check the engine oil level and, if necessary, enter the level recording in the flight log book and the oil quantity added. A copy of the sheet of the flight log book is sent to the engineering department and the data is recorded on the appropriate form. Operating time is also recorded together with the amount of added oil and the date on which the oil was added.

Operating time, in hours and minutes, after the oil tank is refilled is taken into consideration. Two consecutive refills are combined each time the oil consumption is computed. The oil consumption is then plotted on a diagram. The trend of oil consumption can be accentuated by smoothing the data. To compute the smoothed values the following formula is used:

$$S_n = S_{n-1} + 0.2(E_n - S_{n-1}) \tag{10.17}$$

where:　　S_n = smoothed nth reading,

S_{n-1} = smoothed $(n-1)$th reading,

E_n = unsmoothed nth reading,

0.2 = damping factor.

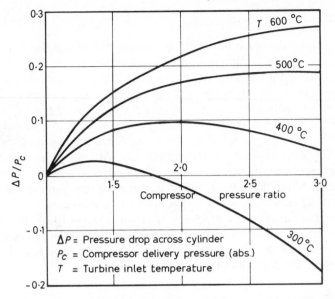

Figure 10.13 *Compressor pressure ratio versus* $\Delta P/P_c$ *for a turbo-charger with variable inlet temperature (60% overall efficiency)*

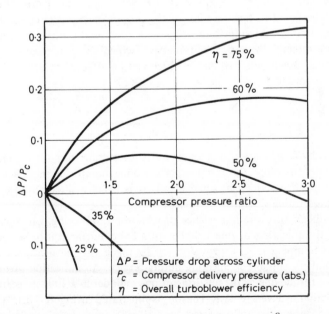

Figure 10.14 *Compressor pressure ratio versus* $\Delta P/P_c$ *for 500° C inlet temperature and different turbo-charger efficiencies*

10.2.7 Charge air pressure ratios

For some engines the exhaust-driven turbo-charger may be regarded as a separately-driven open cycle gas turbine in which the prime mover replaces the combustion chamber. Sufficient power must be developed to 'drive the compressor in order to' supercharge without impairing through-scavenge by creating a high back-pressure.

Relationships derived from continuity and thermodynamic analysis are given in Figure 10.13. This shows that with reducing compressor pressure ratio the pressure drop ratio across the cylinder (P/P_c) decreases. There is a maximum value of the pressure drop ratio for each exhaust temperature. At low pressure charging levels and low exhaust temperature the pressure drop ratio becomes so small that the pressure drop across the cylinder is less than the pressure losses in the system so that there could even be a reversal of flow instead of a scavenging.

The effect of turbo-charger efficiency is shown in Figure 10.14. This shows that a deterioration in turbo-charger efficiency can have a cumulative effect on the pressure drop ratio and for efficiencies of 35 and 25% the flow reversal indicated by negative values of P/P_c is a distinct possibility.

10.2.8 Temperature/expired life integration

Metallurgical changes resulting from prolonged operation are monitored in relation to gas turbines. A value for the expired life of an engine can be obtained by accumulating data on the time spent at each elevated operating temperature, the integral being weighed to take account of the proportionally greater ageing effect of higher temperatures. Attempts are being made at several centres to measure blade temperature directly by optical pyrometry. Two-colour pyrometry may be more successful than brightness pyrometry which can be affected by attenuation caused by particulate matter.

10.2.9 Vibration monitoring

Vibration is by classification, a secondary effect resulting from the operation of the system. While, therefore, this is not a 'performance monitor' in the strict sense (and has already been described in some detail) it is often included in integrated performance monitoring systems.

As an indication of the importance of vibration monitoring, where calculations show that the misfiring of one cylinder of a diesel engine can be serious, Lloyds Register of Shipping require provisions to be made for the measurement of the vibration and for suitable monitoring during service. An indication of the severity of vibratory torque due to the misfiring of one cylinder in a 400 h.p. geared propulsion installation as given by A.R. Hinson, is illustrated in Figure 10.15 (discussion of [6]). The measured vibratory stress for a 10 800 h.p. directly coupled installation is given in Figure 10.16.

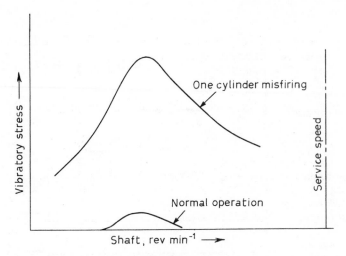

Figure 10.15 *Calculated vibratory stress (torque) effects for a 4000 h.p. geared propulsion diesel installation with one cylinder misfiring [6]*

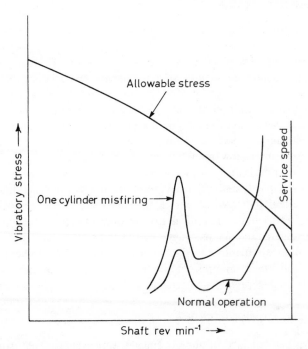

Figure 10.16 *Measured vibratory stress (torque) variations for a 10 800 h.p. direct coupled diesel installation with one cylinder misfiring [6]*

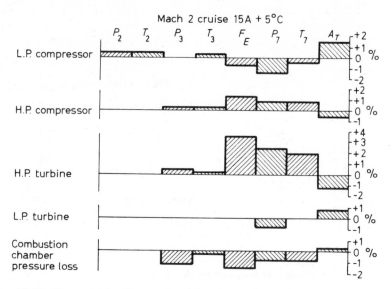

Figure 10.17 *Theoretical effects of performance changes (Based on 1% loss in module efficiencies – Olympus 593-602 engine) [7]*

10.3 Performance trend analysis

Monitoring performance has become an established procedure in the aircraft industry. The decision-problem of establishing deterioration limits beyond which engines should not be allowed to operate has been studied by Davis and Newman [7].

An attempt to provide an indication of parameter deviations associated with component efficiency changes can be derived theoretically, as shown in Figure 10.17. These provide a useful indication of the best parameters to select for trend monitoring, but can be misleading if used as 'failure signatures' for physical damage to main gas stream components. Thus the combined effects of LP compressor efficiency loss and mass flow reduction on engine fuel flow (F_E), jet pipe pressure (P_7) and primary nozzle area (A_J) would mean that a damaged compress would lose both mass flow capacity and efficiency, resulting in a reduced fuel flow. But P_7 and A_J could be either increased or decreased, depending on the relative changes in mass flow and efficiency. Theoretical investigations of 'failure signatures' appear to have limited value and actual recordings before and after failures must be obtained. Rolls-Royce BED are therefore setting up procedures for the retrospective examination of data from engine failures/deteriorations on test beds in order to provide the necessary performance trend data correlation.

10.3.1 Boeing 747/JT9D power plant performance monitoring

Trends in the performance of aircraft engines were systematically studied for the JT9D powerplant in the Boeing 747 and the outcome reported [5]. Information obtained included the aircraft environment, engine operating level and such

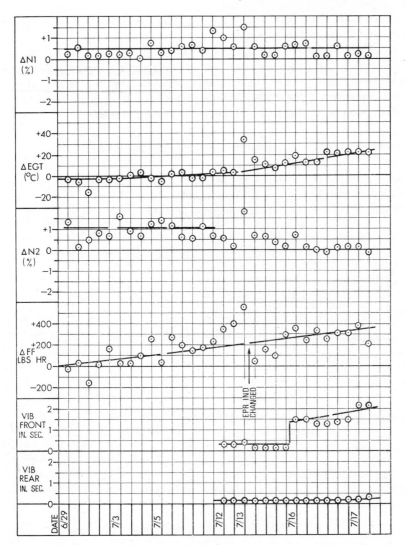

Figure 10.18 *First stage turbine blade failures*

mechanical parameters as oil temperature, oil pressure, breather pressure, vibration (inlet and turbine) and throttle position. The information was collected daily and processed by means of an electronic desk computer Olivetti 101 to compute the deviation of the recorded data from the standard gas generator allowing for both the bleed flow/engine and the primary airflow. This new data was then compared with the previously recorded information to produce a chronological plot. Maintenance action data can be collected and used to detect malfunctions and to determine the degree of overall gas path deterioration.

Failures which have been detected include: bleed air losses, variable stator vane problems, outer liner distress, diffuser case failures, high and low turbine blades, nozzle guide vanes and turbine seals.

One failure of the first stage turbine blades was primarily detected on the indication of a trend plot. As a result of the increased reading in aircraft vibration monitor (AVM) shown in Figure 10.18, a turbine borescope inspection showed first stage turbine blades in a damaged condition apparently resulting from the failure of the blade-retaining rivets. The trends of the performance curves were:

Rotor speed (N_1)	no significant change
Gas temperature (EGT)	$+ 20°C$ rise
Rotor speed (N_2)	-0.7% (decrease)
Fuel flow rate (EF)	$+ 200$ pph (increase)
Vibration	sudden increase on 16/7

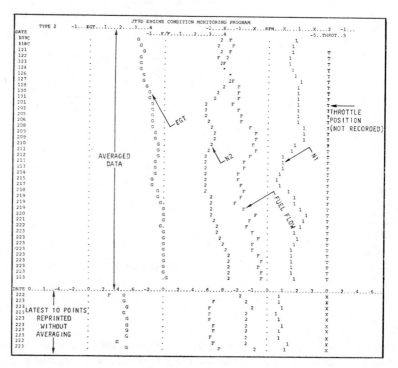

Figure 10.19 *First stage turbine blade-to-outer airseal – excessive clearance*

Excessive clearance of the first stage turbine blade-to-outer airseal for another aircraft was detected from the trend in the plot of Figure 10.19. Trend plots showed increasing gas temperature (EGT) and fuel flow rate (EF) of 2.0 and 2.5% respectively with no significant changes in rotor speeds N_1, N_2.

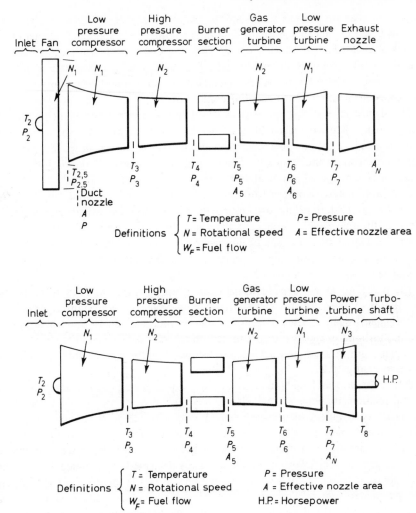

Figure 10.20 *(a) Generalized turbojet engine (b) Generalized turboshaft engine*

10.4 Turbine gas path performance monitoring thermodynamics

As indicated in the introductory 10.1, the numerous independent and dependent parameters involved in a system create a chain of sensitivity arising from the malfunction of individual components. With the advent of the computer, it is possible to manipulate and inter-relate thermodynamic quantities with greater ease and to evaluate the consequences of degressions in particular parameters.

10.4.1 Hamilton Aircraft electronic engine condition analysis system

Through the capability of a digital computer to perform logic decisions based on the outputs of relevant sensors the Hamilton electronic engine condition analysis system (EECAS) [8] accepts gas path data under both steady-state and transient conditions associated with starting up, deceleration and shut-down, then applies an analysis of (a) absolute value of the data, (b) rate of change of the data with time, (c) increments (deltas) in the values calculated by comparing the existing data (suitably adjusted) with reference base-line conditions.

The significance of this approach is based on the interface between components and gas path. Thus, damaged blades are indicated both by gas path thermo-flow dynamics and vibration, fuel control malfunction can be discerned by gas path analysis or position indicators, bearing defects can be detected from both the lubricating oil system and vibration analysis.

Gas path analysis employed in EECAS is a proprietary method based on the parameter inter-relationships established by Urban [9], it considers the engine in terms of the measurable dependent data and the independent performance parameters calculated by a mathematical model based on engine thermodynamics. Typical parameters for various engine arrangements are given in Table 10.1 based on the generalized layouts in Figure 10.20.

The independent and dependent parameters represent the variables in various engine thermodynamic relationships. These thermodynamic equations cover all gas path elements such as the compressor, main burner, gas generator turbine, power turbine, turbojet exhaust, turbojet thrust, horsepower balance and the very important engine air flow versus speed relationships. The concept of this modelling is shown in Figure 10.21.

Differentiation and manipulation of these equations allows the derivation of a general relationship between each change in a dependent parameter and its resulting effects on each independent parameter in turn with all other variables held constant. A matrix may then be formed using these coefficient relationships by superposition of the effect of each dependent variable on each independent parameter.

These general coefficients can then be readily evaluated at an engine design operating point. Thus, constant coefficients relating all of the independent and dependent parameters can be derived for a particular engine operating around a design point. The pertinent independent parameters have to be ascertained for the particular engine while an equal number of dependent parameters must be measured from the engine.

A simplified illustration of gas path analysis applied to a single spool compressor is shown in Figure 10.22. This analysis only covers one portion of the gas path and would have to be expanded in terms of dependent and independent parameters to include all components in the gas path. In this case, a four element matrix is formed in which the independent parameters are turbine inlet temperature (T_4), speed (N), compressor efficiency (η_c) and compressor air pumping capability (T_c). The dependent parameters are compressor discharge pressure (P_3), compressor discharge temperature (T_3), fuel flow (W_F), and jet nozzle exhaust area (A_N).

The variables in this four element matrix are related by the constant coefficients a_1 through a_{16}. This matrix will be evaluated as a third order array by holding the

Table 10.1. *Typical independent and dependent parameters*

Single Spool Turbojet		Twin Spool Turbofan	
Independent	*Dependent*	*Independent*	*Dependent*
Γ_4	Γ_3	Γ_5	P_D
η_C	P_3	η_F	Γ_3
Γ_C	W_F	Γ_F	P_3
η_B	P_5	η_{CL}	Γ_4
A_4	Γ_5	Γ_{CL}	P_4
η^+	N	η_{CH}	N_1
A_N		Γ_{CH}	N_2
		η_B	W_1
		A_5	Γ_6
		$\eta + H$	P_6
		A_6	Γ_7
		$\eta + L$	P_7
		A_{JD}	

Single Spool with Free Power Turbine		Twin Spool with Free Power Turbine	
Independent	*Dependent*	*Independent*	*Dependent*
Γ_4	N	Γ_5	Γ_3
η_C	Γ_3	η_{CL}	P_3
Γ_C	P_3	Γ_{CL}	Γ_4
η_B	W_F	η_{CH}	P_4
A_4	Γ_5	Γ_{CH}	N_1
η^+	P_5	η_B	N_2
A_5	H_P	A_5	W_F
η_{Pt}	Γ_6	$\eta + H$	Γ_6
		A_6	P_6
		$\eta + L$	Γ_7
		A_η	P_7
		η_{Pt}	Γ_8
			H_P

Definitions: η = efficiency
$\qquad\qquad\Gamma$ = air pumping capability
(all parameters corrected to inlet standard day conditions)

speed constant and realizing that the change in A_N can be calculated once the other independent changes are known. In step (B), actual parameter measurements are made for P_c, T_c, and W_F. The corrected values are then used to calculate the deltas shown in step (C) at the given speed with reference to the baseline curves.

Knowing the respective dependent parameter changes, the equations of step (D) can be formed and evaluated by the step (E) matrix solution to derive relative changes in Δ_4, η_c, and T_c. the relative change in A_N can now be calculated.

Finally, the model results are analysed for diagnoses of engine faults. If all the calculated performance parameter relative changes are equal to or very nearly zero, there has been no performance change, step (F). If, as in step (G), a compressor has been degraded for some reason, there will be downward shifts in compressor

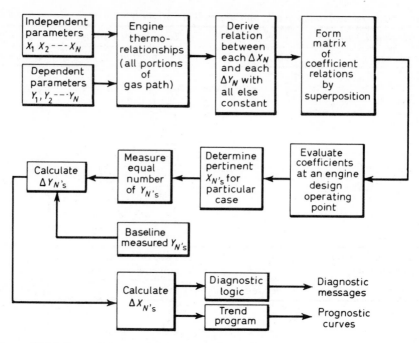

Figure 10.21 *Gas path prognostic/diagnostic procedure*

efficiency and air pumping capacity, accompanied by an upward shift in turbine inlet temperature, while the exhaust nozzle area will remain constant. Consequently, a component fault has been isolated to the compressor and suggested diagnostic actions are shown.

10.4.2 Olympus 593 trend monitoring procedure

Data collected during flight performance monitoring with the Olympus 593 engine was corrected to allow for the effects of variations between different flights of the changing engine inlet conditions and engine running conditions [7]. These corrections are indicated in Table 10.2.

Variations in engine inlet total pressure (P_1) and total temperature (T_1) are taken into account by the usual 'non-dimensional' relationships, as illustrated for N_1, N_2 and F_E in Table 10.2 (P_1 is not measured directly and therefore calculated from pressure height, Mach number, and an intake efficiency relationship). The wide range of T_1 encountered by the Concorde aircraft would require a correction for specific heat level (C_p) if a common datum condition such as Sea Level Static ISA were used for all flight cases. Separate datums were therefore defined.

Engines in current airline service, for which fuel flow rate is the only primary independent variable, require only one parameter such as engine pressure ratio (EPR) to define the non-dimensional running point. Recorded thermodynamic parameters

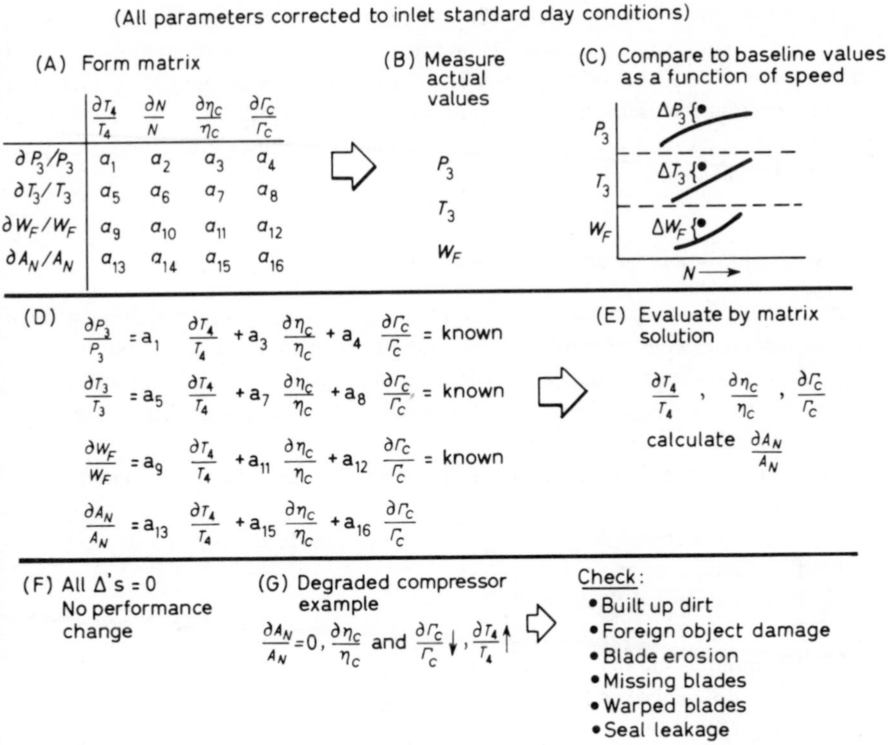

(All parameters corrected to inlet standard day conditions)

(A) Form matrix

	$\frac{\partial T_4}{T_4}$	$\frac{\partial N}{N}$	$\frac{\partial \eta_c}{\eta_c}$	$\frac{\partial r_c}{r_c}$
$\partial P_3/P_3$	a_1	a_2	a_3	a_4
$\partial T_3/T_3$	a_5	a_6	a_7	a_8
$\partial W_F/W_F$	a_9	a_{10}	a_{11}	a_{12}
$\partial A_N/A_N$	a_{13}	a_{14}	a_{15}	a_{16}

$\Rightarrow$

(B) Measure actual values

P_3

T_3

W_F

(C) Compare to baseline values as a function of speed

(D)

$$\frac{\partial P_3}{P_3} = a_1 \frac{\partial T_4}{T_4} + a_3 \frac{\partial \eta_c}{\eta_c} + a_4 \frac{\partial r_c}{r_c} = \text{known}$$

$$\frac{\partial T_3}{T_3} = a_5 \frac{\partial T_4}{T_4} + a_7 \frac{\partial \eta_c}{\eta_c} + a_8 \frac{\partial r_c}{r_c} = \text{known}$$

$$\frac{\partial W_F}{W_F} = a_9 \frac{\partial T_4}{T_4} + a_{11} \frac{\partial \eta_c}{\eta_c} + a_{12} \frac{\partial r_c}{r_c} = \text{known}$$

$$\frac{\partial A_N}{A_N} = a_{13} \frac{\partial T_4}{T_4} + a_{15} \frac{\partial \eta_c}{\eta_c} + a_{16} \frac{\partial r_c}{r_c}$$

$\Rightarrow$

(E) Evaluate by matrix solution

$\frac{\partial T_4}{T_4}$, $\frac{\partial \eta_c}{\eta_c}$, $\frac{\partial r_c}{r_c}$

calculate $\frac{\partial A_N}{A_N}$

(F) All Δ's = 0
No performance change

(G) Degraded compressor example

$\frac{\partial A_N}{A_N} = 0, \frac{\partial \eta_c}{\eta_c}$ and $\frac{\partial r_c}{r_c} \downarrow, \frac{\partial T_4}{T_4} \uparrow$

$\Rightarrow$

Check:
- Built up dirt
- Foreign object damage
- Blade erosion
- Missing blades
- Warped blades
- Seal leakage

Figure 10.22 *Simplified single spool compressor gas path analysis (all parameters corrected to inlet standard day conditions)*

Table 10.2 *Correction procedure for thermodynamic trend parameters*

Trend parameters:
$P_{2STAT}, T_2, P_{3STAT}, T_3, P_{7STAT}, P_{7TOT}, T_7, F_E, A_J$

Mach 2 datum values:

$T_1°K$	$P_1 \text{ lb in}^{-2}$	$N_1 \text{ revs/min}$	$N_2 \text{ revs/min}$
4000.0	10.85	6526	8873

Correction steps:
1. Correct thermodynamic trend parameters to Datum P_1 and T_1
2. Correct N_1 and N_2 to datum T_1.
3. Correct N_2 for 'P_1 effect'
4. Evaluate revs/min deviations from datum N_1 and N_2
5. Correct thermodynamic trend parameters to datum N_1 and N_2.

corrected for P_1 and T_1 can therfore be compared with values given by 'standard gas generator curves' of the particular parameters versus EPR. The differences between recorded and standard gas generator values are then used to produce trend plots.

The Olympus 593, with variable primary nozzle in addition to variable fuel flow, requires two parameters to define its non-dimensional running point and two spool speeds were selected for this purpose. The datum flight conditions define T_1, so the datum revs/min represent in effect datum values of N/T_1. The thermodynamic parameters to be plotted are corrected for deviations from these datum revs/min by referring to small change effect curves.

10.4.3 Axial-flow gas turbine loss parameters

A study made by Timms and Beachley [10] used computerized three-dimensional techniques on a wide range of turbine configurations and operating conditions to compare the Ainley and Mathieson loss data [11] with the Balje and Binsley parameters [12]. Little significant difference was found in the results obtained by using either approach.

For compressible flow the aerodynamic losses are functions of such parameters as:

(1) Reynolds number, R_e,
(2) Blade incidence angle, i,
(3) Blade spacing/cord length, P/C,
(4) Blade height/cord length, B/C,
(5) Maximum blade thickness/cord length, t_{max}/C,
(6) Entry boundary layer thickness/cord length, δ_c/C,
(7) Stagger angle, τ,
(8) Exit Mach number, M_2,
(9) Specific gas constant, R,
(10) Blade height/mean radius, B/Ω_m,
(11) Profile camber.

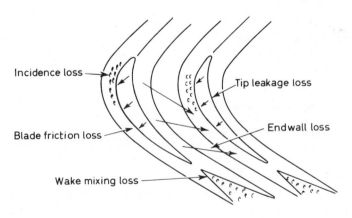

Figure 10.23 *Blading losses*

Equations for pressure loss coefficient [11] and enthalpy loss coefficient [12] omitted several of the foregoing aerodynamic loss parameters. To investigate the significance of these omissions Timm [13] broke the total loss down into component losses. The most important variables can then be considered separately for each component. For ease and accuracy of calculation, the following components are used (Figure 10.23):

(1) The profile loss associated with boundary-layer growth around the blade profile (including the wake mixing loss);

(2) Endwall loss which includes both the annulus friction loss associated with the turning of the boundary layer on the end walls along with the associated separation of flow;

(3) Tip clearance loss associated with the leakage flow across the gap between the rotor blade and the endwall;

(4) Disc viscous loss associated with the viscous friction of the rotor moving relative to the gas flow.

The losses associated with bearings and seals subtract work from the output shaft but do not affect the gas flow characteristics. These losses are not considered here, since they can be added after the aerodynamic analysis is complete.

The comparison of loss methods must be made by reviewing the final calculated turbine performance. One loss coefficient cannot be compared simply to another. Therefore, for a valid comparison: (a) a method of analysis must be selected, and (b) a common measure of performance must be used.

The approach taken to study the Ainley and Mathieson loss data is summarized by the flow chart, (Figure 10.24). The complete computer software package, consisting of one main program and 35 subprograms, requires 2.3 to 2.5 min on the Univac 1108 for compilation. The execution time required to obtain one performance point using reasonably accurate guesses is only 15 to 20 s on the 1108. If the turbine is near choking conditions, however, this time can easily triple, because the sensitivity of the calculations makes convergence difficult to achieve.

Balje and Binsley data had to be modified to allow for off-design performance. The incidence loss ratio and deviation angles of the Ainley and Mathieson data were used and disc friction losses obtained from the work of Daily and Nece [14]. The method of combining radial equilibrium theory and the composite loss method to predict off-design turbine performance is given by the flow chart of Figure 10.25.

10.4.4 Effects of changes in blade profile

There are virtually two categories of blade profile change:

(1) Pure roughening of the blade surface with an almost constant blade profile;

(2) Rough or almost smooth blade surface with a change in profile form which is chiefly marked by a uniform thinning or thickening of the profiles.

Impurities in the working medium attack the surface of gas turbine blades, particularly in open-cycle gas turbine, causing erosion, corrosion and contamination. These factors have been investigated by at the Institute for Turbomachinery and

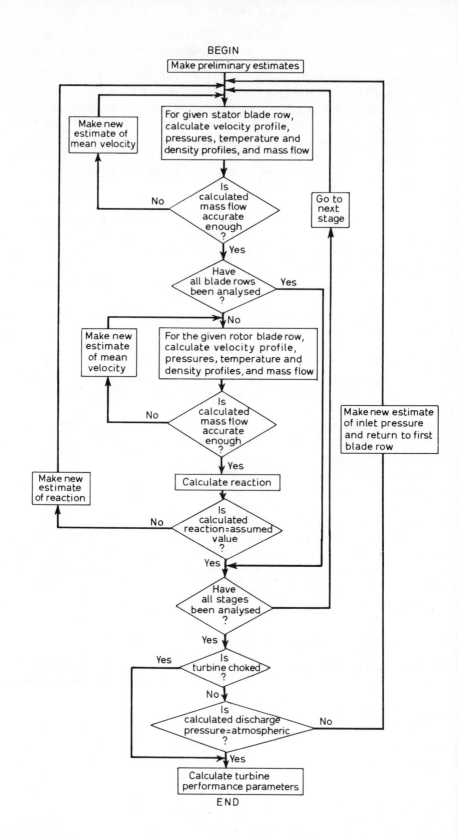

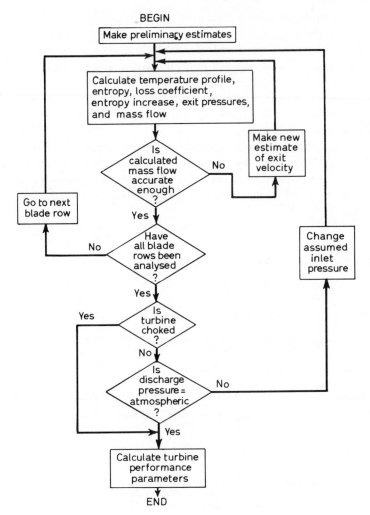

Figure 10.25 *Flow chart of computer program for radial equilibrium analysis with composite loss data*

Gas Dynamics, University of Hanover [15–17]. In steam turbines, salt deposits and water drops have similar effects. Experience shows that in many cases the blades of the first stages are thinned by erosion, the last stages are thickened by deposits. The blade surfaces are roughened throughout all stages [18]. Deposits stick according to the temperature of the medium and the aggregation of deposit materials [19].

Corrosion of the thermally highly stressed first turbine stages causes deposits to

Figure 10.24 *Flow chart of computer program for radial equilibrium analysis with the Ainley and Mathieson loss data*

stick initially to the blade surface according to the temperature and the state of the aggregation of the deposit materials. This results in either the profile surfaces being just roughened or partly thickened; the partly cemented deposits split off under thermal shocks during load changes causing thin oxide layers of blade material to be entrained and the blades gradually thinned.

These changes in blade profiles cause (a) stage efficiency to decrease, (b) turbine loading coefficient ψ to increase, (c) optimum stage efficiency to decrease and occur at a reduced velocity ratio.

The influences of such profile changes on the performance loss on a four-stage air turbine were investigated by Bammert and Sandstede [15]. Figure 10.26 shows the effect of blade roughness on mass flow coefficient. At a constant mass flow co-efficient the turbine loading coefficient is higher for greater roughness, in which:

$$\text{turbine loading coefficient, } \psi_e = \frac{h_e}{(u_t^2/2)} \tag{10.18}$$

where: h_e = effective specific enthalpy change,

u_t = circumferential turbine speed.

$$\text{Mass flow coefficient } \theta_{tot\ E} = \frac{C_{zE}}{\alpha_{tot\ E}} \tag{10.19}$$

where: C_{zE} = gas absolute axial velocity at inlet,

$\alpha_{tot\ E}$ = sonic velocity of gas at inlet.

These curves also show that the effective efficiency η_e is reduced by roughening of the surface and that the effective value of the mass flow coefficient is lowered, in which

$$\text{effective efficiency } \eta_e = \frac{h_e}{h_{sm}} \tag{10.20}$$

where: h_e = effective specific enthalpy change,

h_{sm} = mean isentropic specific enthalpy change.

Thickening of the blade coating produces an almost linear change in turbine ef-ficiency ranging from $\Delta\eta = -1.5\%$ for a thickness of 0.25 mm to $\Delta\eta = -4.2\%$ for a thickness of 0.7 mm.

10.5 Steam turbine performance analysis

Despite the preponderance of material which must be available very little data appears to have been published. It is believed that power stations have maintained performance logs consistently for many years and have built up considerable exper-tise but at the time of writing little of this information appears to have been co-ordinated. Similar considerations apply to other major users such as the marine transportation industry.

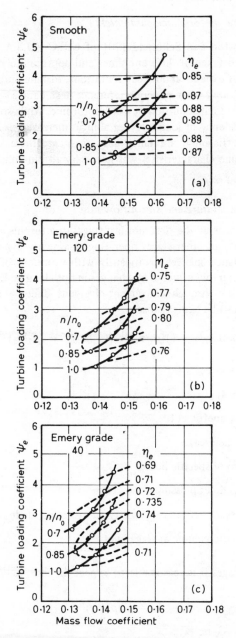

Figure 10.26 *Characteristic curves for a 4-stage air turbine with changed blading for smooth and roughened surfaces* ——— $n/n_0 = constant$ – – – – $\eta_3 = constant$ [25]

10.5.1 Steam turbine performance parameters

It was held by Cotton and Schofield [20] of G.E.C. (U.S.A.) Schnectady, New York, that test data taken at different times and appropriately corrected for such parameters as initial exhaust and reheat conditions, turbine output, steam flow rate etc. could be compared with previous reference test data to provide valuable information on the internal condition of a steam turbine. While some test parameters, which indicate deficiency in load-carrying ability, increased packing leakage or decrease in section efficiency, are fairly easy to locate these authors note that a knowledge of turbine characteristics was necessary to pin-point the physical areas of distress.

10.5.2 Pressure, temperature and steam flow rate

Only the performances of the first and last stages of a turbine are significant under changing conditions, all other stages operating at a nearly constant pressure ratio. First stage performance changes significantly with variations of load; the last stage performance is measurably affected by changes in both load and condenser pressure.

For subsonic flow through convergent-divergent ducts, flow rate through all stages of a turbine can be expressed as

$$W = K[(2\gamma/\gamma - 1)(p_1/v_1)]^{1/2}[(p_2/p_1)^{2/\gamma} - (p_2/p_1)^{\gamma + 1/\gamma}]^{1/2} \qquad (10.21)$$

where: W = flow rate,

$K = C_q \cdot A_n$,

C_q = flow coefficient,

A_n = nozzle area,

γ = ratio of specific heats,

p_1 = stage inlet pressure,

p_2 = pressure between stationary and rotating blade rows,

v_1 = specific volume at stage inlet.

For intermediate stages it is reasonable to expect the ratio (p_2/p_1) and γ to be constant so that:

$$W = C(p_1/v_1)^{1/2} \qquad (10.22)$$

i.e. $W = C \cdot p_1(RT_1)^{-1/2}$,

where: C = constant,

R = universal gas constant,

T_1 = absolute temperature at stage inlet.

0.5.3 Effect of throttle pressure variations

The last stage of a condensing turbine is affected by an increasing throttle pressure. All preceding turbine stages are not affected. With a constant valve setting an increase in throttle pressure results in a directly proportional increase in flow — as demonstrated by the equation, $W = C \cdot p_1(RT_1)^{-1/2}$ — since the pressure entering the second stage increases by the same percentage as that of the throttle pressure. This applies to all succeeding stages except the last stage of a condensing turbine.

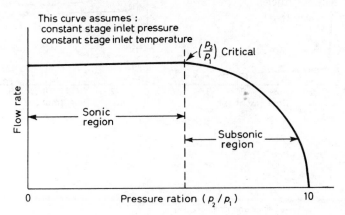

Figure 10.27 *Condensing turbine last-stage steam flow versus stage pressure ratio*

The exhaust pressure of the last stage remains constant, it is dependent on the condensing characteristic of the discharge. The last stage of most turbines is designed with near-critical pressure ratios across the nozzle. As shown in Figure 10.27 the last stage flow becomes sensitive to changes in pressure ratio. Flow velocities through the last stage become very unstable, the thermodynamic performances accordingly impaired and the stage subject to dynamic influences.

0.5.4 Effect of reheat temperature

With the throttle steam condition held constant the equation $W = C \cdot p_1 \cdot (RT^1)^{-1/2}$ shows that an increase in reheat temperature results in an increase in reheat pressure. This effect is reflected in all stages downstream of the reheat inlet so that the total pressure drop decreases across the high pressure turbine.

0.6 Case studies in performance monitoring

As mentioned in Section 10.5, published information is scanty, the following cases were quoted by Cotton and Schofield [20].

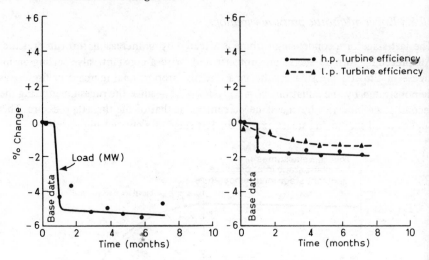

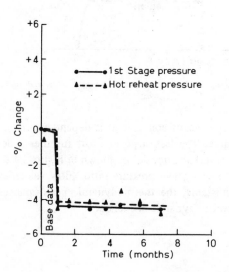

Figure 10.28 *(a) % change in load versus time for Case 1*
(b) % change in h.p-l.p. turbine efficiency versus time for Case 1
(c) % change in turbine pressure versus time for Case 1

10.6.1 Case Study No. 1 – flow restriction (steam turbine)

Symptoms

The load capacity of a turbo-generator suddenly fell by 4% as shown in Figure
10.28 together with associated changes in first stage pressure and turbine efficiency
a summary of the observed performance changes being;

Load change − 4.9%
Feedwater flow rate − 3.6%
First-stage pressure − 4.2%
Intercept valve pressure − 4.4%
High pressure stage efficiency − 1.8%
Intermediate pressure stage efficiency − 0.4%

Deductions.

(1) Drop in first stage pressure and intercept valve pressure suggests a restriction in turbine flow-passing capability;

(2) This is substantiated by the associated 3.6% reduction in feedwater flow;

(3) Causes for the loss in flow-passing capability relates to changes in stage pressure;

(4) Since all stages other than the first and last operate at a constant pressure ratio, change in pressure upstream will be reflected in an equal percentage change in any of the pressure changes downstream;

(5) In this case there is a similar percentage change in pressure of the first stage and the intercept valve;

(6) It was concluded that the loss of flow arose from a restriction at the front end of the turbine either in the first stage or at the control valve.

Inspection revealed that the stem of one of the four control valves was broken so the sequence of opening was interrrupted.

0.6.2 Case study No. 2 − turbine shell distortion

Symptoms

The efficiency of the intermediate pressure turbine in a plant was found, from an enthalpy drop efficiency test, to be 6.9% less than that measured during an earlier test, although no problem was apparent. Further data then acquired showed the following time-compound changes:

Load change −2.0%
First stage pressure + 0.3%
Hot reheat pressure + 6.3%
Low-pressure bowl pressure + 2.2%
High-pressure stage efficiency − 1.3%

Deductions

(1) Increase in the hot reheat pressure at the constant − control − valve setting suggests an area restriction either in the intercept valves or in the stages of the intermediate-pressure turbine;

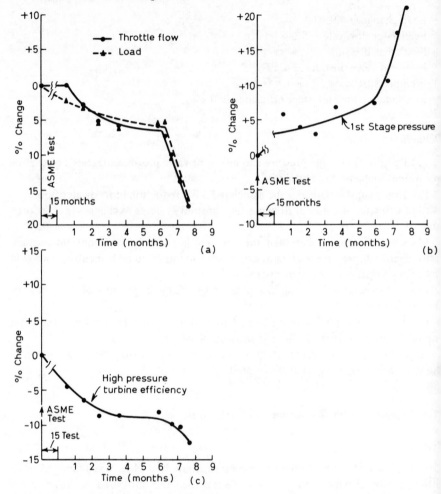

Figure 10.29 *(a) % change in load and throttle flow versus time for Case 3*
(b) % change in 1st-stage pressure versus time for Case 3
(c) % change in high-pressure turbine efficiency versus time for Case 3

(2) The data suggests the cause might be (i) congestion of the intercept valve screens, (ii) severe deposits in the turbine, (iii) mechanical damage causing a reduction in nozzle area;

(3) Congestion of the intercept valve screens was not occurring as the pressure drop was small;

(4) Sudden excessive build-up of deposits in the time-period under review (8 months) was unlikely;

(5) Mechanical damage in the stages of the intermediate-pressure turbine was therefore a likely cause of congestion.

Inspection revealed that the turbine end of the double flow intermediate pressure section was badly damaged due to failure of a number of the 9th stage blades. Additional damage arose from impact by broken blades lodging in the diaphragm. The original defect arose from distortion of the turbine shell causing abrasion of the blade tenon root fixings; the shell distortion arose from the entry of water into the turbine causing shock chilling of the shell in local areas.

10.6.3 Case No. 3 – blade and nozzle deposits

Symptoms

A turbo-generator unit operating at $3500\,\text{lbf}\,\text{m}^{-2}$ suddenly started to lose output at the rate of 1.2 MW each day after operating for 21 months. Readings showed that the performance had gradually deteriorated over the whole period as shown in Figure 10.29. 4(a), (b), (c), the changes being:

Throttle flow rate − 17.2%
Load − 16.5%
First-stage pressure + 21.2%
High pressure unit efficiency − 12.2%

Deductions

(1) The 21.2% increase in the first-stage pressure could be due to either (i) increased flow through the turbine, (ii) a reduction in area of the second or subsequent stages of the high pressure unit;

(2) Increase in flow through the turbine is not applicable as the throttle flow rate decreased 17.2%;

(3) Restriction in the second or subsequent stages was the probable source of trouble and this could arise from either (i) mechanical damage (ii) deposits;

(4) Mechanical damage usually causes a sudden change in performance whereas the observed changes had been gradual and consistent.

Inspection revealed severe copper deposits throughout the entire high pressure turbine, the thickness being measured as follows:

Nozzles:	1st stage	0.041 in
	7th stage	0.093 in
Blades:	1st stage	0.010 in
	4th stage	0.060 in
	7th stage	0.029 in

10.6.4 Case No. 4. – boiler debris

Symptoms

A turbo-generator returned to service after the replacement of boiler tubes had a maximum capacity 9.1% lower than prior to the boiler re-tubing, the comparable

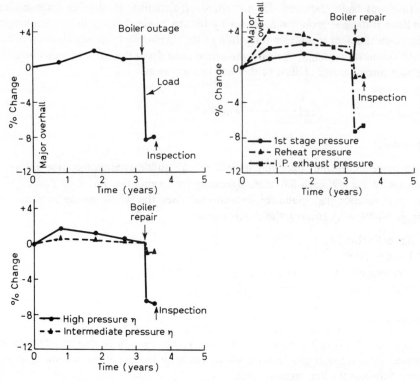

Figure 10.30 *(a) % change in load versus time for Case 4*
(b) % change in turbine pressure versus time for Case 4
(c) % change in h.p.-l.p. efficiency versus time for Case 4

performance results shown in Figure 10.30 4(a), (b), (c) were as follows:

Load − 9.1%
First-stage pressure + 2.1%
Hot reheat pressure − 3.3%
Low-pressure turbine inlet pressure − 9.7%
High-pressure stage efficiency − 8.9%
Intermediate-pressure stage efficiency − 1.4%

Deductions

(1) Increase in first-stage pressure could be caused by (i) increase in flow rate, (ii) increase in first-stage area, (iii) reduction in area of the 2nd stage;

(2) An increase in flow rate would increase the pressure throughout the turbine stages − this does not occur as the hot reheat pressure and low- pressure inlet pressure both decreased;

(3) Similarly, an increase in first-stage area should cause a transmitted pressure increase, which did not occur;

(4) A restriction in flow through the 2nd or subsequent stages of the high-pressure turbine is therefore the most likely cause of trouble and would cause the decrease in the hot reheat pressure;

(5) The further decrease in low-pressure turbine inlet pressure suggests a restriction in flow in the intermediate-pressure turbine as shown by the reduction in efficiency;

(6) That restrictions exist in both the h.p. and i.p. sections suggests that foreign metal particles from the boiler repairs have been transported through to the turbine.

Inspection revealed solid particle erosion of the first stage nozzles and buckets, severe battering of the second stage and similar defects in the intermediate section of the turbine. Fine mesh strainers were not installed (a grave omission) and many pieces of cutting debris was found jammed in the coarse main-stream strainers.

10.6.5 Case No. 5 – nozzle erosion

Symptoms

A turbine experienced an increase in output over a 3-year operating period at a given valve setting. A summary of the trend data over this period was:

Load + 11.0%
First stage pressure + 11.0%
Hot reheat pressure + 10.2%
High-pressure efficiency − 1.8%

Special tests performed on the high pressure section of this unit indicated an increase in output over the entire range of valve points tested. Figure 10.31(a) shows that flow at each valve lift as compared with design, Figure 10.31(b) gives the h.p. efficiency for each exhaust/throttle pressure setting over the whole range as compared with a previous test run.

Deductions

The stage pressures indicate a general flow increase through the turbine. An increase in first stage area resulting in an increased flow at all valve positions could be due to such factors as:

(1) Erosion of the first stage nozzles, probably most on the first and second admission areas of 8-valve machines; this was seen to be the effect as most of the flow increase had occurred before the opening of the third valve,

(2) Damage to the first stage – this would have impaired the h.p. turbine efficiency but such impairment did not occur;

(3) Flow leakage around first-stage nozzles – this would have a much greater influence on the turbine efficiency than actually occurred;

(4) The shape of the high pressure efficiency curve compared with the design curve provides a diagnostic clue, if the 1st stage efficiency falls due to mechanical

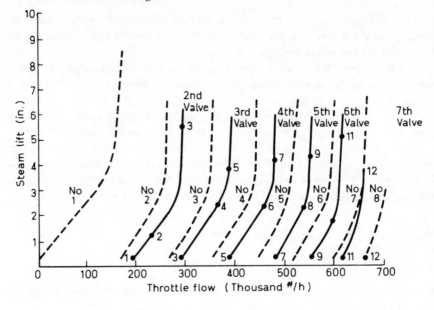

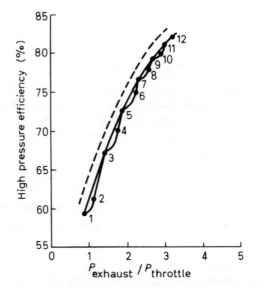

Figure 10.31 *(a) Control-valve lift versus throttle flow for Case 5*
(b) High-pressure turbine efficiency for Case 5

damage the h.p. efficiency would decrease, however, in this case as the load is decreased more energy would be transferred at the first stage while the energy transfer across the other stages would remain constant;

(5) A steep h.p. efficiency curve indicates trouble in the first stage;

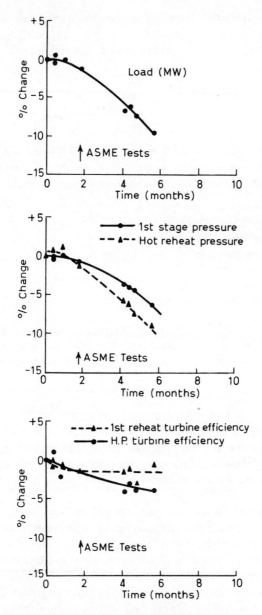

Figure 10.32 *(a) % change in load versus time for Case 6*
   (b) % change in turbine pressure versus time for Case 6
   (c) % change in h.p. and 1st reheat turbine efficiencies versus time for Case 6

(6) With the turbine at full load the efficiency change was − 1.8% (reduction), at low load the change was − 5.9% (reduction);

(7) It was concluded that there was an increase in first stage nozzle area at the 1st and 2nd nozzle areas.

Inspection revealed serious erosion of the 1st, 2nd and 3rd nozzle areas.

10.6.6 Case 6 – foreign bodies

Symptoms

The capacity of a turbine fell 9.5% during the first 6 months of operation with the first stage efficiency and capacity at the start lower than that which had been expected. Some important changes during the 6-month period were:

Load — 9.5%
Feed water flow rate — 8.0%
First stage pressure — 6.4%
First reheat pressure — 9.2%
h.p. efficiency — 4.0%
First reheat efficiency — 1.4%
Second reheat efficiency – no change

The variations are plotted in Figures 10.32(a), (b), (c), (data was not available for a 2-month period hence it is not possible to tell categorically whether the change was sudden but the subsequent data suggest that the deterioration was gradually progressive).

Deductions

(1) The 5.4% reduction in first stage pressure indicates a front end flow restriction due to either (i) control valve malfunction, or (ii) front-end area closure;

(2) The 9.2% decrease in first reheat pressure indicates that in addition to the front end closure there is a flow restriction in the 2nd or subsequent stages of the h.p. turbine;

(3) Since the load loss occurred gradually and the flow restriction is not centred on any one part of the h.p. turbine the original substandard performance possibly arose from the collection of deposits in the h.p. turbine;

(4) The main accumulation of deposits may be in the first stage with a smaller accumulation in the second and subsequent stages of the h.p. section.

Inspection revealed two major problems, (1) heavy deposits located throughout the whole of the h.p. turbine, (2) mechanical damage through the loss of a number of shroudings over the first stage blades.

The deposits were reddish-brown in colour, analysis revealed them to comprise aluminium with small amounts of iron. Nozzle deposits were extremely hard and difficult to remove. Deposits on the latter stages of the h.p. turbine were softer and showed evidence of build-up followed by spall-off.

Mechanical damage from the blade shrouds extended throughout the whole h.p.

turbine, some large pieces were lodged between diaphragm partitions — first, second and third stage nozzle partitions were extensively damaged.

10.7 Performance monitoring systems

Aircraft monitoring systems generally described as AIDS (Aircraft Integrated Data Systems) together with marine power plant monitoring systems have been introduced by a number of organizations. A number of these are reviewed in Section 10.18. A summary of the salient features is given in the following paragraphs.

10.7.1 Performance data recording

Small installations should establish at least a manual record of all measurable quantities. Sensible selection of the frequency of recording should prevent this from becoming a useless clerical chore, although it does have the added operational value of ensuring that each of the measuring points are regularly inspected and accordingly contributes to effective personnel management.

Automatic data-logging has been introduced to many control installations and provides a frequent print-out of a whole range of instruments. So comprehensive and formidable does much of this material become that apart from its use as a spot read-out and the basis of a subsequent post-mortem it frequently has to be discarded. This is regrettable as by establishing a suitable display on a histographical basis the trends could be indicated and recorded in a more compact form.

10.7.2 Data-log information analysis

Data from running logs needs in principle to be related to ideal conditions such as those defined by theory or design with due consideration for the significance of the input parameters.

Thus, in dealing with thermodynamic performance, either as a means of energy conversion or as a source of heat transfer for chemical or physical—chemical processes, deviations from the ideal can be obtained from theoretical considerations. Such calculations must be evaluated to establish the effects of deviations from the design conditions and thus provide a basis for comparison. Typical of such techniques are the comparison of actual and theoretical indicator diagrams in reciprocating internal combustion engines, the temperature gradients in heat transfer equipment and pressure differentials affecting flow through pipes.

Similarly, mechanical effects arising from friction and wear can be analysed by studying the torque and/or speed, together with induced thermal effects such as rise in bearing temperature.

To provide for input parameters it seems more reasonable that instead of analysis being made by comparing direct ratios the comparisons should be based on the ratio of ratios, the ratios being those of the point or output parameter to the appropriate input parameter.

It is possible that comparable data might be based on dimensionless quantities such as the values of Reynold's Number which relate to streamline or turbulent flow. Alternatively, some dimensional analysis such as that used to relate scale effects but which use time as a variable quantity might provide the basis for condition analysis comparison — particularly linked to reliability theory [21].

10.7.3 Engine performance recording

Facilities for the monitoring and recording of flight data were originally built into aircraft as a safety analysis device and have since been extended as a means of providing health monitoring information [22—24].

Typical recorded parameters include gas pressures and pressure ratios, gas temperatures compressor speed, oil pressure and temperature and fuel flow. There have been numerous developments in AIDS for assessing the significance of the total information (perhaps with the addition of a few special sensors, e.g. vibration sensors) in terms of engine health. An airborne computer or airborne tape recorder for use with a ground computer is used to analyse the voluminous information.

10.7.4 DATACHIEF monitoring/maintenance computer

The DATACHIEF computer system introduced by Norcontrol, a division of Noratom-Norcontrol A/S, Norway, provides a comprehensive marine engine-room facility which has been installed in 122 000 dwt OBO ships built for the Swedish Gotaverken Group at Oresund.

Functions of the complete computer system include engine operation instrumentation of an unmanned engine room, control of the power supply, remote control of the main engine from the bridge as well as extensive monitoring of the conditions in the main engine.

A substantial part of the system is constructed in order to reduce total operating costs on the main engine. This part of the system, called Data Trend makes it possible to monitor the combustion process in the engine. It also affords the engineer the possibility of continuously controlling fuel injection and thereby the combustion rate.

An engine equipped with this monitoring system is expected to burn 1 to 2% less fuel than an engine without such monitoring. In addition, the engineer receives other information which, correctly used, can optimize maintenance work.

The most important function claimed for this system in the engine room is the provision of information on the condition of the machinery. This function is performed by the Data Trend subsystem. Information necessary for the engineer's evaluation of when and where maintenance ought to be performed is presented on the basis of continual measurements and computation by the equipment.

Maintenance can thus be performed according to need and not in response to a time-table. The improved routines in the engine room reduce the risk of engine break-down and make significant savings possible. Closer control of engine operating conditions, together with more efficient maintenance, means less wear on the engine. This, along with better control of the combustion process, makes for lower fuel consumption.

Investigation undertaken by NSFI (Norwegian Ship Research Institution) indicate a gradual increase in engine fuel consumption of up to 3 to 4% in comparison to a new engine's performance, and one of Data Trend's most important functions is to reduce this increase in fuel consumption. The system cannot prevent a fault occuring in, for example, fuel injection, but it can notify the engineer about it so that he can act to adjust conditions as soon as possible. Data Trend is fed with data on the various diesel engine components when the engine is new, and with the help of a mathematical model describes the engine 'normal condition'. Greater or less deviations between the engine's 'normal condition' and the 'measured conditions at any one time' will therefore be an expression for a change in operating conditions which has occurred and which demands the engineer's attention. The system has an automatic check-program for the essential parameters of the diesel engine, and this gives an immediate alarm if the measured deviation goes over a set limit. Such an alarm system operates for the metal temperature of the cylinder head and cylinder linings and for the proper sealing of the piston rings. These measured parameters allow the computer system to present 'ideal operating condition', 'present operating condition', and 'relative deviation'. This form of presentation is a valuable aid to the engineer in his control, adjustment and maintenance work.

The system also has a simulation program which makes it possible to estimate the need for maintenance work before it is done. Finally, the system makes trend analyses of the accumulated data on operating conditions. Based on this, it is possible to predict when a certain condition is likely to occur. This is new and useful information for the engineer, who thus gets a more accurate picture for evaluating and planning future maintenance.

In addition to the Data Trend sub-system, Data Chief also includes DATASAFE, an advanced engine operation instrumentation system for alarm and control of a partially unmanned engine room. DATAPOWER is another system for the automation of the electricity supply on board.

10.7.5 LMT-1000 trend analysis system

Continuously-performed steam plant efficiency calculations by this system from Litton Systems Inc., 5500 Canoga Avenue, Woodland Hills, California establishes machinery performance trends, alarms excessive rates-of-change of machinery parameters and indicates potential action.

From the pressure, temperature, torque, velocity, revs/min, and vibration transducers, data are automatically collected and processed. The data being processed are standardized by correcting ambient pressures and temperatures, and different API gravities to a set of standard conditions. When initially installed aboard the ship, baseline data is collected at several propulsion power levels and stored permanently. Power plant losses not otherwise measured or monitored (radiation, turbine

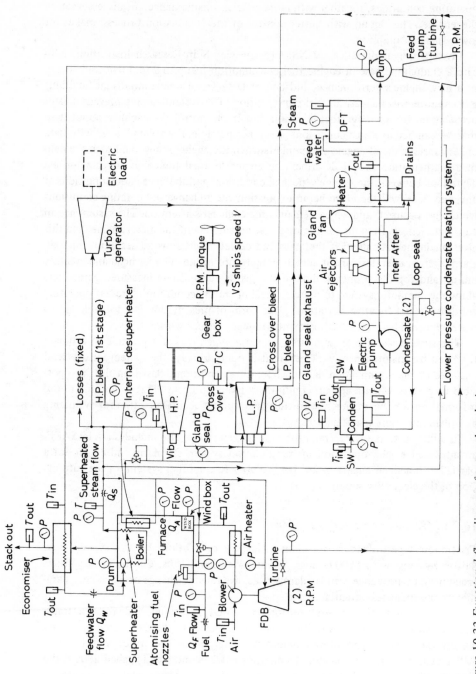

Figure 10.33 *Functional flow diagram for trend analysis*

expansion loss, miscellaneous mechanical, etc.) are derived from these baseline data.

Heat balance equations are used to compute the thermal effectivities or efficiencies of the steam plant and its elements. If a shaft torque meter is installed, it and the shaft rev/min meter are used to compute the mechanical efficiency of the propulsion unit. Deviations from normal temperatures and pressures are indicative of changes in operating points, equipment degradation, or efficiencies. By monitoring and recording these deviations, the system provided data to the ship's engineer permitting him to identify present and potential problems and to take corrective action.

The specific fuel consumption rate can be calculated. Using the output of the ship speed log, the fuel consumed per nautical mile is calculated. These calculations can be repeated at several power levels and ship velocities. Comparison of the measured mechanical shaft horse-power or horse-power calculated from thermodynamic data with attained ship speed (with reference to ship's trials data), indirectly evaluates the level of the hull resistance. The gradual increase in hull resistance due to the accumulation of fouling marine growth is monitored. Mechanical damage to propellor, rudder or other extended hull structure is reflected by a jump in hull resistance following the defect.

A typical ships steam plant showing the minimum number of instruments required for the minimum basic trend monitoring is shown in Figure 10.33.

These minimum basic trend computations are:
(1) Air heater effectiveness;
(2) Boiler efficiency;
(3) Fraction of total power developed in the h.p. turbine;
(4) Main condenser efficiency;
(5) Air ejectors/l.p. consensate heater system efficiency;
(6) De-aerating feed tank efficiency;
(7) Feed pump efficiency;
(8) h.p. heater heat transfer coefficient;
(9) h.p. heater efficiency;
(10) Total fuel flow rate;
(11) Specific fuel consumption rate.

Air heater effectiveness, for a given discharge pressure and rev/min of the forced draft blower is computed from a comparison of the air temperature rise across the heater to the steam pressure drop across the turbine. Fuel, air, water and steam flow rates are used to compute boiler efficiency for each measured output steam pressure and temperature.

The thermal efficiencies of the several turbines are calculated from steam flow rate, inlet temperature and pressure, exit temperature and pressure. Compensation is made for bleed or extraction steam flows. Using the cross-over temperature and pressure of the illustrated cross-compound turbine the relative ratio of high pressure to low pressure turbine power outputs is evaluated. Comparison of computed turbine power output with shaft power computed from measured shaft torque and speed reveals gear reduction box losses.

Main condenser efficiency is calculated, for each sea water cooling flow rate, by use of inlet steam and exit condensate temperatures at the measured condenser vacuum pressure.

The air ejector/lower pressure condensate heater system efficiency is computed from measurements of ejector steam pressure, and condensate flow rate and temperature rise. Pressure and temperature within the de-aerating feed tank, at a given level of feedwater in the tank, is used to calculate the efficiency of the de-aerating feed tank.

The ratio of feedwater flow rate to steam pressure drop across the feedwater pump's turbine, at a particular revs/min and pump outlet pressure, is a measure of pump efficiency for a non-recirculating feedwater pump.

The heat transfer coefficient and efficiency of the economizer/high pressure feedwater heater is computed from measurements of temperature rise in the feedwater and temperature in the stack at a measured gas pressure drop from furnace to uptake stack.

The following ambient parameter measurements are used to normalize the above computations to a fixed set of standard conditions:

Atmospheric pressure,
Ambient air temperature,
Sea water temperature,
Fuel oil API gravity (manual input).

Other parameters measured and monitored by the trend system are not used in the above computer calculations. Deterioration in these values, however, are indicative of changes in the physical plant which are affecting either the effectivity of the particular elements listed above or the mechanical integrity of elements for which efficiencies are not continuously calculated. The auxiliary measurements include the pressures and temperature of the lubricating oil supplied to the propulsion turbine, reduction gear, shaft, fuel oil pumps, condensate and feedwater pumps, forced draft blowers, and secondary steam turbogenerators. Pressure drops across fuel oil, lubricating oil, and steam filters and strainers are also monitored. Excess turbine radial shaft displacement, over a broad range of rotary speeds, indicates turbine vibration above normal and this parameter is continuously monitored.

A number of discrete element status indicators are used by the LMT–1000 to stop or alter the efficiency computations when there are changes in the steam plant which render incorrect the basic programmed equations. These operate until the plant has restabilized. Examples are turning on the boiler tube-soot blower; or closing the turbine interstage steam extraction valves.

References

1   Rogers, G.F.C. and Mayhew, Y.R. (1958), *Engineering Thermodynamics, Work and Heat Transfer*, Longmans Green & Company.
2   Spalding, D.B. and Cole, E.H. (1959), *Engineering Thermodynamics*, Edward Arnold (Publishers) Ltd.

3 Haywood, R.W. *Analysis of engineering cycles*, Pergamon Press (Commonwealth & International Library)

4 Short, K.I. (1972), 'Operational experience with medium speed diesel engines', *Trans. I. Mar. E.* **84**, 37–72.

5 Short, K.I. (1972), 'Engine performance trend analysis', *Aircraft Engineering*, July, 23–25.

6 Glew, C.A.W. (1974), 'The effectiveness of vibration analysis as a maintenance tool', *Trans. I. Mar. E.* **86**, 29–50.

7 Davies, A.E. and Newman, H.L. 'Diagnostics and engine condition monitoring – the relative role of engine monitoring programmes during development and service phases', *AGARD*, 43rd. PEP Meeting.

8 Moses, H.J. (1971), 'Electronic engine condition analysis system', SAE Paper 710448, May.

9 Urban, L.A. (1969), 'Gas turbine engine parameter inter-relationships', Hamilton Standard Division, United Aircraft Corporation, Windsor Locks, Connecticut, USA, Report.

10 Timm, R.T. and Beachley, N.H. (1973), 'Comparative analysis of turbine loss parameters', A.S.M.E. Paper 73-GT-91, April.

11 Ainley, D.G. and Mathieson, G.C.R. (1951), 'A method of performance estimation for axial flow turbines', R & M 294, Aeronautical Research Council.

12 Balje O.E. and Binsley, F.L. (1968), 'Axial turbine performance evaluation – part A, Loss geometry relationships', A.S.M.E. Paper 68-GT-13.

13 Timm, R.T. (1972), 'A comparison study of 3-dimensional analysis techniques and loss calculations for axial flow turbines', Ph.D. Thesis, University of Wisconsin.

14 Daily, J.W. and Nece, R.E. (1960), 'Chamber dimension effects on induced flow and frictional resistance of enclosed rotating discs', *J. Basic Engineering*, **82**, 217–232.

15 Bammert, K. and Standstede, H. (1972), 'Measurements concerning the influence of surface roughness and profile changes on the performance of gas turbines', *Trans. A.S.M.E., J. Eng. for Power*, **94**, July 207–213.

16 Bammert, K. and Stobbe, H. (1970), 'Results of experiments for determining the influence of blade profile changes and manufacturing tolerances on the efficiency, the enthalpy drop and the mass flow of multistage axial turbines', A.S.M.E Paper No. 70-WA/GT-4.

17 Bammert, K. and Stobbe, H. (1970), 'The effect of corrosion, erosion and contamination on the operational behaviour of turbine and circuit of gas turbine plant', *Archiv. F. Eissenhuttenwes*, **41**, No. 11 1055–1068.

18 Forster, V.T. (1966–67), 'Performance loss of modern steam turbine plant due to surface roughness', *Proc. I. Mech. E.* **181**, Pt 1 391–406.

19 Strub, R.A. (1965), 'Field experience on industrial gas turbine installations', *7th. International Congress on Combustion Engines*, London, C.I.M.A.C. Part B.12, 915–939.

20 Cotton, K.C. and Scofield, O. (1971), 'Analysis of changes in the performance characteristics of steam turbines', *A.S.M.E. J. Eng. for Power*, April 225–237.
21 Collacott, R.A. (1975), 'Component life concepts related to a theory of whole life expectancy', *Quality Assurance, J. Qual. Ass. Inst.*, December.
22 Cline, E.L. (1972). 'Shortening the road from subjective to automated vehicle inspection and diagnosis', Paper No. 72017, *1st International Congress on Automotive Safety*, San Francisco, Cal. July.
23 Jacobus, J.L. (1972), 'Automated diagnostic inspection systems', Paper No. 72001, *1st International Congress on Automotive Safety*, San Francisco, Cal. July.
24 Cassentino, F.J. (1971), 'Readymaids Mark II automatic test and diagnostic system – engineering design model for multi-fuel engines', Memorandum Report M-71-15-1, March (Frankfurt Arsenal, Philadelphia, Pa.).
25 Bammert, K. and Sandstede, M. (1972), 'Measurements concerning the influence of surface roughness and profile changes on the performance of gas turbines', *Trans. A.S.M.E. J.Eng. Power*. 72-GT-34, May.

11 Static testing

11.1 Introduction

Defects in components may be detected by the application of physical principles without impairing the usefulness of the components. This technique is more usually called 'non-destructive testing' which is applied not only to the detection and location of flaws but also to the anticipation of variations or non-uniformities in properties which can be tolerated during service.

Three groups of defects can be identified by means of static or non-destructive testing, namely:

(1) Inherent defects which arise from deficiencies in the initial production of the base or raw material;

(2) Processing defects introduced during the conversion of the base material into a manufactured component;

(3) Service defects arising from the operating conditions imposed on the component.

The types of defect or structural variation which can be determined by these test methods accordingly include: blow-holes; chemical composition variations; cracks (surface and sub-surface); dross shrinkage; flakes; grain size variations; heat treatment deviations; inclusions; laminations; laps; machining defects; pits; plating defects; porosity; rolling defects; seams; segregation; tears; weld penetration deficiencies.

The range of devices available is so wide that only a few of the available methods will be described more comprehensive information can be obtained from the NDT centres [1].

11.2 Visual Testing

Experience inspectors can visually detect many defects such as surface cracks and their orientation, oxide films, weld defects and the presence of potential sources of weakness such as sharp notches or misalignment. Vision varies considerably between individuals and for one individual alone may vary according to his alertness or fatigue, enthusiasm or boredom.

Aids to visual inspection include mirrors, lenses, microscopes, periscopes, telescopes, enlarging projectors, comparators, borescopes, photoelectric systems, fibre optics, image intensifiers and closed circuit television (C.C.T.V.).

11.2.1 Borescopes

These are instruments designed to enable an observer to inspect the inside of a narrow tube, bore or chamber. They are precision-built optical systems with arrangements of prisms, achromatic and plain lenses to provide light with the maximum efficiency. Instruments derived from medical sources are commonly called 'cystoscopes', others which embody telescopic features for the examination of tubes may be called 'tuboscopes', flexible instruments of the fibre-optics variety are called 'endoscopes'.

The gas path from the end of the low pressure compressor to the output end of the turbine in a JT9D engine fitted in a Boeing 747 aircraft can be examined from any of 21 borescope inspection ports [2]. In addition, 8 ports are located around the circumference of the diffuser case to permit complete inspection of the combustion section including fuel nozzles, burner liner and first stage turbine nozzle vanes.

The equipment used for this inspection was made in Germany by Avia-Mess-Engineers in co-operation with the Deutsche Lufthansa. The borescope set consists of two main units: one called a chamberscope with a camera attachment for inspecting the whole combustion chamber and its related parts (NGV, fuel nozzles, etc.); the second, a universal unit is for performing the compressor and turbine inspection. The main features of these equipments are: external light source with fibre optic bundle for light transmission, multiple connection for two or more borescopes on the light source and very high light intensity (300 W) which can be converted into a single fibre optic bundle allowing a strong light condensation for the taking of photographs.

11.2.2 Fibre-optic scanners

Many viewing problems can be solved by the use of fibre optic light pipes. Fibre optics can be broadly divided into two types:

(1) Coherent devices (image carrying) which are primarily used for military and medical purposes; in the flexible form also used for applications in connections with quality control inspection;

(2) Non-coherent devices (light piping) which are used for purposes of illumination. Industrial light pipes are most commonly available in either glass or polymer (plastic) materials which are generally drawn into flexible fibres, the fibres are then bundled together to form flexible light pipes of any size and diameter.

Optical fibres transmit light by the phenomenon of total internal reflection, light rays which strike the core/sheath interface at angles within the acceptance angle are reflected back into the core then travel along the fibre by a zig-zag path of successive reflections. Constructionally, an optical fibre consists of a core of high refractive index material surrounded by a sheath of material of lower refractive index, the refractive index (N) being a constant for any particular material,

$$N = (\sin i)/(\sin r) = 1 \text{ for air} \tag{11.1}$$

For the core-sheath arrangement of a fibre optic shown in Figures 11.1 and 11.2

$$N_{air} < N_1 < N_2. \tag{11.2}$$

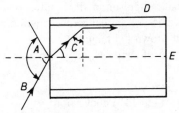

A = Acceptance angle 60°
B = Light ray
C = Critical angle
D = Sheath (N_2)
E = Core (N_1)

Figure 11.1 *Single-fibre*

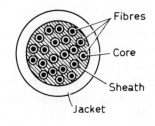

Fibres

Core

Sheath

Jacket

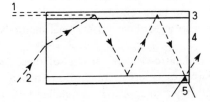

1 = Zone of penetration
2 = Light ray
3 = Sheath (N2)
4 = Core (N1)
5 = Scratch which reaches
　　penetration zone, allowing
　　light to escape and hence
　　loss of efficiency

Figure 11.2 *Fibre bundle*

The sheath is used to provide optical insulation and permits the optical fibres to be embedded in other absorbing materials without loss of light from the core. A flexible pipe may consist of a number of glass fibres of size 0.001 to 0.003 in diameter, bundled in random order in a common jacket (which is made from PVC for general purposes, or from stainless steel for high temperature applications).

For purposes of fault monitoring fibre optic, scanning tubes can be used to examine inaccessible areas or in dangerous environments, other possible applications quoted by Hird-Brown [3] include:

(1) Position indication of switches, relay contacts, valves located in inaccessible cabinets;

(2) Instrument illumination;

(3) Circular to straight-line light conversion;

(4) Annular to digital conversion – if a light bulb or source is flashed in some known manner a digital signal can be conveyed to photo cells;

(5) Registration mark detection in connection with high speed paper webs for cut-off and other machine control functions by the use of micro-optic photo-electric control with a fibre optic scanner;

(6) Computer card and tape readers.

11.2.3 Cold light rigid probes

Working on the image transfer principle and transmitting cold light these instruments use sophisticated coherent fibre bundling techniques to produce slim instruments down to 2 mm diameter, [4]. White light of high intensity is channelled from a 150 W light source unit into a flexible fibre cable through which it is transmitted by total internal reflection into a rigid endoscope. This contains a lens relay system sheathed by glass fibres through which the light passes to the working tip. No light is wasted and no heat is emitted. Thus these instruments are particularly suitable for use in combustible and heat sensitive areas. This technique provides a light intensity 30 to 40 times stronger than conventional methods using miniature lamp illumination.

Forward, sideways and retro-viewing versions of these probes are available for small bores, tubes, gearboxes, internal combustion engines, car chassis members, hydraulic actuators and other components. A detailed look at surface finish in an otherwise inaccessible area can be obtained without dismantling.

The makers report that one of these instruments was used to great effect after an air crash when it was suspected that a contributory cause of the crash was a jammed horizon bar in the pilot's artificial horizon. Dismantling the gyro assembly in the conventional way was thought likely to result in the obstructing object rolling away unobserved and so a viewing probe was inserted through a small hole in the case, to ascertain the cause of the jamming with the minimum disturbance. It was subsequently found that the horizon bar was jammed by a small screw and photographs taken through the endoscope were important exhibits at the Court of Enquiry.

11.2.4 Deep-probe endoscopes

Using the principles of fibre optics these are special modular endoscopes which are available in lengths of up to 21 m (70 feet) for the inspection of pipework in boilers and heat exchangers; radiation-proof versions are widely used for the direct observation of the loading of atomic reactors.

The high precision stainless steel components of the system screw together to provide a rugged viewing system capable of penetrating long bores even when access to the bore entry is severely limited. The option of high intensity quartz halogen illumination available with many of the larger systems affords users a standard of image brightness previously unobtainable with long probes. The standard monocular attachment accepts 6×, 8×, 10×, or 12× wide field oculars; the binocular attachment accepts field oculars of similar magnifications and a special right-angled ocular attachment can also be used. Viewing heads provide 180° direct viewing with 55° field of view for the inspection of blind ends, or 90° lateral view with 55° field of view, or 110° or 135° fore-oblique viewing, or 60° retro-viewing, or 180° wide-angle panoramic 170° field of view.

11.2.5 Pan view fibrescopes

Detailed examination and photography of wall defects in tubes, machinery and engine internals usually demands the use of a lateral viewing fibrescope but rapid and certain steering of the instruments is greatly assisted when one is using a forward viewing instrument for rapid detection and location of defects. This conflict of ideals has been solved by the introduction of the Pan view fibrescope which, by means of a remotely controllable rotating prism built into its tip, can be made to view forwards or sideways as required. This means that the instrument can be inserted using forward viewing and can be stopped to take a detailed sideways look at any passing defect simply by rotating a control knob built into the side of the eyepiece [4].

It is possible to adapt the fibrescope for purposes of photography by the use of the Kowa SQ11 Scope Camera which was designed exclusively for fibrescope photography and incorporates automatic exposure control (with manual over-ride and a full range of shutter speeds up to 1/250 s), adjustable view finder, an automatic lock-on device using locating bayonets pins and device for camera rotation.

11.2.6 Miniature C.C.T.V./endoscope

A technique of inspection of turbine blades and other components within turbojet, turbofan and turboshaft engines for damage produced by the ingestion of foreign objects from fatigue, overheating and similar causes is a routine part of preventive maintenance. Direct viewing through endoscopes (boroscopes) inserted through ports in the engine housing requires prolonged inspection often under difficult and uncomfortable conditions. An improvement has been the introduction of miniature closed circuit television camera equipment coupled to endoscope and light sources appropriate to the type of engine; the results may also be transmitted to remote picture monitors of video-recorded, (Figure 11.3).

Tests with leading airline [5] and engine makers have shown that detection

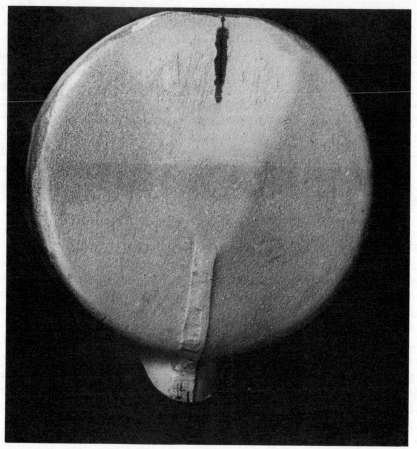

Figure 11.3 *Severe crack in alloy casting shown by the Checkmor Dye Penetrant process*

and assessment of damage can be successfully achieved by the use of TV camera/ endoscope technique in as little as a quarter of the time required for direct viewing, with greatly reduced operator fatigue and improved confidence of results.

11.2.7 Electron fractography

Fracture surface impressions can identify fracture mechanics involved in the failure of a component. The characteristics of the fracture surfaces of metal components can be examined so that the causes and circumstances of failure can be established with a high degree of certainty. Electron fractography as applied by the National Research Council of Canada [6] is the direct application of a sophisticated electronic device to problems arising from the premature failure of components in aircraft, motor vehicles and industrial equipment. Such fractographic findings frequently point toward specific remedial action. An estimate can also be made of the service life spans of particular metal components. These have resulted in recommendations for the scheduling of periodic examinations for early detection of cracks, before complete failure of the components occurs. The estimation of fatigue crack growth rates in aircraft wheels and wing spar components represents a particularly significant contribution of fractography to flight safety.

Each fracture surface has its own 'fingerprint'. In other words, the history of the fracture process is imprinted on the fracture surface, and with the aid of the electron microscope, it is frequently possible to trace the sequence visually. Thus, in general, the path of the fracture passes through the region of the weakest material (in a microscopic sense) just as a chain fails at its weakest link.

Using a two-stage method which does not damage the surface of the fracture, a replica of the failed component is made. A visual inspection of the fracture surface usually pinpoints the area from which the fracture evolved. A strip of cellulose acetate softened with acetone to a sticky consistency is pressed firmly into the fracture area and left there for about an hour. The strip, which has been moulded to the same configurations as the fracture surface is then removed. This provides considerable fidelity of reproduction.

The strip is placed on a glass slide and introduced into a vacuum evaporator, where chromium and carbon are applied to the surface in very thin layers, the chromium at a $45°$ angle (to produce contrast) and carbon at a $90°$ angle.

A small part of this treated strip is then cut away in the area of the fracture surface in which the fracture began and soaked in a water-acetone solution to dissolve the cellulose acetate. An exact replica of the fracture surface in the form of the carbon-chromium film then remains and it is this that is placed in the electron microscope for analysis and 'reading'. Here the dimples, striations and other tell-tale patterns are recognized and the mode of the fracture is determined.

11.3 Liquid penetrant inspection

Discontinuities are delineated by the use of penetrants which serve as an extension of visual inspection. All metals, glazed ceramics, plastics and other non-porous materials can be inspected by the use of liquid penetrants. The defects must behave as capillaries to draw in the penetrant and retain it after excess material has been removed.

11.3.1 Oil and whiting test

This is the simplest and oldest penetrant test. Paraffin is applied to the test surface and, after a penetration time, excess paraffin is removed and the surface dried. A thin coating of whiting (calcium carbonate) is then applied to the surface so that the paraffin seeps out of the crack and stains the white surface.

Another penetrant can be made from a mixture of red alizarin dye, oleic acid and diesel fuel. This is applied, allowed to penetrate and excess material removed.

Defects then appear as surface stains by the use of dessicated magnesium or talc in alcohol as a developer.

A commercial range of dye penetrants is available typical of which is the checkmor 200 Red Dye Penetrant Process [7] this is applied to the thoroughly cleaned and degreased component and left for a maximum of 30 minutes penetration time. The sensitivity of the process is then influenced by selecting the appropriate removal method appropriate to the surface texture and inspection location. The properties of these products are given in Table 11.1.

Table 11.1 *Properties of a dye penetrant and washes approved to DTD929*

Checkmore 200 dye penetrant	Contains no low boiling components; no characteristic odour; self-emulsifiable, non-corrosive to metals; $102°C$ flash point.
Water wash	May disperse dye causing pink background
Quickmore wash	Petroleum-based solvent with coupling agents; sulphur and chlorine free; non-corrosive to metals; $83°C$ flash point'
Quickmore 403	Blend of surfactants; slight odour; sulphur and chlorine free; non-corrosive to metals; $83°C$ flash point.
Quickmore 10(S)	Blend of non-ionic surfactants inhibitors and couplers; non-flammable; sulphur and chlorine free. non-corrosive to metals.

Typical results to be obtained from this penetrant are the cracks shown in Figure 11.3.

11.3.2 Fluorescent penetrant

Greater sensitivity to crack detection can be achieved by the use of a suitable penetrant in conjunction with ultra-violet illumination. The penetrant dyes give off a green fluorescence when activated by U.V. light which makes it possible to identify quite fine cracks as shown in Figure 11.4.

11.4 Thermal methods

Flaws in a component alter the temperature distribution and may therefore be detected by heating the component. Heating is carried out by either induction, infrared or electric radiation and detecting the surface temperatures can be done by wax, stearine, temperature sensitive phosphors, oil film colours etc. This technique is particularly applicable to the examination of bonds between materials.

11.5 X-ray photography

Radiological examination is based on the fact that X-rays and γ-rays can pass

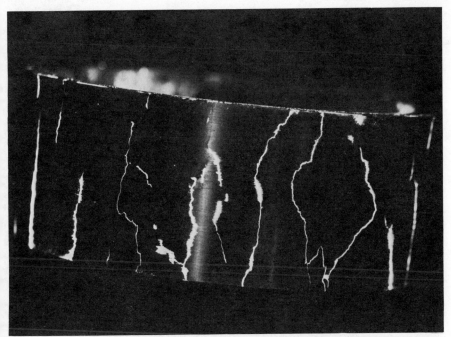

Figure 11.4 *Cracks revealed on a machined magnesium ring by Britemore Fluorescent Penetrant process and inspected under ultra-violet light*

through materials which are optically opaque. The absorption of the initial X-ray depends on (1) thickness, (2) nature of the material, (3) intensity of the initial radiation, all factors which affect the transmitted radiation. The primary methods for detecting the transmitted radiation make use of (a) a fluorescent screen (fluoroscopy), (b) a photographic image (radiography).

With fluoroscopic inspection the transmitted radiation produces a fluorescence of varying intensity on the coated screen. The image is positive so that the brightness of the image is proportional to the intensity of the transmitted radiation; it is low in cost, produces speedy results and has a scanning capability. Disadvantages are (1) lack of a record, (2) generally inferior image quality.

Radiography provides a permanent record when sensitized film is exposed to X-rays. Flaws and voids, being hollow, show up as dark areas, whereas refractory inclusions appear as light areas since they absorb more radiation.

Specifications relevant to radiological practice are listed in Table 11.2.

11.5.1 Radiographic examination of faulty composite materials

An evaluation of the results of radiographic examination of fibreglass, boron-fibre and graphite-fibre epoxy resin composite panels was made [8] using specimens which had been fabricated with known flaws. A 50 kV X-ray machine with 0.5 mm effective focal spot and a beryllium window X-ray tube was used with a

Table 11.2

B.S.2600/1962	General recommendations for the radiolographic examination of fusion-welded joints in thickness of steel up to 2 inches.
B.S.2737/1956	Terminology of internal defects in castings as revealed by radiography.
B.S.2910/1962	General recommendations for the radiographic examination of fusion-welded circumferential butt joints in steel pipes.
B.S.3451/1962	Testing fusion welds in Al and Al-alloys.
B.S.3971/1966	Image quality indicators (I.Q.I) and recommendations for their use.
B.S.4080/1966	Methods for NDT of steel castings.
B.S.4097/1966	γ-ray exposure containers for industrial purposes.
B.S.499/1965 Part 3	Terminology for fusion-welding imperfections as revealed by radiography.
B.S.1500/1958	Fusion welded pressure vessels for use in chemical, petroleum and allied industries.
B.S.2633/1966	Class 1 metal arc welding of steel pipelines for carrying fluids.
B.S.3351/1961	Piping systems for the petroleum industry.
B.S.4206/1967	Methods of testing fusion-welds in Cu and Cu-alloys.

single-emulsion, fine-grain film to avoid the parallex effects of standard double-emulsion film for subsequent advantage in projection or enlargement. Effective and less costly examination of the raw material was conducted by back-lighted tape viewed through a pedestal magnifier. Inspection could be done more quickly by employing an X-ray image intensifier, or, for images up to 35x magnification, an X-ray vidicon [9].

Radiographs illustrate the ability to determine fibre orientation and separation, wrinkles, and kinks, foreign objects or inclusions can also be detected. Water intrusion and entrapment in honeycomb aircraft structures has been detected by radiography. Because the X-ray absorption of resin and graphite-fibres is similar, the fibre orientation is not clearly discernible. Perry [10] used a few tracer threads of high density, such as lead-silicate fibre or metal wires, to outline fibre orientation. The addition of a small amount of Sb_4O_3 to adhesive, in order to make it more radiopaque than the surrounding silica—phenolic composite has been advocated D. Hagemaier [11].

Cinefluorography of small ablative thrust chambers during hot-firing, for evaluating failure modes, also performed by Hagemaier [12] showed that the joints separated during the cooling-down period between hot-firings because of a differential in thermal expansion of the components. As a result of these findings, the thrust chambers were redesigned.

Neutron radiography for engineering evaluation of composite structures used thermal neutrons which are highly attenuated by boron and hydrogen atoms [13, 14]. When an adhesive-bonded honeycomb specimen is neutron radiographed, the hydrocarbon adhesive becomes very apparent because of its high neutron capture. This condition is reversed for X-ray radiography where the metal components are high attenuators. A high density boron-fibre composite becomes completely opaque to a thermal-neutron beam.

11.5.2 Radiological assessment of aero-engine integrity

A linear accelerator with a great output of high energy X-rays has been used by the Non-destructive Testing Centre, Harwell in conjunction with Rolls Royce (1971) Ltd., Bristol Engine Division to produce quality radiographs showing rotating and static components within an engine under running conditions. Accurate measurements can be made, as for example, the clearances between seals, in the interpretation of these radiographs to assess the condition within engines [15].

Current radiographical methods do not involve 'stop-motion' techniques but, by selecting suitable viewing positions, exposure times ranging from 0.3 s (used for acceleration studies) to 2 minutes can be used. The linear accelerator yields 1500 to 2000 rad min^{-1} at an energy level of about 8 MV emitted from as a pulse train at a frequency of 50 to 500 p.p.s. with durations of 2–3 s.

Typical radiographs, Figure 11.5, show that the events occuring within an operational engine can be accurately revealed and discrepancies identified. Further references to this technique and its applications are [16–21].

11.5.3 Proton scattering radiography

Intense beams of energetic protons show sharp shadow cast by the edges of the massive steel bending and focussing magnets when the beams are incorrectly adjusted. Even very thin objects placed in the beam often showed up quite sharply defined. This phenomenon was first applied when considering the possible use of the 160 MV protons from the Harwell Synchrocyclotron for radiography and led to a new type of radiography using protons [22].

Radiograph production depends on the multiple coulomb scattering of energetic charged particles as they pass through the electron clouds round the atoms. Repeated small deflections produce a small net deflection of an individual particle emerging from any material. Particles are not absorbed by this process hence the selective absorption involved with X-rays or neutrons does not apply. Intensity changes at an edge of the material.

Figures 11.6(a) and (b) the intensity distribution from a square cut edge of material perpendicular to the plane of the diagram and illuminated with a parallel beam of protons. When added the spiked intensity distribution has a sudden intensity jump at the geometrical 'shadow' of the edge increasing from the normal intensity. This intensity distribution is highly visible and outlines an edge with

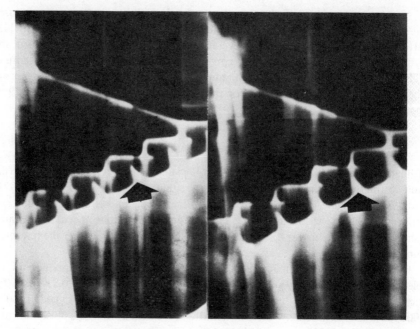

Figure 11.5 *These sections from two radiographs show a labyrinth seal. The static component has the stepped profile and the rotating component has the sloping profile with projecting fins. The relative displacement to the left in view A as compared to view B is clearly visible*

double light and dark bands sharply separated from each other and rapidly faded off away from the 'shadow' of the edge. High proton radiographs show the object as an outline whenever there is a rapid change of thickness. This is in contrast to an X-radiograph (recorded on film) where the different thicknesses of the object produce a different shading of the areas. These two representations of an object are however accepted with equal plausibility by the mind. The difference is that between a pencil drawing and a charcoal sketch.

Xero-radiography in which a charged selenium plate uses the modified charge distribution rendered visible by dusting with a fine powder has many of the general features of proton radiography; whereas the edge pattern is fundamental to proton scattering, it is entirely a property of the recording device with xero-radiography. Details of proton radiography based on the theory of multiple scattering are described by West and Sherwood [23, 24].

One aspect of proton scattering radiography and of xero-radiography is the wide range of detail seen on a single radiograph as compared with that in an X-radiograph recorded on film.

A second feature is the ability to show up cylindrical or spherical edges at the same time as other detail. The 'burning out' of the cylindrical parts with X-rays does not occur with protons and the other details are equally well shown with

Figure 11.6(a) *Wrist watch. Proton scattering radiograph, 160 MeV, face in contact with film (D. West and A.C. Sherwood, AERE R7190 1972, and SPECTRUM 1974 and ATOM April 1975)*

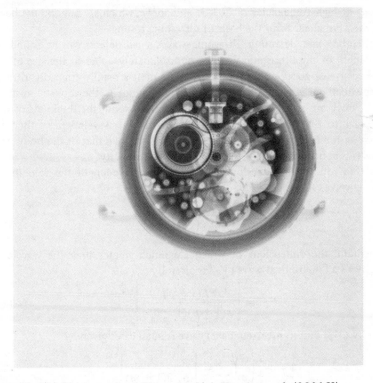

Figure 11.6(b) *Wrist watch of Figure 11.6(a). X-radiograph (280 kV)*

protons as with X-rays. Proton scattering radiography may be of use in the selective examination of certain planes in the object similar to that which can be obtained with X-rays (tomography) by careful coupled motion of the film and X-ray tube during exposure. A further advantage of proton scattering radiography which it shares with xero-radiography is the facility of printing the radiographs without loss of tonal range. The maximum intensity ratio on a proton scattering radiograph is $3:1$ so that printing on to paper which can reproduce a $10:1$ tonal range does not detract from the radiograph.

11.6 Sonics

Use of the 'ring' of metal when struck is an old technique for detecting gross 'flaws'. A steel specimen containing a flaw is dull and harsh compared with the note emitted from a flawless material. Devices such as stethoscopes, microphones and electronic amplifiers have been used to improve the listening sensitivity of such tests. While these tests provide reasonable results with welds and similar joins they are of limited value with more complicated faults.

11.7 Ultrasonics

Excitations at high frequencies of $20\,\mathrm{kHz}$ and above, which are beyond the hearing range of the normal ear form the basis of ultrasonic testing.

By generating and detecting ultrasonic waves small defects can be found since the wavelength of the ultrasonic waves is approximately equal to the size of many defects. Because of good elastic properties most metals readily transmit ultrasonic vibrations which are scattered or reflected by a flaw due to the acoustic mismatch.

Waves, which are propagated over the surface of a solid, the thickness of which perpendicular to the surface is large compared with the wavelength of the waves, are known as Rayleigh waves and have a velocity less than that of the body waves; for metals it is approximately 90% of the shear wave velocity.

For a thin rod of diameter much less than the wave-length the velocity of a longitudinal wave (V_L) is given by

$$V_L = \left(\frac{E}{\rho}\right)^{1/2}. \tag{11.3}$$

For a medium, the dimensions of which are much greater than the wavelength, the velocity of a longitudinal wave (V_L) is given by

$$V_L = \left[\frac{E(1-\rho)}{\rho(1+\rho)(1-2\rho)}\right]^{1/2}. \tag{11.4}$$

The velocity (V_T) of a transverse (shear) wave is in all cases given by

$$V_T = \left(\frac{G}{\rho}\right)^{1/2} = \left[\frac{1}{\rho}\cdot\frac{E}{2(1+m)}\right]^{1/2}. \tag{11.5}$$

The velocity (V_R) of a Rayleigh surface wave is given by

$$V_R = KV_T \qquad (11.6)$$

where: E = modulus of elasticity;

ρ = density;

m = Poisson ratio;

G = modulus of rigidity;

K = bulk modulus.

Typical values for ultrasonic velocities are given in Table 11.3.

Table 11.3

Material	Velocity, $m\ s^{-1} \times 10^2$		
	V_L	V_T	V_R
Aluminium	62.7	31.0	28.0
Brass	47.0	21.4	19.3
Copper	46.3	21.3	19.1
Lead	19.5	6.4	5.8
Steel	57.5	30.9	27.9

11.7.1 Ultrasonic testing technique

Techniques available for electronic testing include pulse echo, transmission, resonance, frequency modulation and acoustic imaging.

11.7.2 Pulse echo technique

A piezoelectric crystal coupled to a surface by means of a liquid medium is commonly used to generate ultrasonic waves. Three main types of ultrasonic wave may be generated, namely (a) longitudinal (b) shear (or transverse) (c) surface wave. Longitudinal waves arise when the crystal is applied parallel to the surface, if the angle of incidence of the crystal is 30° about 98% of the energy becomes shear.

The basic testing technique is that of pulse-echo. The transmitter emits an ultrasonic pulse; stops; receives the echo displayed on a cathode ray tube. The echo is a combination of return pulse either from the opposite side of the specimen or from an intervening defect. The time travel of the reflected wave is determined and appears as a pulse to the side of the original pulse and its position determines the location of a defect. An approximate concept of the size, shape and orientation of the defect can be determined by checking the specimen from another location or surface.

Figure 11.7 illustrates a typical instrument in which the transmitted pulse is directed into the specimen and the echo amplified and displayed on the cathode

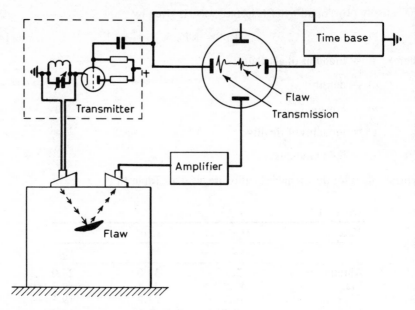

Figure 11.7 *Typical pulse-echo ultrasonic tester*

ray tube. The time base is adjusted to keep the transmission pulse in the same position on the display.

11.7.3 Transmission technique

This technique uses continuous waves from one transducer which are passed right through the test piece. Flaws tend to reduce the amount of energy reaching the receiver and by this means their presence can be detected. This was the original ultrasonic technique but has been replaced owing to problems of modulation associated with standing waves which can be created in the test piece causing false readings to be obtained.

11.7.4 Resonance technique

This consists of moving the transmitter over the surface and observing the transmitted signal. Resonance in the absence of flaws will keep the transmitted signal high. Flaws will cause the transmitter signal to reduce or disappear. Such a method is highly suitable for the testing of thin surfaces as a development of the transmission technique and is widely used as a means of testing the bond between thin surfaces.

11.7.5 Frequency modulation technique

Only one transducer is used to send ultrasonic waves continuously at changing radio requencies. An echo returns at the frequency of the early original transmission and thus interrupts the new changed frequency. By measuring the phase between frequencies the location of the defect can be measured.

1.7.6 Acoustic imaging

Prior to the introduction of holography, techniques were evolved to produce an optical image from the ultrasonic energy. These generally relied on thermal variations in the presence of flaws producing optical effects on chemicals or temperature sensitive phosphors.

1.7.7 Boiler tube thickness probe (ultrasonic)

Instead of evaluating metal wastage in marine boiler tubes by removing selected tubes and assessing condition from the frequency and size of the pits, an ultrasonic probe technique has been developed by Peters, O'Malie and Greenwood of the Royal Canadian Navy Defence Research Establishment [25]. This method was developed specially to deal with the superheater tubes of the Babcock and Wilcox X100 boiler which are difficult to reach from the combustion area of the boiler.

Two ultrasonic units were built, one utilizing a contact probe and one an immersion probe. The ultrasonic contact probe used a 5 MHz transducer (Automation Industries SFZ 57A2216 5.0/.375) incorporated into the probe, this transducer was chosen to suit a Sperry UCD battery operated reflectoscope. A lucite or nylon shoe which tends to wear into the profile of the tube, and can easily be replaced when it wears out, was fitted onto the transducer.

Alcohol coupling was maintained through a small white plastic tube and the alcohol inserted by means of a large hypodermic syringe. The alcohol supply could be attached to either of two receptacles on the probe so that the couplant runs down onto the probe-tube interface when used in either the port or starboard boilers.

The special feature of the probe is that it can measure the tube wall thicknesses at the bends and around the bends, as well as along the straight lengths. A flexible steel spring, to which the probe is attached, keeps the transducer at right angles to the tube wall at all locations. The probe is attached to a long flexible steel wand which enables the operator to manipulate the probe from the end; that is, the probe can be rotated, pushed or pulled while in the tube.

1.7.8 Ultrasonic contour scanning

The contour of jet engine main shaft forgings can be scanned by ultrasonic probes using an immersion technique to rapidly determine whether defects exist – used as flaw detection method to avoid costly machinery of defective forgings. A helical scan path ensures quick, comprehensive surface scanning.

The testing probe is fed forward under the action of a selsyn motor until rollers on both sides of the probe carrier meet the component. When the rollers touch the component, internal buffer springs cause a microswitch to operate and stop the feed. Shafts are mounted vertically, with the largest diameter at the bottom. The probe is started at the largest diameter and moved vertically at a steplessly adjustable feed of between 0 and $\frac{1}{4}$ in per revolution of the turntable. Forward motion of the probe assembly automatically follow the contour of the engineer shaft. A joy stick enables the probe arm movement to be manually over-riden at any time. Separate switches select vertical movement and movement of the arm by the joy stick. The turntable can also be inched by a pushbutton. Electromagnetic brakes on turntable and probe arm allow the operator to quickly re-examine a certain position and prevent undue overrun when changing from manual to auto or auto to manual.

The position of the flaw is determined by three parameters: the height, from a scale on the main column; the angular position taken from vernier on the turntable and the flaw depth below the surface by direct reading from the cathode ray tube of the ultrasonic tester. A television camera with videotape recording of the trace and a commentary give positions of any indicated defects.

11.7.9 Ultrasonic triangulation fault location

In contrast to existing ultrasonic techniques which require the transducer be located fairly near the suspected crack, triangulation can detect a flaw as long as it is within the area being scanned by the sound signals. A system devised by the George C. Marshall Centre, Huntsville, Alabama, U.S.A. was used to detect flaws on the Saturn rocket case.

Four barium titanate crystals, one of which is used as a transmitter, were arranged in a retangular pattern near the weld area being examined Figure 11.8. Each crystal was able to move around a quadrant which couples a wave-guide to the plate being tested. This, gives a null and well-defined ultrasonic beam which radiate through the metal. The beam is resonant in the thickness of the plate, and thus is capable of travelling through it, as if the plate were a waveguide. Reflected energy from any flaws was picked up by one or more wave-guide detectors. The return signal was then processed by a computer which directed the sensor beams so that they intersected at the flaw. This point was either displayed on a cathode-ray tube or on a digital readout. The amplitude of the return signal determines whether the flaw should be passed or failed: this was done with automatic signalling equipment.

11.8 Stress wave emission

The phenomenon of 'tin cry' has been known for centuries, but the mechanism responsible for this audible manifestation of crystallographic changes in tin and tin-rich alloys was only adequately explained recently. In 1950, Kaiser, [26] observed stress wave emissions from a range of metals, and from dry spruce wood and reported that if a metal was strained to a given level, burst emissions of noise were observed, and if the metal was then unloaded and re-loaded, no fresh noise bursts occurred until the previous maximum stress level had been exceeded.

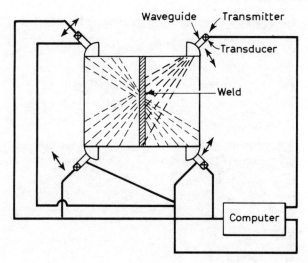

Figure 11.8 *Ultrasonic method used by NASA engineers to test rocket cases works by pulsing high-frequency sound and measuring the return signal*

A dramatic research breakthrough occurred, by fortunate accident, when a 6.5 m diameter motor case tested by the Thiokol Corporation, failed under hydrotest at 56% of proof pressure [27]. Almost by chance, SWE transducers had been placed upon this vessel to record information during what was expected to be a normal and satisfactory pressure test. By the time engineers had put the fragments together two things had happened.

(1) The SWE had been analysed and used to indicate the origin of failure.

(2) The critical defect size had been calculated by linear elastic fracture mechanics.

Both predictions were remarkably accurate and the mishap provided a stimulus to both SWE research and to the development and application of fracture mechanics throughout US and later UK industry.

A typical response characteristic of a sensitive piezoelectric element to a stress wave is shown in Figure 11.9. This is a rapidly rising burst followed by a relatively slow decay. The amplitude of such stress waves is related to the energy of the event that produces them and thus, if they can be detected and measured, an estimate can be made of the extent of material breakdown.

11.8.1 Simple stress wave emission monitoring system

A simple SWE monitoring equipment is illustrated in Figure 11.10. It is not always easy to measure pulse amplitudes without sophisticated equipment but a compromise is to feed such a signal through an electronic counter set to count any signals which just exceed the background level. Provided one is using a single-frequency resonant transducer, a low-noise amplifier and a low cut-off filter (to reduce

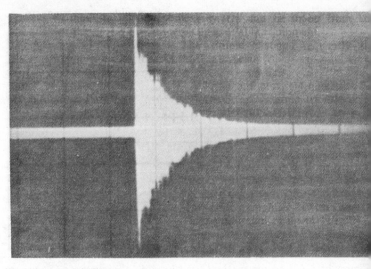

Sensor voltage 50µ V/Div

TIME (I msec/ Div)

Figure 11.9 *Typical stress wave emission sensor response*

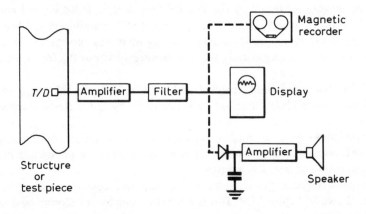

Figure 11.10 *Simple stress wave analysis equipment*

extraneous noise) to produce the signals, the 'count' will be an approximate measure of the burst energy. (The accuracy tends to deteriorate when pulses of widely different amplitudes are being counted or when the counter trigger level is much above the background noise level.)

For laboratory studies of mechanisms of failure etc., such a count can be expressed as total count (or total energy release) or as an energy release rate and a convenient system can be made from commercially available equipment.

The simple system used by the Admiralty Materials Laboratory [28], consists of

(1) A 5 mm × 5 mm × 4 mm thick LZT piezoelectric ceramic transducer, resonant in thickness mode at ∼ 300 kHz and fixed to the specimen with a viscous silicone grease;

(2) The electrical output from the transducer can be fed along $< \frac{1}{2}$ metre coaxial cable to a low-noise amplifier (with switched filters and gain of up to 110 dB);

(3) The amplifier output can then be fed to a Hewlett Packard type 5326C frequency counter (or similar) which (with slight modification) feeds either total count or count rate into a compatible digital/analogue converter for presentation on any potentiometric chart recorder or X–Y plotter.

11.9 Magnetic testing methods

Inhomogeneities such as blowholes, cracks and inclusions in a magnetic material produce a distortion in an induced magnetic field. These inhomogeneities have different magnetic properties from that of the surrounding material which produces a distortion of the magnetic field. Magnetic testing methods use this leakage flux to magnetise random particles so as to reveal the presence of the defects.

There are five basic methods of magnetisation for purposes of magnetic flaw testing [29]:

(1) Current flow method;
(2) Induction, magnetic flow method;
(3) Induction, threading bar method;
(4) Induction, magnetizing coil method;
(5) Induced current flow method.

11.9.1 Current flow magnetisation

Current flow machines normally provide a sustained current through the specimen, ink being applied while current flows. The specimen is usually clamped between contact pads on a static machine, but portable units are available in which the contacts take the form of hand-held prods, and these are often used for checking components which are difficult to mount in a static machine. Good electrical contact is essential, and the contacts are usually provided with copper gauze pads, sufficient pressure being used to prevent arcing between the pads and the specimens.

For testing purposes it is usual to apply a sufficiently heavy current to give a satisfactory magnetic flux in the specimen, and to use a low voltage to safeguard the operator. As a rough guide, most steels can be satisfactorily tested using an alternating current of 500 A rms in^{-1} diameter or, for specimens of irregular shape, 150 A rms in^{-1} of periphery.

11.9.2 Induction, magnetic flow method

Figure 11.11 shows the arrangement of a typical magnetic flow machine, the

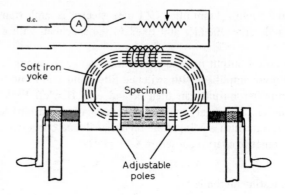

Figure 11.11 *Magnetic flow machine*

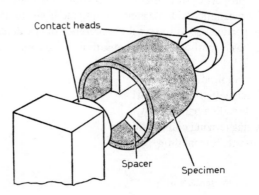

Figure 11.12 *Threading bar method*

specimen being clamped between adjustable poles in the magnetic circuit of a powerful electromagnet. Good contact between the poles and specimen is essential, otherwise a marked lowering of the field strength will result. Laminated pole pieces are often used to ensure that good contact is maintained with specimens of curved or irregular shape, and in some portable equipments which employ a permanent magnet, contact is obtained through a number of spring-loaded pins.

11.9.3 Induction, threading bar method

This method is used for testing rings and tubes, and is illustrated in Figure 11.12. A current flow machine is used, and a conductor connected between the contact heads of the machine. Current flowing through the conductor induces a magnetic flux in the specimen at 90° to the direction of current flow; this flux may be used to reveal defects in line with the axis on the specimen.

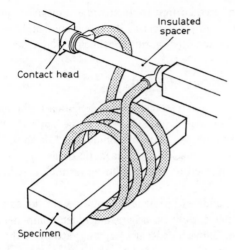

Figure 11.13 *Magnetizing coil method*

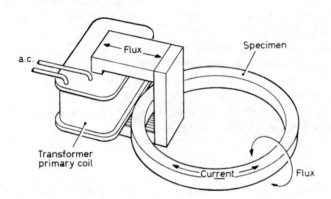

Figure 11.14 *Induced current flow method*

11.9.4 Induction, Magnetizing coil method

A current flow machine is also used for the magnetizing coil method. An insulated heavy gauge copper wire or strip is connected between the contact heads of the machine as shown in Figure 11.13 and formed into a coil; a.c. coils have $2\frac{1}{2}$ to 4 turns and d.c. coils 6 to 10 turns, the space between turns being less than the cross-sectional diameter of the wire in order to minimize the flux leakage. The magnetic lines of force resulting from passing current through the coil, will induce a magnetic flux in the specimen, in the direction of the coil axis.

11.9.5 Induced current flow method

Figure 11.14 shows the coil arrangements for this method, in which current is induced to flow through the specimen by the action of the primary coil of a transformer. The induced current itself provides a magnetic field within the specimen, which may be used for detecting defects lying mainly in a longitudinal direction. This method is often used on ring specimens of large diameter.

11.9.6 Magnetic particle inspection inks

The magnetic particles used to reveal defects are either in the form of a dry powder, or suspended in a suitable liquid. They may be applied by spray, pouring, or immersion, depending on the type of component. Magnetic flaw detection 'inks' consist of finely divided black or red magnetic oxides of low coercivity (i.e. they will not retain the magnetism induced during testing), suspended in a liquid (normally paraffin).

Pigments may be added to provide a contrast with the surface of the specimen. Black inks are suitable for use on bright, machined components, but red inks may be more suitable for unmachined parts or, alternatively, a thin coat of white paint or strippable lacquer may be added to the component before carrying out the test. The solids concentration in inks manufactured to BS 4069 should be 0.8 to 3.2% by volume. Methods of determining the solids content of magnetic inks are detailed in BS 4069.

Fluorescent inks are also widely used and are often specified where high sensitivity is required. Inspection of a component to which fluorescent ink has been applied, should be carried out under u.v. light. The solids content of fluorescent inks is approximately one-tenth that of pigmented inks.

Figures 11.15(a) and (b) show the way in which magnetization can reveal cracks in a material which are not visible to the eye.

11.9.7 Strippable magnetic film

Cracks and other defects on the surface of magnetic materials can be located by use of a self-curing silicone rubber containing a dispersion of iron oxide particles from which a counterpart of the original surface can be prepared [30].

The rubber solution is poured into or onto the area under inspection and a magnetic field induced by a permanent or electromagnet around the area. Under its influence magnetic particles in the solution migrate to cracks. A catalyst is added to the rubber solution before pouring in order to control the rate of curing. After curing, a rubber plug can be withdrawn from holes or a coating removed easily as silicones exhibit an inherent ability to release from a surface. Permanent or electromagnets magnetize the poured solution as it is curing. Cracks can then be seen on the cured rubber as intense black lines. Investigation of small cracks may need a low powered magnifying system.

This technique has been effectively used to examine areas with limited visual access such as straight, tapered and threaded holes from $\frac{1}{8}$ in (3 mm) diameter; metal under coatings such as paint, electroplate and flame sprayed metal surfaces; complex shapes and poor surface conditions; defects which need magnification for

Figure 11.15(a) *A group of ferro-magnetic components before magnetic testing*

Figure 11.15(b) *The same group as figure 11.15(a) after magnetization and the application of Supramor Grade 4 Black showing grinding and hardening cracks*

detection and interpretation. Defects only 0.002 in (0.05 mm) long on surfaces of 40 rms have been identified.

11.10 Electrical NDT techniques

Methods using electrical effects to test for the presence of flaws may be classified as:
    (1) Electrical resistance measurement;
    (2) Triboelectric effect sensing;
    (3) Thermoelectric effect sensing;
    (4) Static field marking.
    Electrical NDT methods are useful for locating faults in the adhesion of the lining of metal bearings, splits in bimetallic strips, defective insulation and similar defects.

11.10.1 Electrical resistance measurement

Electrical resistivity in the neighbourhood of a defect differs from that in a sound material so that the presence of a defect alters the potential difference between two electrodes as shown in Figure 11.16(a) and (b).

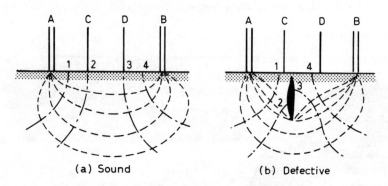

Figure 11.16 *Effect of a flaw on lines of electrical potential*

    In a typical apparatus the specimen forms part of one arm of either a Wheatstone bridge or similar circuit such as Kelvin bridge and the resistance of the specimen between two constant-distance probes acts as the basis for test. As the probes are moved along the specimen the resistance remains constant until a fault is located between the probes.

11.10.2 Triboelectric effect sensing

A voltage is produced if two surfaces of dissimilar metals rub against each other. Typical instruments using this principle as a means of flaw detection include a tool to reciprocate the surfaces together and a microvoltmeter which is read after the reciprocating motion has ceased. This test is very sensitive to segregation of alloys, particularly variations in the amounts of copper.

11.10.3 Thermoelectric effect sensing

The thermal e.m.f. effect when the junctions of dissimilar metals are at different temperatures has been exploited to detect variations in metallurgical structure, heat treatment and chemical composition. Such an instrument uses two probes one of which is heated to a fixed temperature while the other is at room temperature. When the probes are in contact with a specimen the e.m.f. depends upon the thickness and thermal conductivity of the specimen and can be calibrated in terms of coating thickness to sense variations over the surface of the specimen.

11.10.4 Static field marking

Similar in principle to magnetic particle marking, this method is suitable for testing non-conducting materials such as porcelain insulators. Fine calcium carbonate powder is sprayed onto the surface by an air jet and becomes positively charged. They are attracted to cracks which become delineated by the powder.

11.11 Eddy current testing

Coatings over ferrous substrates can be measured by using a magnetic circuit set up by a constant energizing signal and observing magnetic flux changes brought about by changes in reluctance (magnetic resistance) in the circuit due to the coating.

The probe comprises an exciter cell involving encapsulated coil windings which are energized by a low frequency generator. The twin contacts of the probe establish a strong localized electric field which influences the magnetic flux through the pole and therefore the e.m.f. reading of the amplifying windings of the detector cell. Pre-calibration against the bare substrate material makes it possible for the detector reading to be expressed in terms of coating thickness.

11.12 NDT selection

The methods of static (non-destructive) testing appropriate to a range of circumstances have been evaluated by a number of authorities.

11.12.1 Structural validation

This is a term given by Birchon [31] to the assessment of the health of a structure in the light of the duties it has to perform. It is an overall approach intended to observe only those aspects of flaw behaviour relevant to a structure's service duties and to assess their significance, either to establish the minimum extent of repair, or

Table 11.4

Technique	Principle	Advantages	Limitations	Coarse filter potential
Radiography X-ray γ-ray Neutron	Absorption and scattering of penetrating radiation.	Versatile, well understood.	Sensitivity often low for crack-like defects. Two-sided access essential.	C
Visual inspection Borescopes	Lens or fibre optic examination. assessment prior to using	Allows an important visual assessment prior to using further NDT if necessary.	Surface inspection only. Lens systems relatively inflexible. Fibre optics provide flexibility but limit resolution.	B
Holography	Wave front reconstruction. 3-D imaging of Fresnel diffraction patterns.	Wide range of interference techniques, makes possible detection of very small strain and dimensional changes, vibration modes etc.	Surface inspection only.	A
Dye penetrant	Entrapment of visible or fluorescent chemicals.	Low cost, can be very sensitive.	Surface-breaking defects only.	C
Thermal Thermographic compounds	Critical temperature sensitive selective reflection of light from mesomorphic compounds.	A picture of thermal patterns enables rapid detection of discontinuities. Not affected by emissivity changes.	Expensive, Rather limited temperature range covered by these compounds at present.	B
Infra-red	Detection and recording of IR transmission, absorption, spectrometry and generation.	As above. Real time CRT displays are available.	Real time equipment is bulky and expensive. All methods upset by emissivity variations and heat generation and dissipation conditions.	
Surface impedance	Relation of heat flow to temperature in low frequency sinusoidal thermal waves.	One side access only: will detect sub-surface defects.	Relatively slow in operation.	
Electrical Corona discharge	Generation of electrical discharge in gas pores under high	Can detect sub-surface cavities as well as surface flaws.	Only of value in electrical insulators. Require bulky equipment.	C

Table 11.4 (Continued)

Technique	Principle	Advantages	Limitations	Coarse filter potential
Resistivity	d.c. and a.c. resistance determination.	Simple, sensitive and fairly easily interpreted.	Highly conductive materials and debris in cracks give problems. Surface defects only.	C
Capacity	Dielectric variation detection.	Follows dimensional (t) variations.	Poor resolution. Primarily for non-metallic materials only. Conductive backing necessary.	C
Eddy current	Electromagnetic induction effects.	Applicable to a wide range of conducting materials. Can work without surface preparation. One-sided technique.	Few or no polymer applications. Limited to surface defects.	C
Microwave	Transmission refraction polarization and scatter of waves in IR/RADIO region.	A non-contacting inspection technique for a wide range of non-conducting solids, liquids and gases.	For non-metallics only. Poor resolution at present	C
Acoustics and ultrasonics				
Vibration analysis	Spectrum analysis comparisons during life of components and machinery. (Operational or induced vibration).	Self-emitted noise fairly easily detected. Gives prior warning of failure.	Analysis problems may be complex.	A
Holography	Imaging of ultrasound by holographic reconstruction in light.	Pictorial indication. The holograph can be depth scanned.	The technique is still in its infancy. No universal agreement on test method, or reconstruction at present.	A
Ultrasonic	Transmission, velocity, reflection attenuation, phase change, resonance and spectrum analysis of all forms of high frequency sound waves.	Some single-sided techniques. Applicable to majority of materials.	Some difficulties are experienced on differentiation between types of defects.	C
Stress wave emission	The detection and processing of stress waves emitted from lightly strained materials.	Single-sided access can show both severity and location of defects in a very wide range of materials and structures.	Still in its infancy, requires specialized techniques.	A

Table 11.4 (Continued)

Technique	Principle	Advantages	Limitations	Coarse filter potential
Magnetic				
Magnetic particle	The physical migration of solid magnetic particles under influence of a magnetic field.	Low cost, reliable and sensitive. Very widely used.	Primarily for surface breaking defects, rather time-consuming and contaminates clean surfaces.	C
Flux sensors	The detection of residual or induced flux perturbations by flux-sensitive devices.	Does not involve direct contact surfaces.	Mainly for surface breaking defects. Object or head has to be rotated. As above for ferromagnetic materials only.	C
*Magneprint	The recording of magnetic perturbations by magnetic sensing paint forming a strippable film.	A pictorial system, clean and capable of high resolution with none of the disadvantages of MCD (Magnetic Crack Detection).	Limited to surface breaking defects in ferromagnetic materials.	C
Miscellaneous				
Contamination/analysis (e.g. SOAP)	Detection of traces of wear products in oils, vapours, coolants, etc from machinery.	Can be developed to produce prior warning of failure.	The technique involves impeccable administration and regular sampling as well as expensive and sophisticated equipment.	A
Photoelectron/emission	Detection of spontaneous electron emission from plastically deformed surfaces.	Single-sided technique.	As yet for surface effects only – not well understood.	C
β-ray backscatter	The detection of reflected radiation. (β-emitting isotopes used).	Single-sided technique suitable for layer thickness determinations and aggregate composition assessment.	Limited depth of penetration relies on layers or aggregates of notably different atomic number.	C
Displacement gauges	Strained displacement measurement by rugged sensitive LVDT (Linear Variable Differential Transformer) devices.	Senses change in structural behaviour from serious defects in any material. Can be used for degradation during service.	Requires prior knowledge of defect areas.	C

* A patented process developed by AML
Key – A possible, B limited, C unsuitable

to establish confidence in operating the structure with a known remanent strength. It differs from established methods of NDT because it is concerned with the positive function of establishing 'fitness for purpose', rather than the negative activity of 'searching for defects'.

Some form of 'inspection for defects' is a necessary part of the quality control essential in manufacturing and maintenance to ensure 'good workmanship', but there is scope for improvement in methods employed to establish defect rejection limits, and in restricting the amount of nugatory work so often conducted.

The principal characteristics of NDT techniques which have been evaluated by the Admiralty Materials Laboratory (AML) in relation to structural valiadation are presented in Table 11.4.

Table 11.5 *Non-destructive testing techniques applied to corrosion evaluation*

Technique Corrosion type	Radiography	Ultrasonic	Eddy current	Magnetic particle	Dye penetrant
Hidden wastage of unknown mechanism	Best general technique	Limited	Limited	N.A.	N.A.
General wastage	Poor	Best general technique	Limited	N.A.	N.A.
Pitting corrosion	Best general technique	Can detect	Good method	N.A.	N.A.
Intergranular corrosion	Can detect	Best general technique	Good method	Can detect	Can detect
Dezincification	Can detect	Best general technique	N.A.	N.A.	N.A.
Corrosion fatigue	Best general technique	Good method	Good method	Good method	Can detect
Stress corrosion cracking	Good method	Best general technique	Good method	Good method	Good method
Hydrogen embrittlement cracking	Can detect	Best general technique	Good method	Good method	Good method

11.12.2 Corrosion evaluation by non-destructive testing

Some wastage mechanisms such as corrosion— erosion and high temperature oxi-ation result in a general loss of metal. Visual examinations of components suffer-ing from external attack by such mechanisms may readily indicate the nature of the attack, but the extent of the damage may not be apparent. In other instances, where the attack is internal, the wastage may be detected only by utilizing non-destructive tests. The extent to which such techniques can be used were reviewed by Peters [32] and summarized in Table 11.5.

11.12.3 Recommended NDT techniques for composite structures

A survey by Hagemaier, McFaul and Parks [8] following experimental investi-gations of different NDT techniques led them to make the following recommen-dations:

(1) Radiography: This can be used to reveal delaminations, resin-rich and resin-starved areas, porosity, bond-joint separations, cracks, fabrication fit-up, inclusions, fibre orientation and crushed honeycomb core;

(2) Ultrasonics: These can be used to reveal disbonds, delaminations, porosity and resin content; also for thickness gauging and, under controlled conditions, a correlation between physical and mechanical properties;

(3) Liquid Penetrants: Can be used for finding discontinuities open to the sur-face such as cracks, porosity, disbonds and delaminations;

(4) Thermal or Infrared Testing: This can be used mainly for detecting disbonds in laminate or honeycomb structures.

The applications and limitations of nondestructive test methods for fibre glass, boron and graphite fibre composites, based on test results, are abstracted in Table 11.6.

The whole range of methods for measuring the thickness of surface coatings may be summarised as follows:

(a) Chemical dissolution;

(b) 'Strip and weigh' (weight difference technique);

(c) Mechanical displacement;

(d) Microscopy (including interferometry);

(e) Coulometric;

(f) Thermoelectric;

(g) X-ray fluorescence;

(h) β-backscatter and transmission;

(i) γ-absorption

(j) H.F. and L.F. eddy current;

(k) Ultrasonic;

(l) Electro-magnetic induction.

Methods (f) to (l) inclusive are the non-destructive techniques used. The relev-ance of these techniques are given in Table 11.7.

Table 11.6 *Applications and limitations of nondestructive test methods to composite materials*

Defect	Fibreglass	Boron fibre	Graphite fibre
Unbonds	Sonic Ultrasonic Thermal (thin sections only)	Sonic Ultrasonic Thermal (thin sections only) Eddy-Sonic	Sonic Ultrasonic Eddy-Sonic
Delaminations	Sonic Ultrasonic	Sonic Ultrasonic Thermal (thin sections only)	Sonic Ultrasonic
Fibre-orientation (Tape)	Back-lighting (magnified) X-ray (low kV)	Back-lighting (magnified) X-ray (low kV)	X-ray (low kV)
Fibre-orientation (Laminate)	X-ray Microscope (edge)	X-ray Microscope (edge)	Microscope (edge)
Inclusions	X-ray Ultrasonic	X-ray Ultrasonic	X-ray Ultrasonic
Crushed honeycomb	X-ray Ultrasonic (through transmission)	X-ray Ultrasonic (through transmission) Eddy-Sonic	X-ray Ultrasonic (through transmission) Eddy-Sonic
Resin-rich and resin-starved areas	X-ray Ultrasonic	X-ray Ultrasonic	X-ray Ultrasonic
Micromechanical studies	Acoustic emission	Acoustic emission	Acoustic emission
Thickness gauging	Micrometer Ultrasonic	Micrometer Ultrasonic	Micrometer Ultrasonic
Porosity and cracks (internal)	Ultrasonic X-ray	Ultrasonic X-ray (thin sections only)	Ultrasonic X-ray (thin sections only)
Porosity and cracks (external	Penetrant	Penetrant	Penetrant

Table 11.7 Due to the large number of coating/substrate combinations and the infinite variety of items requiring measurement a range of instrument techniques may apply as follows [33]

Coating	Substrate								
	Zinc	Stainless steel <5% ferric	Steel	Silver	Plastics Glass Ceramics	Nickel	Magnesium and alloys	Copper and alloys	Aluminium and alloys
Anodic	–	–	–	–	–	–	CB	–	CB
Brass	X	(T)	S	B	BT	X	BC	T	B
Cadmium	B	T	SB	T	B	B	B	BT	B
Chromium	B	C	SB	B	B	B	B	BC	B
Gold	B	B	SB	B	BT	B	B	B	B
Lead	B	B	SB	BT	BT	B	B	BT	B
Nickel	N	N	N	N	N	N	N	N	N
Electronless Nickel	T	T	SB	TB	B	S	B	B	B
Palladium	B	B	SB	(T)	B	B	B	B(T)	B
Paint, plastics, rubber, etc.	CB	CB	SB	CB	B	ST	CB	CB	CB
Rhodium	B	B	SB	T	B	B	B	BT	B
Silver	B	B	SB	BT	BT	B	B	B	B
Tin	B	B	SB	(T)	BT	B	B	B	B
Zinc	T	B	S	B	B	B	B	B	B

B = Beta backscatter, N = Modified electromagnetic, C = H.F. eddy current, S = Electromagnetic, T = Specialized eddy current, ( ) = Measurement possible with difficulty or some modification, X = Not possible.

References

1 N.D.T. Centre, Harwell *Quality Technology Handbook* IPS Press, Guildford.
2 ——— (1972), 'Engine performance trend analysis' *Aircraft Engineering*, July, 23–25.
3 ——— *Fibre-optic scanners,* Hird-Brown Ltd, Lever Street, Bolton BL3 6BJ.
4 ——— *PanView Fibrescope*, Specfield Limited, 12 Park Street, Windsor, Berks.
5 Maddox, V. (1973), 'How to measure piston rod motion in reciprocating compressors' *Hydrocarbon Processing,* January.
6 Wiebe, W. (1974), 'Diagnosing fractured components', *Science Dimension,* **6**, No. 5.
7 Vagg, V.A. 'Penetrant inspection processes', Burmah-Castrol Industrial Limited, Swindon, SN3 1RE.
8 Hagemaier, D.J., Mcfaul, H.J. and Parks, J.T. (1970), 'Nondestructive testing techniques for fibreglass, graphite fiber and boron fiber composite aircraft structures', **Materials Evaluation,** September.
9 Baldanza, N. (1965), 'A review of nondestructive testing for plastics; methods and applications', Plastic Report No. 22, Picatinny Arsenal, Dover N.J.
10 Perry, H. (1962), 'Tracer radiography of glass fiber reinforced plastics', *Proc. 17th Annual Meeting,* Reinforced Plastic Division, Society of the Plastics Industry, Chicago.
11 Hagemaier, D.J. (1966), 'Nondestructive testing of small ablative thrust chambers', AFML-TR-66274, *Testing Techniques for Filament Reinforced Plastics*, Dayton, Ohio, September.
12 Hagemaier, D.J. 'Cinefluorography of small ablative thrust chambers during hot firing', *Materials Evaluation,* April.
13 Tomlinson, R. and Underhill, P. (1969), 'Production neutron radiography facility for routine NDT inspection of special aerospace components', A.S.N.T. Spring Conference, Los Angeles, California, March 10–13.
14 Haskins, J.J. and Wilkinson, C.D. (1969), 'Neutron radiography; some applications for NDT' A.S.N.T. Spring Conference, Los Angeles, California March 10–13.
15 Pullen, D.A.W. (1974), 'High energy radiography', Atom (UKAEA), **212**, June.
16 Pullen, D.A.W. (1971), 'Radiography applied to determining dynamic conditions inside aero gas turbines' *British Journal of Non-Destructive Testing*, **13**, 42–45.
17 Stewart, P.A.E. (1972), 'Radiography of gas turbines in motion', *Chartered Mechanical Engineer,* **19**, May 65–67.
18 Ratcliffe, B.J. (1970), 'Metal intensifying screen for radiography', *British Journal of Non-Destructive Testing,* **13**, March 55–59.
19 Callister, W.C. (1970), 'Profile radiography by gamma rays', *Non-Destructive Testing,* **5**, August 214–219.

20 Mitchell, C. and Pulk, R.A. (1956), 'Applications and problems in stroboradiography', *Non-destructive Testing* (USA) **14**, Sept/Oct.
21 Vincent, B.J. (1960), 'Stroboscope radiography with a linear accelerator', British Journal of Applied Physics, **II**, March.
22 West, D. (1974), 'A general description of proton-scattering radiography', A.E.R.E. Harwell, Didcot, Berks, Report R 7757. June.
23 West, D. and Sherwood, A.C. (1972), *Nature,* **239**, 157–159.
24 West, D. and Sherwood, A.C. (1973), *Non-Destructive Testing,* **6**, 247–257.
25 Peters, B.F. O'Malia, J.A. and Greenwood, B.W. (1969), 'Ultrasonic evaluation of tube wastage in marine boilers', *Trans. I.E.R.E.* London, Joint Conference on Industrial Ultrasonics 177–184.
26 Kaiser, J. (1950), Ph.D. Thesis, Technical Hochschule, Munich.
27 Stanley, J.E. and Esgar, J.B. (1966), 'Investigation of hydro-test failure of Thiokol Chemical Corporation 260-inch diameter SL-1 motor case', NASA Technical Memo X-1194, January.
28 Birchon, D. and Warren, R.H. (1971), 'Stress Wave Analysis', *J. Royal Naval Scientific Service,* **27**, No. 2.
29 Birchon, D. and Warren, R.H. (1974), 'Basic non-destructive examinations – magnetic flaw detection', Report BL/8, Issue 2, Civil Aviation Authority, P.O. Box 41, Cheltenham, June.
30 ——— (1974), 'Eddyprobe Mark II', Inspection Instruments (NDT) Ltd, 32, Duncan Terrace, London N1 8BR.
31 Birchon, D., Warren, R.H. and Wingfield, P. (1974), 'Structural validation', *Chartered Mechanical Engineer,* June.
32 Peters, B.F. (1972), 'Non-destructive testing for corrosion evaluation', *Journal of the Engineering Institute of Canada,* March.
33 Latter, T.D.T. (1970), 'Non-destructive coating thickness measurement', Fischer Instrumentation (GB) Ltd., Newbury, Berks.

12 Monitoring systems in operation

12.1 Introduction

Monitoring systems are already in use in a number of industries. Such industries are usually capital-intensive and considerable financial penalties may arise from undetected malfunctions or may involve exceptional hazards such that, for reasons of safety, high levels of operational reliability are of equal or superior importance to considerations of financial obligations.

Active measures are being taken to miniaturize and rationalize many of these systems and the other devices described in previous chapters. When this has been achieved, new technologies will be available and the use of monitoring systems extend to normal consumer equipment and thus improve the availability and safety of such equipment.

12.2 Marine monitoring systems

Despite the traditional conservatism of the marine engineering industry, the introduction of very complex equipment coupled with problems of manning and pressures of commerce have compelled some shipping companies to introduce some measure of condition monitoring. A number of systems available to both mercantile and naval vessels are briefly described in the following paragraphs.

12.2.1 French marine engine room surveillance

An engine room surveillance system with computer was installed in the Shell Tanker *Dolabella* in 1966 and also used for such calculations as boiler, turbine and propeller efficiences [1], the installation was primarily associated with the development of an unmanned ship system (U.M.S.). Another installation in 1967 produced by the French automation company, Compagnie d'Etudes et de Realisation de Cybernetique Industrielle (CERCI) for the M.V. *Aquilon,* a refrigerated cargo ship included some refrigerated cargo logging and machinery surveillance capabilities [2].

12.2.2 West German marine process control systems

A refrigerated cargo ship the M.V. *Polar Ecuador* was fitted with a digital computer when it was built in 1967 by Blohm and Voss [3]. The operational specification calls for alarm surveillance, automatic logging and centralized monitoring of engine room machinery and refrigerated cargo plant. The computer handles approximately 380 parameters.

12.2.3 Italian marine surveillance equipment

Following a pilot project on the M.V. *Esquilino* in 1969 a significant computerized surveillance system has been installed in the container ship S.S. *Lloydiana* [4]. The twin computer installation comprises an IBM 1130 plus a system 7 with AEG machinery performance monitoring software and can include the following capabilities:

(1) Electrical power plant monitoring;
(2) Refrigerated cargo monitoring;
(3) Ship condition calculations;
(4) Main propulsion machinery (32 500 shp) performance monitoring.

Participants in this project were the Engineering Faculty of Genoa University, C.E.T.E.N.A. (Genoa's centre for marine research), Lloyd Triestino, Italcantieri, IBM Italy and Registro Italiano Navale.

12.2.4 Japanese marine monitoring systems

Studies by the Ship Bureau of the Japanese Ministry of Transport resulted in the development of six highly automated ships built between 1969 and 1972, namely:

(A) M.V. *Seiko Maru* [5]
(B) M.V. *Mitsuminesan Maru* [6]
(C) S.S. *Kinki Maru* [7]
(D) S.S. *Tottori Maru* [8]
(E) M.V. *Ohtsukama Maru* [9]
(F) M.V. *Yamazuru Maru* [10]

A summary of the condition monitoring and related facilities included in the systems embodied in these vessels is given in Table 12.1

Table 12.1

	A	B	C	D	E	F
Alarm surveillannce	X	X	X	ʌ		X
Data logging						
Alarm diagnosis and failure analysis	X	X	X	X		X
Condition monitoring	X	X	X			
Boiler performance monitoring			X	X		
Abnormality detection				X		

12.2.5 Spanish marine condition monitoring system

A computer installation using an IBM 1800 and System 7 has been used on the bulk carrier *Castillo de la Motta* [8].

The Spanish marine organisation E.L.C.A.N.O. is involved in collaboration with

C.E.T.A.N.A. the Italian organization managing the *Lloydiana* project with which it has many similarities.

12.2.6 Swedish marine process-control computers

Kockums of Malmo with the Swedish Ship Research Foundation initiated developments in 1967 leading to the development by A.S.E.A. of the TURBODAC for steam turbine installations and DIESELDAC for diesel installations. They provide for the following controls:
  (1) Alarm monitoring of machinery including variable alarm limits;
  (2) Trend analysis and recording;
  (3) Hull condition monitoring;
  (4) Machinery performance monitoring.

12.2.7 U.K. marine condition monitoring systems

A digital computer system was installed on board the *Queen Elizabeth II* following a feasibility study by the British Ship Research Association partly financed by the National Research and Development Corporation [11]. The following facilities are provided:
  (1) Alarm surveillance and automatic logging of engineering plant;
  (2) Main condensor vacuum control;
  (3) Optimization of utilization of fresh water plant;
  (4) Weather routeing;
  (5) Hotel services stock recording.

In 1969 a computer-based alarm surveillance and monitoring system developed by the English Electric Company (now G.E.C.) Power and Marine Division was fitted on the board the *Glen Avon*, a 1700 dirt sludge carrier owned by Bristol Corporation [12]. This led to the development of the G.E.C. 2112 automatic watchkeeping system which has the largest sales of any computer-based machinery space surveillance system [13].

12.2.8 U.S. marine condition monitoring systems

America has a comparatively small merchant fleet for a nation of its size and has not previously been deeply involved in sophisticated marine technology until recently when a study was made of methods to improve marine technology through aerospace electronics [14]. A scheme, spectrum 75 entitled the 'Integrated Standardized Shipboard Automation System' has been submitted by the Honeywell Marine System Centre to the U.S. Maritime Administration Office of Research Development (M.A.R.A.D.) may establish a break-through in this field for U.S. technology.

12.3 Marine condition monitoring requirements

A summary of methods which have been successfully used in connection with various types of propulsion machinery is given in the following table. Most effort has been directed towards systems for large-bore diesel engines and steam turbine machinery but such systems are becoming available for geared medium speed engines.

Table 12.2 *Summary of condition monitoring techniques applied to marine engines*

(1) Slow running diesel engines

Component	Method
Turbochargers and charge air coolers	Calculation of efficiencies, air delivery and heat transfer coefficients, vibrations
Piston rings, cylindrer liner	Piston ring transducer, analysis of liner metal temperatures, liner surface temperatures, liner wear sensor
Combustion	Analysis of cylinder pressure

(2) Steam turbine machinery

Fuel injection System	Analysis of fuel oil pressure during injection
Main steam boilers	Calculation of corrected fuel gas temperatures and heat transfer coefficients
Main steam turbines	Calculation of corrected pressure ratios and turbine efficiencies, bearing metal or discharge oil temperature, rotor vibration

(3) Auxiliaries

Main reduction gears	Bearing metal analysis or discharge oil temperature
Stern tube bearing	Metal analysis or discharge oil temperatures, wear
Rotating Machinery	Measurement and analysis of vibrations
Heat exchangers	Calculation of heat transfer coefficients

In addition to the main propulsion and auxiliary machinery attention is also being given to the structural condition of ships and other marine constructions such as oil rigs.

12.3.1 In-water hull inspection robot scanner

An extension of the visual examination of components is reported [15] in the development of a robot submissible vehicle for inspecting the underside of the hulls of V.L.C.C.s and other large marine vessels by using remote control and without the need for drydocking.

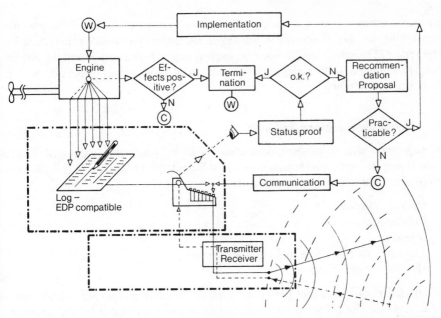

Figure 12.1 *Data acquisition entered in a computer-compatible log book and radioed to a land-based centre (Norddeich)*

Known as SCAN, the remote-controlled vehicle traverses the hull and relays information via T.V. cameras to surveyors on the surface. This robot, is saucer-shaped, some 5 ft in diameter and adjusted to neutral buoyancy in sea water. In operation, it is manoevered by divers to the starting position under the tanker, and compressed air forced into the central buoyancy chamber; this gives a strong up-ward thrust against the vessel's bottom plates. Control is then taken over by the operator on a tender or jetty beside the tanker.

12.3.2 Hull stress monitor

A device has been developed by Mitsui Shipbuilding & Engineering Co. Ltd. to monitor the stress created by waves on a ships hull, it is of particular significance in preventing possible dangers on very large cargo-carrying vessels. and the first unit has been installed on the 270 000 dwt tanker 'Navigation Monitor' owned by Exxon, U.S.A.

Hazards which a ship may encounter are monitored by a continuous watch on the hull strength under various weather and sea conditions. This is done by a sensor which is affixed to a specific position in the midship section of the hull, where the hull strength as well as degree of strain exerted on the hull are most critical and apparent [16].

Strain detected is converted into a stress value to represent the condition of the hull and shown on a display unit in the wheel house. If any danger is indicated an alarm warns the crew in charge. The device also makes forecasts on possible dangers to the ship if she still continues to sail under the condition perceived by the sensor. The sensor incorporates an explosion-proof stress gauge and the display unit is equipped with a signal change unit, root mean square processor, meter, indicator and printer.

12.4 Marine diesel engine monitoring

A suggestion that ship-board computers might be reasonably employed in the monitoring of diesel performance trends was made by Monceaux [17] of Bureau Veritas in relation to the monitoring of:

Cylinder output,
Combustion,
Injection,
Liner and liner port fouling by various deposits,
Heat exchanger efficiency.

Computation of the output per cylinder using sensors to provide the cylinder pressure, revs/min and injection pump rack position make it possible for the computer to record the cylinder pressures in relation to the crank angle position. This enables the mean indicated pressure and the developed output per cylinder to be calculated. From this information, it is possible to know the total delivered output and also the load sharing of the cylinders i.e. to check the right adjustment of the injection pumps. This output value can be also computed with the torque-meter output value.

12.4.1 Engine cardiograms

Any changes in the functional values and in the existing correlation between them over a period of time indicate changes in the regularity of certain processes and events within the diesel engine operating cycle.

By means of time series analyses it is not only possible to determine the instant when such changes began but also to give a prognosis as to when such developments will lead to a critical condition. The problem requires, to mention only the most important points:

Logging of a multitude of measured values at a frequency which depends on operating conditions and environmental influences,
Calculation of many mathematical relationships where the measured values serve as input variables,
Reducing the above to reference and comparative conditions,
Concentration of the functional values in accordance with a logic that corresponds to specific conditions,
Using the results right away for trend studies to determine optimal overhauling and repair times,
Comparisons between actual and reference values,
Checking data plausibility with a view to pickup and line control.

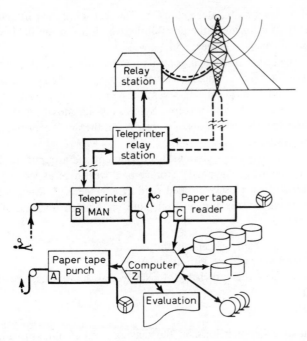

Figure 12.2 *Punched-tape input relay overland by teleprinter to a computer centre for evaluation and results then transmitted to the source-vessel*

These tasks can only be carried out by means of a process control computer. In addition, the aim, i.e. early recognition of eventual failures, can only be achieved if the computer is used in 'ON-LINE' operation.

In order to keep the high expenses involved within acceptable limits, Pauer of M.A.N. Hamburg reported a 'soft-ware project study', chosen as a considerably less expensive solution without, affecting the development of the soft-ware, i.e. the processing logic of the measured values [19]. In lieu of the computer, there is an EDP-compatible log book (Figure 12.1). The measured values are consequently still logged by conventional methods, but are transmitted at regular intervals by radio to the M.A.N. computer. The progress of the data becomes apparent if Figure 12.1 is compared with Figure 12.2. The teletypewriter establishes a recording medium by means of a perforated tape which, by way of a paper tape reader, serves as a data input medium for the M.A.N. computer. Consequently, this computer, in OFF-LINE operation, functions as the later shipboard computer which is to operate ON-LINE. The results of this programmed logic, i.e. a diagnosis of the status of the propulsion system, are output parallel to the clear text at the printer, again in the form of perforated tape, and return to the vessel in the same way as described before, only in the reverse order. This method is at present in operation for test purposes and with a view to later extension. The test solution is, of course, cumbersome and far from utilizing all the possibilities which will be available in the future, the realization of which is, however, being worked at.

The best diagnosis is not much good if there is no therapy. A diagnosis must be based on one or, in most instances, on several symptoms. There is causality between the symptoms and a diagnosis or several diagnoses. Only a knowledge of this causality permits a systematic therapy. Symptoms—diagnoses-relationships can be formulated first verbally by means of a matrix and then in the abstract with the aid of Boolean expressions.

The symptoms—diagnoses-relationships document and accumulate the knowhow and experience of many engineers. Storage of a wealth of experience and knowledge in a matrix (in a closely defined field of engineering), when compared with conventional storage in the human brain and on written material affords several decisive advantages:

> Virtually unlimited storage capacity,
> Unlimited learning capacity without delay,
> Independence from the knowledge of a single person in the case where a decision is required,
> Immediate reproducability.

Incorporation of a decision logic in the diagnosis programme provides up-to-the-minute assistance to the engine control personnel of future vessels in order to optimize availability of the propulsion system and to achieve maximum transport capability.

12.4.2 PAC/DAC engine operation monitors

Signals from monitoring transducers may of course be connected to fairly conventional electronics and displays. With an increasing number of transducers and monitoring functions, the process computer has distinct advantages because of the many functions it can carry out, for example: reading of binary and analogue signals, storing of data in digital form, performing calculations and control functions, presenting processed data, and carrying out various logical sequence.

The M.V. *Taimyr* was one of the first vessels to be fitted with such a computer. The M.V. *Hoegh Multina* uses a more advanced system with a large number of functions for engine monitoring and control together with facilities to analyse engine data for preventive maintenance [20].

The systems consists of two main parts, the *Eo-system* and the *Preventive maintenance system* and is served by two small process computers PAC (Process data Acquisition Computer) and DAC (Data Analysis Computer). PAC performs the monitoring, the process control and a preliminary evaluation of the main engine's condition. Outline diagrams are shown in Figures 12.3 and 12.4.

DAC performs further calculations of process parameters and evaluation of data from special instrumentation. Moreover, DAC acts as a standby unit for the

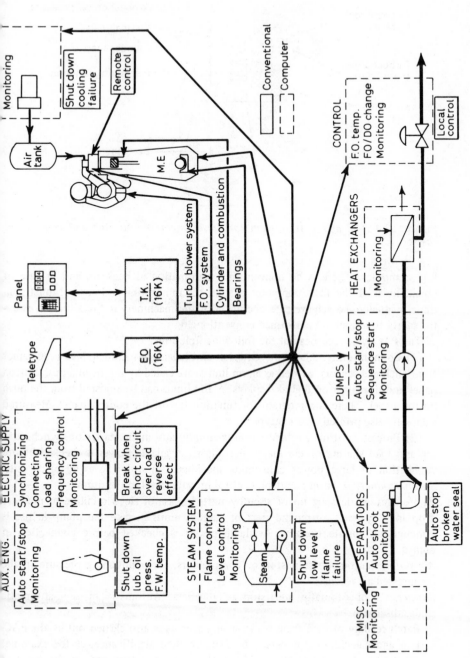

Figure 12.3 *Systems included in the PAC and DAC diesel engine monitoring and control system*

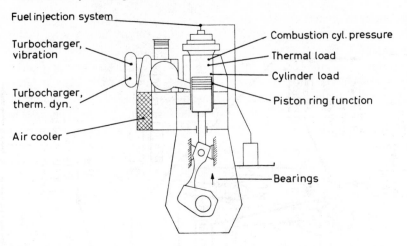

Figure 12.4 *PAC and DAC system for diesel engine monitoring and component check*

PAC-computer. There is no conventional automation as back-up, and in case of break down of the total computer system the monitoring and control must continue manually. To achieve safe operation of the machinery in this situation essential safety functions are performed conventionally.

The PAC-system carries out the following functions:

(1) Eo-monitoring: Alarm presentation and logging is included. Measured values from analogue sensors, as well as alarm limits, are called by means of a push button panel and displayed on digital displays. Alarm limits can be adjusted from the push button panel. There is a program for function test of the control panel. Measured values are also printed on a teletype.

(2) Process control: *Electrical power supply;* automatic start/stop, synchronising and load sharing of the diesel alternators. Available power is checked before connection of large power consuming machinery in order to avoid black outs. *Steam production;* burner control. Control of water level in the drum.

Pumps; automatic start up of standby units in case of electric failure or pressure drop in a system. Automatic start of standby units in case the pump in use is switched off unintentionally. Available power is checked before connection of pumps.

Air supply; automatic start/stop of compressors, for maintaining pressure in the compressed air vessels.

Purifiers; semi-automatic start/stop of purifiers. Automatic or semi-automatic sludge discharge.

Batch reading of data from some analog sensors is also carried out by the PAC-system and calculations processed to evaluate their significance, as for example:

(1) Turbo charger evaluation: The airflow rate is determined on the bases of pressure drop measurement across the inlet duct of the compressors. The

efficiencies of compressor and turbine are evaluated by means of suitable pressure and temperature readings, and thermodynamic theory. The rotors are monitored for vibrations. Evaluation of air coolers is also included (contamination of air and water side, efficiency of heat transfer, water velocity).

(2) Cylinder monitoring: *Piston rings;* the piston ring condition is sensed by means of proximity transducers, which are mounted flush with the running surface of the cylinders. It detects rings not in proper contact with the cylinder as well as broken piston rings. It is shown in principle in Figure 12.5 with the electronic interface producing digital information for further evaluation by the computer. Within a period of time, the computer evaluates the state of the piston rings.

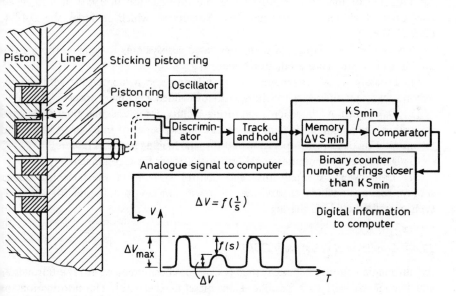

Figure 12.5 *EO-system method for piston-ring condition monitor*

Cylinder lubrication and wear; this function is based upon a surface thermocouple mounted in flush with the running surface of the cylinder. If the lubrication breaks down friction produced temperature pulses develop due to heat when the piston rings pass over the element. These pulses are evaluated by the computer, in terms of wear intensity and lubrication condition. Temperature from ordinary thermocouples in the liners are also monitored, statistical distribution parameters being calculated. These parameters are useful in giving the general state of the cylinder, indicating gas leakage and thermal load.

Combustion; the cylinder pressure is recorded continuously and digitized. The computer calculates mean indicated pressure, cylinder power and rate of heat release. The thermal load is evaluated by means of heat flux recording in the cylinder head.

Analogue signals from the sensors are sampled by the computer and presented graphically on a CRT screen, namely:

(1) Cylinder pressures;

(2) Piston ring sensor signal;

(3) Cylinder surface temperature;

(4) Fuel injection pressure;

(5) Trend diagrams for various engine and component data — three different trend diagrams can be presented simultaneously from a total of 56 trend functions.

12.4.3 DATACHIEF (Diesel)

An outcome of the *Taimyr* project was the introduction in 1970 of a range of computerised shipboard equipment by Norcontrol which includes the *DATA-CHIEF* (Diesel) [21].

This consists of a family of computer-based susbsystems for engine room automation and maintenance prediction comprising:

(1) Datasafe: A watchkeeping and monitoring system complying with classification society requirements for ships with unattended machinery spaces;

(2) Datatrend: A condition monitoring facility for diesel machinery designed to assist in the planning of maintenance. It utilizes a series of transducers developed for monitoring piston ring operation and thermal loading in the cylinders;

(3) Datadiesel: A computer-based bridge controller for diesel machinery;

(4) Datapower: A monitoring and control system for ships electric power supply; features include supervision of power consumption, sequential control, sychronization and load sharing.

12.4.4 Condition check system CC10

An automatic condition check system (CC10) has been prepared for the Burmeister and Wain K-GF series of 2-stroke mains diesel engines [22]. The information for this system is collected from the following engine locations:

(1) Air and gasways, including turbochargers and air coolers;

(2) Combustion system, including injection system, fuel consumption, engine output, load distribution between cylinders, and the speed of the ship;

(3) Cylinder units, including thermal load, wear and piston ring function;

(4) Auxiliary systems, such as water coolers, lubricating oil coolers and pumps.

The majority of pressure and temperature measurements are carried out by conventional pressure transducers and temperature sensors and a number of more special temperature sensors in the cylinder cover and cylinder liner. The data from the sensors is transmitted to a digital computer programmed to process the results from the individual parts of the system and to evaluate these as an integrated part of the complete system. Figure 12.6 shows the system comprising of a diesel engine, micro-computer with sensors, and the operator's console which is placed in the control room.

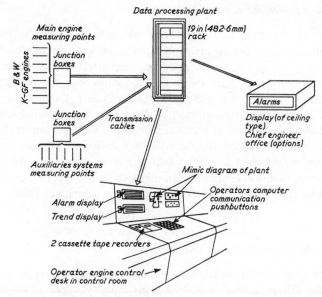

Figure 12.6 *System layout for the automatized CC10 condition check system [22]*

The condition check system is (in principle) based on the manual condition check system and thoroughly tested over several years, by comparison of data model curves with service measurement values recorded manually with fixed, rather considerable time intervals. By recording the deviations in the key diagrams and comparing and evaluating the deviations, a graph of developments in the condition of the actual components is obtained from which diagnosis of the abnormal conditions can be derived.

The automated system CC10 comprises of a higher number of components and gives a more thorough indication of the condition on basis of measurements, calculations, and evaluation. In addition to trends and diagnosis, it also gives an alarm by diagnosis if an abnormal condition arises and develops towards a critical limit. Extrapolation of the trend values in future and the statement of limit for maintenance indicate the time for overhaul work, in other words the required prediction.

As these trend evaluations are performed continuously, the engine is guarded to a degree manual inspections can never do, so that the personnel can concentrate on the rest of the maintenance programme.

12.5 Marine turbine monitoring systems

The numerous auxiliary systems associated with marine turbine systems — particularly steam turbines — considerably complicate the control and monitoring situation. Under these circumstances the computer occupies a central information storage and synthesising role which has been fully exploited by the availability of mini-computer facilities.

12.5.1 Siemens integrated monitoring system

The purpose of the integrated monitoring system is to sound an alarm in the event of disturbances in marine machinery and to indicate the existence of faults. Signals are admitted by binary and analogue transmitters. The binary transmitters (contact transmitters) are connected so that the alarm operates when the contacts open Analogue values from sensors such as resistance thermometers, thermocouples and 20 mA transmitters initiate the alarm when the measured value deviates from an adjustable minimum or maximum limit. For some parameters the mean of a number of measured values can be taken and individual measured values compared with the mean; when deviations from the mean value occur the alarm operates. An internal monitoring circuitry ensures that in the event of interruptions of the transmitter circuits, such as breaks in the wiring or contact faults, an alarm is also initiated.

12.5.2 ASKANIA integrated fault monitoring

A typical fault monitoring installation used for machine purposes, as in the case of steam turbine driven tankers of 300 000 tons capacity, speed 15 knots driven by 36 000 hp turbines, employs an electronic module system with a computerized data store facility and a common technique, sensors and actuators. The control circuits and control loops for a Siemens installation have the same structure and signal language (4–20 mA) and manage: burners; boiler controls; auxiliary control loops; bridge controls; emergency generating plant, etc.

The burner management system consists of two functional groups, (a) sequence control group, (b) safety group. The general circuitry is shown in Figure 12.7 an automatic sequence being initiated at the touch of a button which ensures that all operations are performed in the correct sequence depending on a number of major

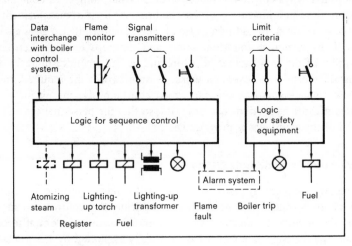

Figure 12.7 *Boiler management system*

interlocks, all of which apply both when flashing up and when shutting down. Safety controls are independent and automatically shut off the fuel supply if an abnormal condition arises – duplicate safety controls are provided.

The core of the slip automation system is formed by a central information and monitoring high level logic system such as the SIMROS(R) 21 produced by Siemens Akt, Erlangen. This keeps a continuous watch on all readings for possible deviations beyond acceptable limits. The nature and location of each fault is immediately indicated by visual and audible alarms in the machinery control room and as group alarms on the bridge and in the engineer's accomodation. All alarms are stored until acknowledged. The block diagram is shown in Figure 12.8. Measuring points are coded and values can be printed out on a strip log either by interrogation from a decade keyboard as required, or, at regular intervals; faults are automatically recorded on the strip log as they arise, the date and time of the fault being printed and the date and time of the correction.

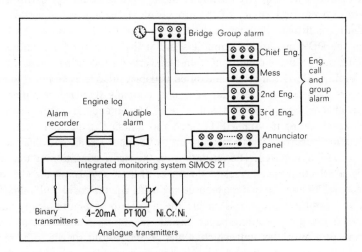

Figure 12.8 *Integrated monitoring system, SIMOS(R) 21*

12.5.3 Main propulsion machinery – condition monitoring

A system designed primarily for gas turbine-powered warships with potential for use in merchant ships using steam, diesel or gas turbine propulsion systems was described by Anderson [23].

It was realised during the development of the propulsion control system for the Royal Navy's type 21 frigate and type 42 destroyer that there would be a need to record permanently certain critical parameters concerned with the ship's propulsion machinery and its associated control system. These recordings could be used during the sea trials period to compare the actual performance of the equipment with the results of the computer simulation carried out during the design stages of the control system. Comparison of the recordings enables the performance of the propulsion machinery to be optimized by adjustment of the propulsion controls and to anticipate any problems which might arise.

The recordings are made by a Dynamic Data Recording (DDR) Equipment produced by Hawker Siddeley Dynamics Engineering Ltd., Hatfield, Herts which, subsequent to the sea trials could be used to monitor machinery performance and to assist in analysing causes of machinery breakdown. The DDR consists of a signal conditioning unit, and a signal selection panel and ultra violet (u.v.) recorder. A magnetic tape recorder and multiplexing unit can also be introduced.

The equipment accepts input signals consisting of any variable quantity capable of being measured by a transducer with an electrical output, or an electrical signal from the control system. Types of transducers commonly used, measure variables of position, rotational speed, pressure, temperature, torsion and fuel flow rate.

The input signals are fed into a signal conditioning unit. This acts as a signal protection device which allows existing control system and instrumentation transducers to be used without any possibility of the DDR affecting the performance of the control system or accuracy of the propulsion machinery instrumentation, even under fault conditions. High value resistors are used for this purpose, both at the inputs of the DDR operational amplifiers and at the outputs of the transducers or signal sources. The signal conditioning unit converts the incoming signals into a d.c. voltage level, proportional to the measured variable, suitable as an input to the u.v. recorder or magnetic tape recorder. The unit contains standard amplifier boards with each channel being provided with a range of gain and bias settings so that the required signal scaling is obtained. The unit also contains a regulated voltage source for calibration purposes, it can measure up to 68 input signals and this can be increased to 114 if required. The outputs from the signal conditioning unit are fed into a signal selection panel. A selected number of these signals can then be fed into a u.v. recorder or a magnetic tape recorder by using a patchboard and jumper leads.

Gain and positioning controls enable the signals to be placed clearly and conveniently on the u.v. recording paper. Signals fed to the magnetic tape recorder go by way of a multiplex unit which allows several output signals to be recorded on a single tape recorder channel. Signals stored on magnetic tape can be fed back into the signal selection panel and then to the u.v. recording paper if required. If a loop tape recorder is used, a large number of propulsion machinery parameters can be recorded covering the last 40 minutes of operation. If a fault should develop, selected parameters can then be fed to the u.v. recorder which will simplify considerably the problem of fault diagnosis. Regular recording will also show any fall-off in the performance of the propulsion machinery.

In gas turbine-powered ships the common practice is to change complete engine modules instead of overhauling them *in situ*. Setting up the new engine module is considerably simplified with the help of the Dynamic Data Recording Equipment since responses can be easily recorded and diagnosed.

12.6 Shipboard vibration monitoring

Systems for the regular monitoring of vibration have been accepted by some marine organisations as a fundamental monitoring procedure. The Canadian Navy has published some accounts of their procedure and successful results from British Navy experience have been recorded by Corben [24].

12.6.1 Canadian shipboard vibration monitoring programme

The Canadian Forces have introduced the use of a portable octave band analyser for a machinery health monitor as part of their preventative maintenance programme on ships [25]. Velocity measurements are taken near each principal bearing on a machine on an overall and octave band basis — the overall velocity level to define whether a machine is in good condition or not, the octave band analysis to determine the nature of the fault.

At specified overhaul shops of the Canadian Navy vibration analysis is applied during the test runs after machinery overhaul. The 44 principal machines on each ship were originally included in the programme but this was extended by the dockyards to include over 100 mechanical and electrical machines. For each type of machine in the programme standardized monitoring points were defined adjacent to each principal bearing. A list of normal vibrational modes and a table of machinery faults and their vibrational effects is prepared together with graphs showing the maximum, average and minimum levels experienced in the fleet at each point on each machine.

Prior to a ship entering for refit, complete octave band surveys are carried out on all machines. These are supplemented by discrete frequency analysis if necessary. From these surveys the ship's maintenance and repair list is suitably revised so that only machinery with unsatisfactory readings is opened up. Post refit surveys are made both as a quality control and to provide datum information for the next refit.

As an extension to this programme, machinery vibration levels were measured monthly by ship's staff on a number of destroyers. Overall levels were taken and if they had risen by a factor of two (6 VdB), octave measurements were then taken to determine the cause.

The principal program is VIBANAL, which checks new data for errors and compares the new data with the average values for the machine. The program output consists of the machine vibration levels and the listing of each point that is more than 6 BdB over the average value. The format is such that the analyst can determine quickly the faults in the machine and to recommend the best refurbishment. Another program, VIBLIST, is usually run in conjunction with VIBANAL, and lists and at each point, and compares the results of the last three vibration surveys, thus expediting the historical comparison of the machinery vibration.

IBM utility programs enter the data into the master storage tape, from which it can be recalled when required by VIBLIST or other comparative work. The data is stored on the master tape so that all the readings from a given point on any machine type are together. This enables the machinery norms to be updated when required.

The program AVLIST takes all the readings between specified time periods and gives an updated set of the norms or of the average values alone, for use by the shipboard operator. For each point the norms consist of:

(1) The maximum values recorded in service;

(2) The minimum values recorded in service, but with a low limit of 70 Vdb (any reading lower than 70 Vdb indicates an instrument defect);

(3) The average of all the readings, but ignoring readings greater than 130 Vdb (which would indicate a seriously defective machine).

At present AVLIST is run every two years. A graphical plotting program, VIB-PLOT, takes the output from AVLIST and draws a new set of norms.

12.7 Spectrometric oil analysis programme – marine

A programme based on the use of flame spectrometric methods has been used for some time by the Royal Canadian Navy [26]. The programme embraces nearly all R.C.N. ships based at H.M.C. Dockyard, Esquimalt, The submarine H.M.C.S. *Grilse* initiated the programme and other vessels were introduced as follows:

H.M.C.S. *Grilse*	May 1966
H.M.C.S. *Qu'Appelle*	June 1966
H.M.C.S. *Yukon*	June 1966
H.M.C.S. *Saskatchewan*	September 1966
H.M.C.S. *Mackenzie*	January 1967
H.M.C.S. *St. Croix*	March 1967
H.M.C.S. *Columbia*	April 1967

With the introduction into the programme of a number of auxiliary and training vessels in March 1967 a total of 250 samples were then being analysed each month. These were analysed quickly with a target of 24 to 48 hours within which to submit a report, when irregular results are observed a check sample is requested and, if the result is confirmed, operating personnel on the vessel are notified by telephone or radio.

12.8 Monitoring integrity verification

The correct functioning of marine electronic control equipment can be checked by means of built-in-test-equipment (BITE) which identifies defective printed circuit modules so that they can be replaced by new modules. A typical BITE is produced by Hawker Siddley Dynamics Engineering Ltd. [27].

12.8.1 Built-in-test-equipment (BITE)

The system to be tested is connected to the BITE through an interface unit. This is essentially a routing system such that the stimulus and measurement devices can be connected to the system under test. The operator controls and monitors the test sequence via the control and display panel. Control signals are fed to the central

test sequence and feeds signals back to the display panel. The operator is informed by the diaplay of the result of the test and, if required, the most probable corrective action.

The BITE system self-test facility may be periodically exercised to ensure that it is functional prior to the testing of a suspect system. The system to be tested is connected to the test system and identified by coding of the connectors. Three lines coded in binary are used for identification. The insertion of the connector mutes the system, taking it off-line.

Once the sytem has been connected, 'System Test' is selected and the test sequence started by pressing the select test button. The central control then sets up the first test in the sequence such that appropriate stimulus and measurement devices are connected to the system under test. The measurement is taken and the result compared to alarm levels and a 'pass' or 'fail' determined and displayed to the operator.

The operator selects the tests in sequence until a failure is indicated or the sequence is completed. When a failure occurs the indicated faulty module is replaced and the test connectors are removed and the system may be re-selected for control.

The control panel is mounted in the machinery control console adjacent to the systems requiring test. The system test socket carries connections such that a system may be fully tested. When a connection is made from this socket to any system or the 'Self-Test' socket, the BITE system will automatically carry out the tests on the selected system or on the test equipment itself dependent upon the test mode selected.

The mode switch is a four-position rotary switch which selects the appropriate mode of operation:

(1) Off,
(2) Self-test 1. Lamps and display checks,
(3) Self-test 2. Stimulus and measurement checks,
(4) Systems test.

With the mode switch in position 2, and the Select Test button depressed, the displays are automatically checked. The test number starts at '00' and runs up to the full test number of count, and any discrepancy will be indicated to the operator. With the mode switch in position 3, a check of the stimulus and measurement devices is carried out. The test sequence is then identical to that for a normal system. The self-test facility is normally concealed under a flap so that operation of the panel is not confused.

With the mode switch in position 4, and with the system to be tested linked to the system test socket, the system test is carried out. When system test is selected, the 'Off-Line' lamp is illuminated confirming that the system under test is muted. The first test is selected by pressing the 'Select Test' button, the result of test (i.e., pass or fail) is displayed and the affected module or minimodule indicated. Testing is continued by pressing the 'Select Test' button again and observing the results.

The ability of a system to drive an actuator is checked by a rotating drum indicator. The direction of rotation is checked by the operator against the 'opening' or 'closing' lamp, the correct sense being illuminated.

Stimuli fall basically into two types, namely analog and digital logic signals. Logic signals are applied to the system under test to simulate the condition of various machinery state inputs. Up to 24 bits may be applied for a given test, or any part of the 24 bit word applied for each test over the sequence. All inputs can be fed with a logic '1' or '0' signal or be left open circuit, depending upon the requirement of the selected test.

A range of fixed frequency signals and a variable frequency generator are available as inputs to the system. Programmable d.c. signals as well as defined d.c. voltage levels can be selected and applied. Signals returned from the system under test are measured to check whether they fall within the programmed limits. Both analog and digital logic signals are processed in parallel and the result displayed as a pass or fail. The type of test is specified, namely, whether an analog voltage, logic level or combination of these is to be toleranced.

Most modifications to existing systems that are subject to test can, in most cases, be confined to 'software', that is, to the program. Alterations can be made by changing the PROM (Programmed Read Only Memory), a replaceable integrated circuit. Extra stimulus and measurement devices may be added to the system to increase its applicability.

12.9 Aircraft condition monitoring

Safety considerations dominate aircraft condition monitoring philosophy although matters of financial viability and operational availability are improved by the use of monitoring techniques.

12.9.1 Olympus 593 performance trend monitors

Olympus 593 engines installed in Concorde development aircraft were used to obtain trend monitoring developing data reduction techniques for producing trend plots of thermodynamic parameters which could be correlated with any engine failures. These have proved a basis for in-service trend analysis [28].

The flight test data recording systems in Concordes 002 and 01 were used and the parameters selected from the comprehensive facilities (Table 12.3).

Trend monitoring was performed at take-off, Mach 1.2, Mach 1.6 and close to Mach 2.0, data being taken over 10 second periods with parameter values at 1 second intervals. Data was initially obtained in printed form and then a system devised for magnetic tape.

A computer extracted the engine trend data from the total flight test recording and output it to an IBM magnetic tape. The computer also applied instrument calibrations and linearly extrapolated parameter values to give values at the precise times required. This data was put into a disc file FLT within a remote access computer system known as AMOS (A Multi-access On-line System).

Table 12.3 *Trend monitoring parameters, Olympus 593/Concorde*

Main gas stream	Mechanical condition
l.p. revs/min (N_1)	Engine vibration
h.p. revs/min (N_2)	Jet pipe vibration
l.p. comp. dely. pressure (P_2)	Eng. diff. oil pressure
l.p. comp. dely. temp. (T_2)	R/H gearbox diff. oil pressure
h.p. comp. dely. pressure (P_3)	Engine oil inlet temperature
h.p. comp. dely. temp. (T_3)	No. 1
Engine fuel flow (F_E)	No. 2 & 3 Bearing oil
Reheat fuel flow (F_R)	No. 4 drain temp.
Jet pipe temp. (T_7)	No. 5
Jet pipe static pressure ($P7_S$)	CSD oil inlet temp.
Jet pipe total pressure ($P7_T$)	CSD oil outlet temp.
Primary nozzle area (A_J)	Main labyrinth vent temp.
	No. 2 & 3 cold vent temp.
	No. 5 cold vent temp.
	Turbine cooling air temp.
	Anti-icing pressure.
	h.p. services bleed pressure.
	Reheat burner pressure.

The first terminal activity is to run the programme INST with the flight data to create a further file PRE, a summary file SUM is also produced for a few parameters to select the most appropriate data within the 10 second period of each flight. The selected data together with an identification of flight number, aircraft number and engine number is made into a data file DAT.

This data is run with the programme TREND which corrects the thermodynamic parameters to datum conditions and creates the results file RES. A further file TEX can be run with the RES file to produce a printed output of the observed and corrected data for the particular flight. The RES file can also be added to an accumulating data bank of RES files from previous flight basis using a computer line printer.

12.9.2 Aircraft flight recorders

From the early days of aviation, when manuscript notes were made by the pilot or navigator, in-flight recording has developed to its present stage involving highly complex instrumentation systems recording in some cases millions of discrete measurements throughout the period of a single flight.

Early records were used for purposes of flight test development and more recently for airworthiness data which subsequently expanded its use into performance monitoring and accident investigation. Legislation requiring the carrying of reliable crash-resistant data recorders on large civil aircraft came into force in the U.S.A. in August 1958, closely followed by similar requirements in France and Australia and in 1965 became mandatory in the U.K. [29].

Experience shows that flight recorders are valuable in the analysis of two very general accident categories:

(1) Where some form of aircraft or system malfunction has been the primary accident cause. This category may extend from, for example, a simple electrical or instrument system failure to the in-flight disintegration of the whole aircraft structure. In such cases wreckage analysis will probably play the principle role in attempting to establish the primary failure. The data recorder serves a very useful purpose in this type of accident, however, by supplying supporting data such as time, height, velocity and acceleration factors.

(2) Where the accident has an operational cause, that is, it is one in which no engineering defect or deterioration in aircraft performance has occurred. Examples of this are certain handling accidents resulting in loss of control and those due to weather or navigational factors. In such an accident wreckage analysis may provide data concerning final configuration and broad details of impact attitude and speed. Beyond this the information derived from the wreckage will probably be largely negative. In this type of accident an accident data recorder may well play the predominant role by providing a time history of the principal manoeuvres throughout the flight.

The common parameters which are recorded are: pressure altitude; indicated airspeed; aircraft heading; normal acceleration.

Other information which is mandatory in some countries is: pitch angle (U.K.); radio beacon interception data (France).

Experience which has been accumulated in the extraction and evaluation of accident data records suggests that more performance monitoring data as a mandatory requirement would assist in accident prevention and improve flight safety. Protected 'voice recorders' have been used in Australia and the U.S.A. for a number of years, they have facilities for bulk erasure following an uneventful flight.

Three basic methods of recording are used in the present accident data recording systems, they are:

(1) Oscillographic engraving on metal foil: In this type of recorder either aluminium or high nickel content steel foil is used as the record medium on which styli engrave, in analogue form, either continuous or sampled traces representing each parameter.

(2) Photographic recording: With this method the displacement of light beams, via galvanometer controlled mirrors, produce traces; may also be in analogue form on photosensitised paper.

(3) Electro-magnetic recording: In these systems both analogue and digital recording techniques are used. Of the former, frequency modulation methods are normally employed using plastic-based magnetic tape as the recording medium. The digital type of recording systems currently uses pulse code modulation techniques to produce digital data on stainless steel wire. Developments in electro-magnetic recording media include the use of stainless steel tape, primarily because of its good survival and data packing properties.

Most of the present types of recorder are of the fixed duration type involving the removal of the record and where required, erasure of the data at specified intervals, a few are of the continuous loop type where a constant amount of data is stored and continuous erasure of the 'first recorded' data takes place on the aircraft. The continuous loop types of recording systems have a facility for parallel recording of the data on a second non-protected recorder from which the recording medium may be readily removed for replay and storage. Means are provided on all types to ensure compliance with statutory data storage periods.

Electromagnetic recording systems generally include such subsystems as: sensors; transducers; signal conditioners; multiplexers; modulators; analogue to digital converters; stabilized power supply; recording stages.

With digital recorders the binary system of digital coding is used although recording techniques vary and differing formats are employed for word sequences and frame synchronization.

12.9.3 Aircraft integrated data system (AIDS)

The efficient routine data processing of aircraft recorded data is only practical with digital systems which can be interfaced with ground electronic data processing systems (E.D.P.). The standard data acquisition system which is capable of interfacing with other AIDS equipment developed by various aircraft companies is characteristically knows as the ARINC 563 system.

The areas in which AIDS data can be used most effectively includes the following:

Flight Safety analysis,
Crew proficiency,
Autoland evaluation and perfection,
Engine health monitoring,
Operations and logistics,
Aircraft and engine performance monitoring,
Ad hoc analysis.

The capacity of AIDS to monitor crew proficiency and the application of operating standards and limitations was originally viewed with a certain suspicion by airline pilot's organizations. Handled properly and with mutual confidence, AIDS benefits both the pilots and the airline, and gradually the fear of a 'big brother' concept has dissolved. In particular, the capability of AIDS to unveil shortcomings in aircraft equipment, ground equipment, procedures and operating instructions enhances pilots' confidence in the system and raises their respect for it. The current installations in civil aircraft and those planned for the aircraft now on order cover most of the objectives. Differences between airlines are more of emphasis than of basic design.

There are, however, other influences that directly reflect themselves in the selected system configuration. Of these, three seem to be most significant. First, airlines that have dispersed facilities tend to use a non-continuous maintenance recorder plus a separate performance recorder. On the other hand, airlines having maintenance and operational facilities in one area tend to use one recorder for both maintenance and performance. Secondly, airlines with a hub-type route system put less emphasis on a capability to transmit recorded maintenance data to their maintenance engineering centre than airlines with other route systems. Thirdly, airlines with a minimum response time requirement look for complete onboard processing, while those with no requirements for quick response prefer large capacity recorders and remove the tape cassette, say, only once a week.

12.9.4 System integrated fault monitor – B.747 AIDS

Aircraft integrated systems (AIDS) have been introduced as part of a programme aimed at reducing engine maintenance and operating costs by applying condition monitored maintenance. Applied by American Airlines, Tulsa, Oklahoma this concept involves the use of a maintenance recorder system the phototype of which was described by Kruckenberg [30] and installed in a BAC 1-11 aircraft along with a performance recorder and designated ASTROLOG.

Experience with maintenance recorders [31] indicated a need for high accuracy with sensor and system accuracies, particularly for the basic parameters of total air temperature, total air pressure, engine fuel flow rate. Experience also indicated that high sampling rates where needed for the parameters especially for engine pressure measurements although not for engine speed.

A B-747 AIDS consists of:

(3) Flight Data Acquisition units (FDAU)
(1) Flight Data Entry Panel (FDEP)
(1) Data Management Unit (DMU)
(1) 4-minute Loop Recorder (LR)
(1) Digital AIDS Recorder (DAR)
(1) Digital Flight Data Recorder (DFDR)

The arrangement of these system components is shown in Figure 12.9.

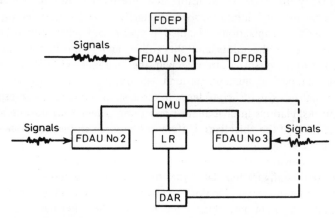

Figure 12.9 *AIDS prototype system schematic arrangement*

The maintenance recorder parameter list provided for processing is given in Table 12.4.

Table 12.4 *Maintenance recorder parameter list*

Parameters	Sampling rate per second	Required accuracy (±)
Aircraft		
calibrated airspeed (CAS)	2	3 knots
altitude (Fine)	2	45 ft.
Greenwich mean time (GMT)	1	1 s
static air temp (SAT)	0.5	0.4%
ram air temp (TAT)	0.5	0.750°C
mach number	1	0.005
pneumatic bleed duct pressure	0.25	0.8 psi
Engine		
Low compressor		
total discharge pressure (P_{13})	2	1%
total discharge temperature (T_{13})	1	1.7°C
speed (N_1)	4	0.2%
High compressor		
static discharge pressure (P_{S4})	2	1%
total discharge temperature (T_{145})	1	6°C
speed (N_2)	4	0.1%
oil in temperature	0.25	3°C
engine pressure ratio (EPR)	2	0.012
oil breather pressure	0.25	0.3 psi
fuel flow	2	45 pph
oil pressure	0.5	2 psi
exhaust gas temp	1	5.5°C
turbine cooling air pressure (P_{s51})	0.25	1%
nacelle temp	0.25	—
variable stator angle (Beta)		1°
bleed flow		7%
power level angle (PLA)	0.5	1°
oil quantity	0.25	
engine vibration (2 Pickups)	1	—
start air pressure	0.25	± 6psi
Discretes		
squat switch	1	—
fuel heater switch	0.25	—
nacelle anti-ice	1	—
high stage bleed valve	1	—
pylon valve	1	—
pressure relief	1	—
wing anti-ice valve	0.25	—
pack valve position	0.25	—

12.9.5 C5A Malfunction detection analysis and recording system

The C5A Malfunction Detection Analysis and Recording System (MADARS) operates on the TF39-GE-1 engine produced by G.E.C. (U.S.A.) for the Lockheed C-5 aircraft. Fault isolation and incipient malfunction detection is performed on a number of line replaceable units (LRU's) including the engine itself and other parts of the propulsion subsystem such as the engine controls, engine overheat warning system including fire extinguisher, engine start system, thrust reverses system, engine instrumentation and constant speed drive, engine lubrication system, engine fuel system, ignition system [32]. The various components (LRU's) which would be included in this surveillance includes:

Engine,
Fuel pump and filter,
Fuel heater,
Main fuel control,
Fuel plumbing,
Fuel nozzles,
Pressurization/dump valve,
Oil tank and cap,
Oil cooler,
Lub/scavenge pump and filter,
Lube plumbing,
Scavenge pump,
Scavenge filter,
External gearboxes,
Compressor inlet temp. sensor,
Variable-stator actuator,
Variable-stator feedback,
Ignition leads,
Igniter plugs,
Fan tach generator,
Fan tach generator cable,
Thermocouple harness,
Thermocouple flex leads,
Main electrical harness,
Anti-icing valve.

The organization of the MADAR system is shown in Figure 12.10 and consists of two subsystems one of which is automatic and the other manual which share the printout unit (POU) and the maintenance data recorder (MDR) for data interface with the flight or ground crew. The automatic portion of the subsystem operates whenever its power switch is on. The digital computer (DCOMP) is an eight thousand word, 24 bit per word, general purpose machine. All MADAR stored programmes are contained in this machine.

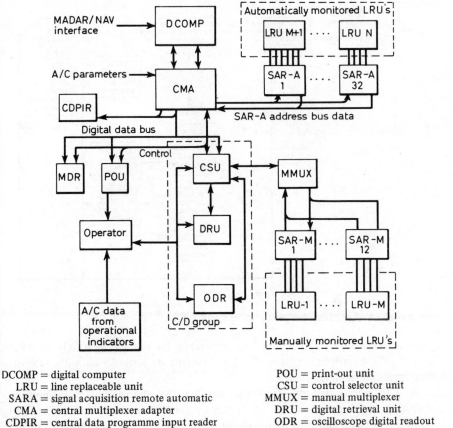

DCOMP = digital computer
LRU = line replaceable unit
SARA = signal acquisition remote automatic
CMA = central multiplexer adapter
CDPIR = central data programme input reader
MDR = magnetic-tape data recorder

POU = print-out unit
CSU = control selector unit
MMUX = manual multiplexer
DRU = digital retrieval unit
ODR = oscilloscope digital readout
C/D group = control/display group

Figure 12.10 *Organizational layout of the MADAR malfunction detection analysis and recording system [32]*

12.9.6 Ignition system truth table (MADARS)

A typical ignition sytem monitoring routine used with the malfunction detection analysis recording system (MADARS– was developed from a truth table for voltage and current limits furnished by the ignition system manufacturer. A typical truth table with appropriate messages is shown in Figure 12.11.

12.9.7 B.E.A. – manual engine data monitoring

British Airways (European Division) formerly B.E.A. have found it valuable to employ manual in-flight logged data to determine engine performance trends and to use these as a means of advance warning of engine failure or malfunctioning. This technique has been adopted as a routine since August 1968. The original objectives were to back up their system of metallic particle monitoring to detect and identify failures and malfunctions in areas other than 'oil-wetted', e.g. turbine and com-

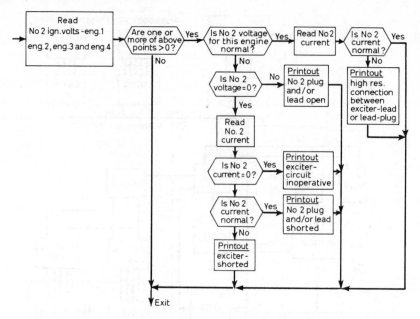

Figure 12.11 *MADARS ignition system truth table*

pressor components, and to use this experience for developing a fully automatic engine data monitoring system for later aircraft in conjunction with automatic data recording equipment.

A perforated tear-off proforma attached to the aircraft supplementary navigation log, is used to enter the relevant parameters once on each flight during stabilized cruise conditions with anticing systems not in use. The start-up or shut-down turbine vibration indication peak reading is recorded because, on the Spey engine, excessive vibration peaking under these conditions is usually indicative of turbine blade failure. In addition start-up turbine gas temperature is recorded, since excessive temperature or repeated higher-than-normal temperatures during the start cycle when rotation and mass airflow are low, can indicate a malfunction which will subject the turbine to thermal stresses and result in early failure or reduced fatigue life.

The completed data strip is telexed directly to the London Airport Engineering Base computer facility where it is processed and a daily printout sent to the Early Failure Detection Centre.

The computer programme converts this data into non-dimensional form and compares the engine performance with that of the ideal or 'brochure' engine, i.e. the manufacturer's test bed performance curves, at the same thrust or EPR (exhaust pressure ratio), then plots the percentage differences or 'deltas'. The first ten 'deltas' are arithmetically averaged, to establish the signature level of the particular engine, subsequent deltas show deviations from this datum. A mathematical smoothing technique is applied to eliminate wild scatter or 'noise' and the smoothed values plotted to give a trend analysis plot of the last 25 flights. In addition a plot is produced of vibration readings directly from the raw data.

Increasing or decreasing trends, away from the datum, are studied to interpret the cause and initiate appropriate actions. A supplementary computer programme produces a printout, twice a week, to include all relevant engine and associated component changes, and this information is carefully weighed during trend analysis.

The success of this technique has enabled detection of incipient failures and system malfunctions well in advance of meaningful symptoms becoming apparent. Linked with metallic particle monitoring the manual in-flight monitoring has substantially contributed to safety, reliability and reduction in repair costs.

12.10 Condition monitoring – generating plant

The net effect of individual component reliabilities generally governs the availability of the plant. Components normally fail through some deficiency, (a) in the design of the plant or component, (b) in the manufacture or erection stage, (c) during operation. Weeks and Hodges [33] reported some of the component failures in generating plant as operated by the Central Electricity Generating Board. A schedule of typical techniques is given in Table 12.5.

12.11 Automotive diagnostic equipment

A significant new breed of test and diagnostic equipment has been developed in recent years, ranging from hand-held units to huge consoles, from 'plug in' diagnostics to equipment which offers similar facilities without any pre-wiring.

A review of existing facilities [34] referred to the Volkswagen computerized diagnostic system, Renault inspection system, the BMW diagnostics system and a diesel diagnostic service offered by Mobil Oil of Belgium. The development of such facilities is spurred by national demands for increased vehicle safety. The known requirements include:

Germany (West)	lights; steering; tyres; suspension; powertrain; body; mirrors; exhaust.
Japan	emission
U.K.	brakes; steering; lights; seat belts. tyres; bodywork.
France	brakes; powertrains; steering.
Sweden	exhaust; brakes; lights; steering; wheels; chassis; body suspension; shock absorbers; engine.
U.S.S.R.	brakes; steering; engine; lights; windscreen wipers.

12.11.1 Military vehicles diagnostic equipment

Diagnostic equipment to aid the more effective servicing of combat and transport vehicles has been developed by the U.S. Army Tank-Automotive Command, Warren

Table 12.5 *Some condition monitoring techniques – electrical power stations*

Plant type	Condition monitoring technique
Rotating machinery in general	(a) Vibration monitoring (b) Performance analysis
Electrical rotating & stationary plant	(a) Insulation condition
Steam turbines	(a) Vibration monitoring (b) Performance analysis
Gas turbines	(a) Comprehensive performance trend analysis (b) Spectrographic analysis of lub. oils (c) Acoustic and vibration monitoring
Alternators	(a) Insulation conditions (b) Leak monitoring from coolant system (c) Vibration monitoring (d) Performance analysis
Steam condensers	(a) Leak monitoring
Boilers & boiler plant	(a) Flame monitoring (b) Mill fire detection (c) Leak monitoring (d) Smoke & flue gas composition monitoring
Reactors	(a) Burst fuel element detection (b) Graphite monitoring (c) Boiler tube leak detection (d) Corrosion monitoring
Transformers	(a) Bucholz gas analysis (b) Transformer oil monitoring (c) Insulation monitoring
Transmission system	(a) Detection and recording of fast transients (b) Infra-red detection of faulty joints

Table 12.6 Typical measurements

Intake manifold pressure	Oil temperature
Engine speed	Starter current
Fuel rate	Battery voltage
Ignition waveform	Battery current
Airflow	Crankshaft position
Exhaust analysis	Fuel pressure
Injector pump pressure	Fuel temperature
Exhaust blowby	Throttle position
Cylinder power drop	Cylinder head vibration
Oil pressure	Water temperature

Table 12.7 *Typical diesel engine, measurements*

Measurement	Sensor location	Engine system analysis
Temperature (26 Points)	(12) Exhaust ports (pipe plug)	Power and injectors lubrication and safety monitor lubrication
	(1) Oil before engine oil gallery	
	(2) Oil cooling line	
	(1) Fuel supply line	Correction factor lubrication
	(1) Fuel return line	and turbo-supercharger
	(2) Oil return line	
	(2) Exhaust outlet	Turbo-supercharger M
	(2) Intake manifold	Turbo-supercharger
	(2) Intake airflow	Correction factor
	(1) Ambient	Correction factor
Pressure (9 points)	(2) Intake manifold (after turbo)	Turbo-supercharger and intake system lubrication and
	(1) Engine oil galleries	safety
	(1) Engine oil	Monitor lubrication
	(2) Ambient pressure	Correction factor
	(2) Injection pump	Injection pump, injector, timing
	(1) Fuel pump	Fuel
Vibration (18 points)	(12) Rocker box (valve) cover	Valve intake and exhaust,
	(6) Engine block vibration	adjustment and timing internal malfunctions (bearings etc.) isolator
Flows (7 points)	(1) Fuel supply line	Intake and fuel
	(1) Fuel return line	Intake and fuel
	(1) Blowby (breathing system)	Engine wear lubrication
	(2) Engine oil	
	(2) Air-intake (supercharger)	Intake
Current (1 point)	Starter	Starter and engine compression
Position (1 point)	Crankshaft	Time base
Speed (3 points)	(2) Engine cooling fan	Cooling system
	(1) Engine under test	Basic control
Torque (1 point)	(1) Dynamometer (test cell 2)	Basic control
Throttle position (1 point)	(1) Engine under test	Basic control
Voltage (1 point)	Magneto primary	Ignition
	starter	Ignition

Michigan in collaboration with Frankfort Arsenal, Philadelphia, Pa. (785). Three generations of USATACOM diagnostics were produced for this project by Dyna-sciences Corp. Chatsworth, California, namely:

(1) First generation: A capability for diagnosing a complete vehicle by using a separate transducer kit and diagnostic computer;

(2) Second generation: In which the transducers are incorporated in the vehicle as a permanent installation;

(3) Third generation: With both transducers and computer built into the vehicle as a permanent installation.

The transducers are used to measure parameters such as those in Table 12.6.

The manner in which these measurements were obtained and the diagnostic significance attributed to each is listed in Table 12.7.

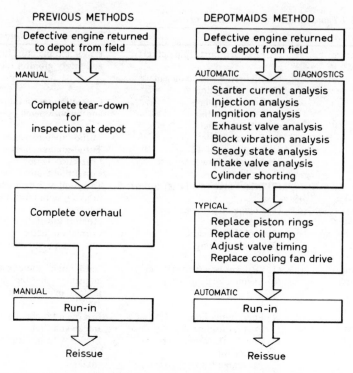

Figure 12.12 *Comparison between depot overhaul procedures*

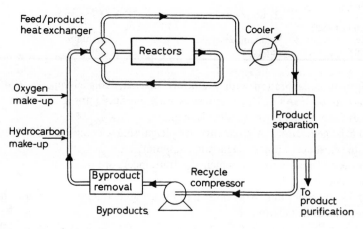

Figure 12.13 *Process diagram*

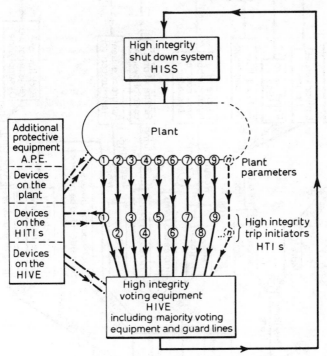

Figure 12.14 *High integrity protective system (HIPS)*

One of the techniques developed in connection with this equipment known as DEPOMAIDS is used at the U.S. Army Letterkenny Depot, Chambersbing, Pa. has been to use two test cells in order to provide a depot-based diagnostic facility. The left cell has provision for monitoring the engine or transmission and absorbing power with two dynamometers. Two dynamometers are used because the combat vehicle cross-drive transmission has two output shafts. The right cell is for engine testing without the transmission and requires the engine to be operational. It contains only one absorption dynamometer. In front of the test cell the control room houses the computer-controlled diagnostic equipment automatically diagnosed before and after the overhaul. A comparison between this technique and the previous method is shown in Figure 12.12.

The third generation with in-vehicle computer equipment is intended as a relatively simple installation capable of indicating (i) a 'go' condition, (ii) a marginal condition which requires maintenance action, (iii) an immediate warning alarm which demands immediate action to prevent a catastrophic failure.

12.12 Systematic fault monitor selection

A systematic technique for choosing fault monitors (trip-initiating parameters) was reported by Stewart and Hensley involving the use of a 'family tree' of fault

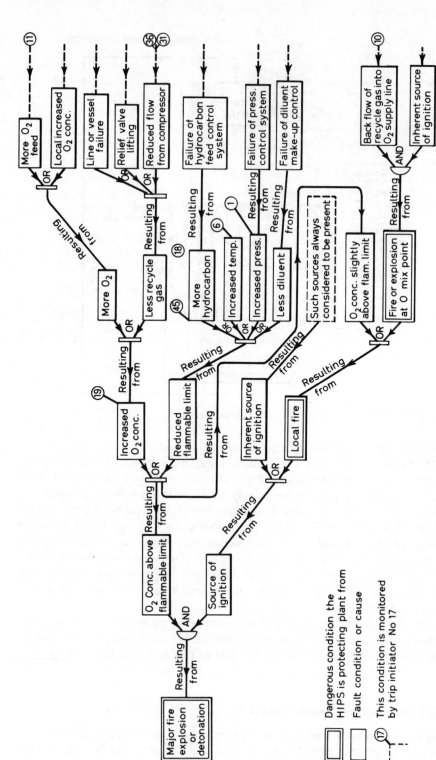

Figure 12.15 *Logic diagram for the design of HIPS*

conditions and causes which could cause a major hazard in a chemical process plant. The process diagram is shown in Figure 12.13 and the high integrity protection system (H.I.P.S.) in Figure 12.14.

A logic diagram was used, a small part of which is shown in Figure 12.15. A parameter was selected to be a trip initiator for each 'line' of logic and it was assumed that there could always be a source of ignition available somewhere in the system. To assist in the design and in the selection of these parameters specialist disciplines were involved:

(1) Electrical engineers designed the majority voting equipment;

(2) Chemists defined and analysed the potentially dangerous conditions, their origins and manifestations;

(3) Chemical engineers defined the effects of those manifestations on parts of the plant;

(4) Mechanical engineers determined the behaviour of the pressure vessels, machines and components, and the effects of the controls and protective devices;

(5) Mathematicians analysed failure probabilities.

As a consequence of the logic diagram analysis a diversity of trip initiating parameters were introduced to provide a broad spectrum of failure sensors to present to the majority voting system.

References

1 Hopkins, R.C. and Chaveret, L., (1971), 'Analysis of results obtained with the Dolabella computer', Association Technique Maritime et Aeronautique.

2 ————— (1969), 'Computerised navigation in one of a series of refrigerated cargo liners from Dunkirk', *Shipbuilding and Shipping Record*, **113**, 7 March 319.

3 ————— (1967), 'Polar Ecuador — lead ship of six refrigerated cargo vessels from Hamburg', *Shipbuilding & Shipping Record*, **111**, 7 December 805.

4 Sitzia, J. and Sartirana, A. (1973), 'Computerised automation of the container-ship *MODIANA*', *Ship Operation Automation, IFAC, IFIP Symposium*, Oslo, July.

5 ————— (1970), '*SEIKI MARU*: extensively automated and Computer controlled tanker', *Shipping World and Shipbuilder*, November.

6 Kobayashi, O. (1972), 'Outline of development of advanced integrated control system for ships', *Sempaku* (Shipping) 4/5, May.

7 Ooe, Y. and Ozawa, N. (1972), 'Fully automated ship:*KINKO MARU*', *Nippon Kokan Technical Report — Overseas*, June.

8 Saito, H. and Okano, S. (1973), 'Computerised super-automation system of the *TOTTORI MARU*' *Ship Operation Automation, IFAC, IFIP Symposium*, Oslo, July.

9 Tamura, I. (1973), 'Some experiences in cargo oil and water ballast handling by shipboard computer', *Ship Operation Automation, IFAC, IFIP Symposium*, Oslo, July.

10 ——————(1972), 'Another super-automated 0/0 carrier delivered', *Zosen*, November.

11 ——————(1972), 'Computer system in the Queen Elizabeth II', *B.S.R.A. Research Report* NS 322.

12 Balme, K.P. (1969), 'Running a ship by computer', *Marine Engineer & Naval Architect,* January.

13 Hatfield, M. (1974), 'User experience of a computer based watchkeeping and control system', Institute of Marine Engineers/Nautical Institute, Joint Automation Conference.

14 Barna, J.D. (1971), 'Study to improve maritime transport through aerospace electronics', Applied Information Industries report for U.S. Maritime Office of Research & Development, June.

15 ——————(1974), 'SCAN in-water hull inspection robot scanner', *Marine Week*, 1, No. 25, 29.

16 ——————(1975), 'Hull stress monitor for VLCCs', *Marine Engineering Review*, June, 66.

17 Monceaux, F. (1974), 'Computer use for monitoring and maintenance improvements of ship propulsive plants', *Bulletin Technique du Bureau Veritas*, June 42—48.

19 Pauer, W. (1974), 'Engine cardiogram; M.A.N. engine diagnosis for marine diesels', *Ship Repair & Maintenance International*, September 43—48.

20 Langballe, M. (1972), 'The computer system for engine monitoring and control aboard *HOEGH MULTINA*', *Shipping World & Shipbuilder*, April 486—488.

21 Sandtor, H. (1973), 'Datatrend; a computerised system for engine condition monitoring and predictive maintenance of large bore diesel engines', *Ship Operation Automation, IFAC, IFIP Symposium,* Oslo, July.

22 Bakke, T. (1975), 'Diesel engine design with a view to reduced maintenance costs', *Trans. I. Mar. E*, 87.

23 Anderson, N.D. (1974), 'Condition monitoring for main propulsion machinery', *Marine Engineering Review,* October.

24 Corben, F. (1972), 'Detection of a defective bearing using the vibration monitoring system', Report LC/501/3909/1a, Electrical Laboratory, Admiralty Engineering Laboratories, West Drayton, Middleses.

25 Glew, C.A.W. (1974), 'The effectiveness of vibration analysis as a maintenance tool', *Trans. I. Mar. E.* 86, 29—50.

26 Waggoner, C.A. and Dominique, H.P. (1968), 'Progress report on the RCN(WC) spectrometric oil analysis program', Materials Report 68-A, Defence Research Establishment Pacific, Victoria B.C. Canada, January.

27 ——————(1974), 'Trouble-shooting marine control systems', *Marine Engineer Review*, November.

28 Davies, A.E. and Newman, H.L. 'Diagnostics and engine condition monitoring — the relative role of engine monitoring programmes during development and service phases', AGARD, 43rd P.E.P. meeting.

29 Feltham, R.G. (1970), 'The role of flight recording in aircraft accident investigation and accident prevention', *J. R. Aero. Soc,* 74, No. 7 573–576.

30. Kruckenberg, H.D. (1972), 'Design and testing of the American Airlines prototype B-747 AID system', *J. Aircraft (U.S.A.),* 9, No. 4 April.

31 Pauliny, S. (1970), 'American Airline maintenance recorder experience', Paper 700317, National Air Transportation Meeting, SAE, April.

32 Horne, E.W., Price, L.R. and Edwards, M.S. (1970), 'Programmed engine maintenance – CSA malfunction detection analysis and recording system (MADRAS)', SAE Paper 700820, October.

33 Weeks, R.J. and Hodges, N.W. (1969–70), 'Component failure in generating plant', Conf. Safety and Failure of Components, *Proc. I. Mech. E,* 84, 3B.

34 Collacott, R.A. (1975), 'Diagnostic plug-ins', *Motor Management,* 10, No. 5, September.

35 Stewart, R.M. and Hensley, G. (1971), 'High integrity protective systems on hazardous chemical plants', Report SRS/COLL/303/2 UKAEA, Systems Reliability Service, Culcheth, Warrington, Lancs. May.

13 Fault analysis planning and system availability

13.1 Introduction

The ability of operators to carry out preventive and corrective maintenance involves a balance between reliability and maintainability associated with the identification of hazard systems together with the provisioning of adequate spares. The assessment of such conditions is a design function involving the study of fault trees with the use of a failure modes, effects and criticality analysis (FMECA), itself an extension of the techniques of quality assurance [1].

13.1.1 Failure modes, effects and criticality analysis (FMECA)

In an illustration of the FMECA principle, Johnston [3], considered a ship's auxiliary system consisting of a piping network taking the form of a main which supplies various users via branches and which has a number of pumps (or compressors) to give a degree of duplication (or 'redundancy') in the equipment producing the fluid flow in the network.

As the flow chart, Figure 13.1 indicates, the first step in the analysis is to annotate the system's arrangement drawing. This is done by allocating a number to each valve to provide reference points to identify the bounds of each functional sub-system. A check list is prepared to ensure that each sub-system is examined in turn, recording the result of the examination, namely — 'sub-system satisfactory' or 'modification proposed'.

13.1.2 Failure effects assessment (FEA)

In a full FMECA it would be necessary to define various failure criteria but, for a FEA of a piping system, a general and somewhat vague class of 'anything causing a sub-system not to function properly' is sufficient. This definition therefore covers fracture of a pipe, excessive leakage at a joint, or seizure of a valve in its open or shut positions. It also provides the basis for the failure effects sheet, which should contain space for comment under the following headings:

  (a) Bounds of sub-system,
  (b) Effect on system of sub-system failure,
  (c) Other systems affected when sub-system fails,
  (d) Proposed modification to sub-system,
  (e) Remarks.

Quality Assurance recommends the following laws as guidelines for the analysis of a piping network.

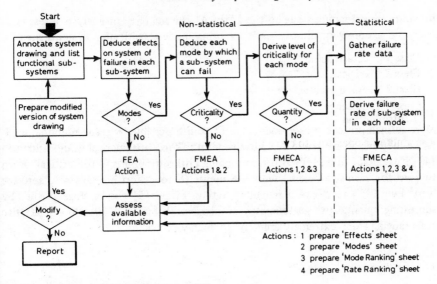

Figure 13.1 *Typical FMECA system [3]*

(1) All branches from the main should have isolating valves at their junctions with the main to better preserve its integrity;

(2) All other isolating valves should be sited at those junctions where they will be most effective in minimizing the effects of pipe failure;

(3) The fluid flow pattern created by the isolation of a failed sub-system should be analysed and a check made that any specified flow limitations are not exceeded;

(4) All essential equipment should be supplied direct from the main;

(5) Where essential equipment is duplicated for standby purposes, at least one of them should have an independent source of supply and discharge;

(6) Supplies to (and discharges from) essential equipment should be kept separate from those of non-essential equipments;

(7) Non-essential users should be grouped into branch networks to minimize the number of connexions taken direct from the main.

13.1.3 Critical areas assessment

The flow chart shows that the next depth of analysis is the FMEA. This, by its study of the modes of failure, can direct the designer's attention to potentially critical areas. It is, however, doubtful whether such a study would be worthwhile for the marine piping network of the example. There are very few modes by which it can fail and, to be meaningful, they need to be allied to failure rate data. A mode analysis is more useful when it is used to assess the design of an automatic control system or of a piece of machinery consisting of many discrete parts, either of which can fail or degrade via a number of modes.

While some good can come from stopping at the FMEA level, its real value is in providing the stepping stone to the FMECA. The majority of analysts assess criticality on a purely qualitative basis by using mode classification systems such as:

Class 1　catastrophic;
Class 2　critical;
Class 3　non-critical;
Class 4　nuisance.

More refined criticality assessments are of course possible by using pseudo-quantitative methods which relate a qualitative term to some arithmetical value. Anderson [4] gives details of a method for combining criticality points (4 for 'critical' down to zero for 'non-significant') with life points (8 for 'short life' down to zero for 'very long life') to derive a reliability rating. Eisner [5] goes a stage further by suggesting 'quantitative guesstimates' of failure probabilities relative to a qualitative scale ranging from 'extremely unlikely' to 'highly probable'.

13.1.4 Failure mode analysis

The close inter-relation between many components in a complex assembly requires high individual component reliability to ensure satisfactory operation of the whole system.

13.1.4.1 Product reliability

Two classes of reliability are:

Inherent reliability: maximum potential reliability of the product. It is the inherent maximum which can be attained if everything were made and functioned as designed – it is a function of design;

Achievable reliability: actual reliability in practice. It is less than the inherent reliability as it includes manufacturing and assembly errors in addition to design imperfections – it is a function of quality control.

13.1.4.2 Fault tree method

An established design is considered and the failure of each component assumed. The interaction of these faults is then analysed.

As an example of this method Green [6] systems engineer, Westinghouse Nuclear Energy Corp, Pittsburg, Pa., U.S.A. included a portion of a fault tree for the failure mode analysis of a car engine (Figure 13.2).

Fault tree methods of failure mode analysis may be used as a design technique to assess inherent reliability, they may be used as a source of clues when a mechanism fails to work correctly.

13.1.4.3 Potential fault analysis

Since a design must be substantially complete before the fault tree can be

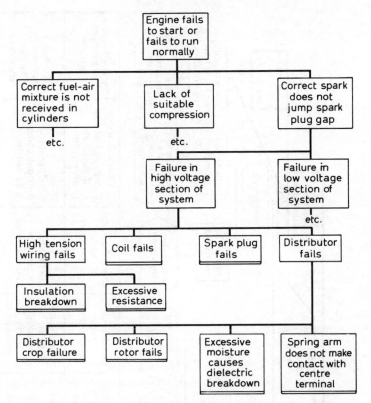

Figure 13.2 *Portion of fault tree for failure mode analysis of car engine*

constructed in detail the 'potential fault analysis' method has been developed as a project control method. The objective of this method is to consider things which could go wrong during production, installation and operation and develop remedies.

Figure 13.3 shows part of a potential fault analysis chart developed for a control drive mechanism for a nuclear power reactor.

The four basic problems:

'Will not be delivered on schedule',
'Cannot be installed',
'Will not operate as designed',
'Will not last for design life'.

These apply to most mechanical components and form the framework for the chart.

13.1.4.4 Potential fault analysis questionaire

A design review procedure mentioned by Green involves the use of a questionaire prepared by some reliability engineering group other than the design group.

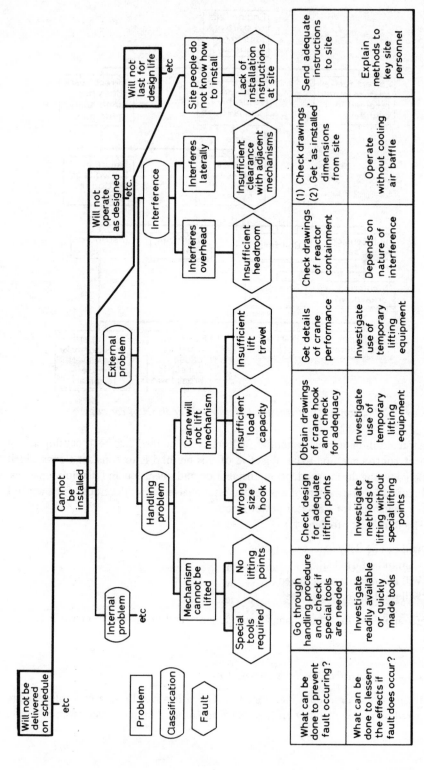

Figure 13.3 *Portion of potential fault analysis chart for control rod drive mechanism*

Design review categories cover:

(1) Design concept,	(12) Ageing
(2) Materials selection,	(13) Interfaces,
(3) Manufacturing,	(14) Stresses,
(4) Bearings,	(15) Thermal effects,
(5) Seals,	(16) Environment,
(6) Rotating parts,	(17) Testing,
(7) Prime movers,	(18) Instrumentation,
(8) Pressure vessels,	(19) Safety considerations,
(9) Surface finishes,	(20) Maintenance,
(10) Dimensional factors,	(21) Operation.
(11) Wear,	

Typical questions might be:

(4) Bearings
Identify any bearings in the device.
For each of these:
> What type of bearing is it?
> What alternatives were considered?
> Why were the alternatives not used?
> Identify the direction of loading.
> What are the effects of misalignment?
> How is the bearing lubricated?
> How is the bearing cooled?
> What is there to prevent debris, dust, grit, etc., from entering the bearing?
> What is the effect of such dust, grit etc.?
> What self-aligning features are present in radial bearings?
> What provisions are made to prevent brinelling of antifriction bearings during shipping?
> Are non-inflammable lubricants used in hazardous environments?

13.1.5 Fault tree analysis

Failure analysis of complex systems may be simplified by the use of fault trees. The method is considered particularly useful as a supplementary analysis to a failure modes and effects analysis. However, fault tree diagrams may also be very useful in the training of operating personnel.

The purpose of a fault tree analysis is usually either:

(1) To determine the possible causes (single or multiple failures) of an end effect as measured at the functional output of a component or a system; or

(2) To determine the possible end effects of a failure causing mutually exclusive sequences of events.

In the first case the approach implies working 'backwards' from the undesired event to its causes, and in the other case 'forwards', i.e. from a cause to its possible end effects.

A fault tree may be considered as a particular type of an 'event logic' diagram. An event logic diagram may be defined as a logical representation of the interrelationship of events. The diagram is based on events interconnected by logic gate determining the relation between 'input events' and 'output events' [7]. A fault tree analysis is a logical event diagram describing the cause—effect relationship of failures as shown in Figure 13.4. The construction usually starts with the definition of the undesired end event, i.e. a plant or system failure. The possible causes are then defined and connected to the 'top' event by logic gates. The procedure is then repeated for each of the causes, and causes of causes, etc., until the desired level of detail is obtained for each particular cause, i.e. an initial and independent event which may be defined as a basic event [8]. The backwards analysis is useful for determining whether all conceivable causes have been considered in a Failure Modes and Effects Analysis, and to identify failure combinations which would otherwise not have been discovered.

13.1.6 Fault logic – lubrication system

A logic chart based on the performance monitoring involved in the diagnosis of a gas turbine lubricating system is shown in Figure 13.5. This analysis depends upon direct measurement of pressures, temperatures, and chip detectors. A logical reasoning process is pursued in which the sensor outputs are sequentially compared to derive conclusions regarding engine condition [9]. The programme is organized to make successful inspections and draw conclusions through logical fault isolation of one type of fault at a time. Complete analysis involves running the entire programme which embraces all of the lubrication system including such components as the bearings, sump, oil breather system, and scavenger pump which are not shown in this simplified diagram.

13.1.7 Fault logic – fuel control system

The use of performance trends in conjunction with a logic chart as a basis for the EECAS [9] is shown in Figure 13.6. This demonstrates diagnostics by interrelationship in this interfacing system and the use of transient condition data with associated rates of time change. For instance, one possible approach would be to store in a diagnostic computer a complete set of curves defining the controller operation. This would be an economically costly trade off. Consequently, the method of logical fault combinational isolation is used because it will enable adequate diagnostics for approximately 80% of possible fuel control problems with very little additional instrumentation cost.

Deterioration of the fuel controller will be evident through four engine malfunctions : (1) too slow or too fast acceleration, (2) speed oscillations, (3) stall resulting in deceleration, (4) over-high exhaust gas temperature.

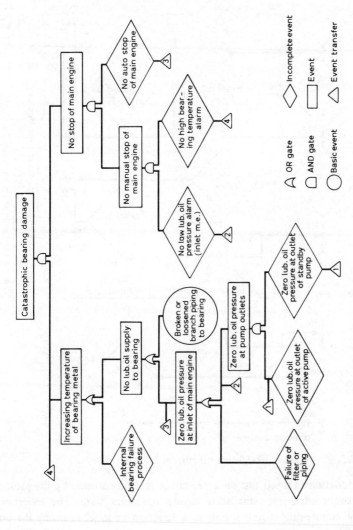

Figure 13.4 Typical fault tree for a bearing failure analysis [10]

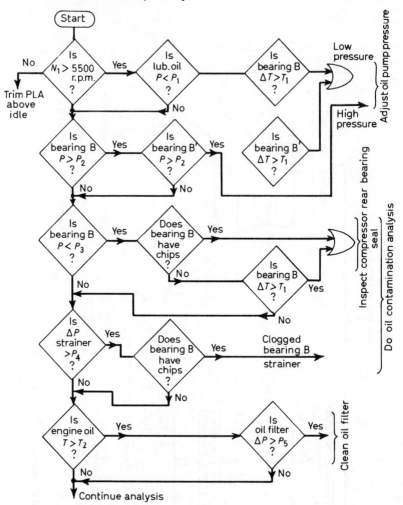

(Assume bearings B and B' at mid and rear engine
shaft locations respectively)

Figure 13.5 *Logic diagram lubricating oil system for a gas turbine*

For malfunction (1), too fast an acceleration is demonstrated by the prolonged
speed overshoot of a pre-established engine operating state. The overshoot is
detected by monitoring engine speeds as a function of time, knowing the power
lever angle (PLA). A data window is also provided such that the analysis will not be
performed unless PLA rate and amount of change falls within specified bands required for meaningful results.

For malfunction (2), engine oscillations can be detected by monitoring the
changes in high rotor speed and compressor discharge pressure. The autopilot
must be off to eliminate any problems in this component and the power lever
angle (PLA) should be stabilized for a sufficiently long period of time.

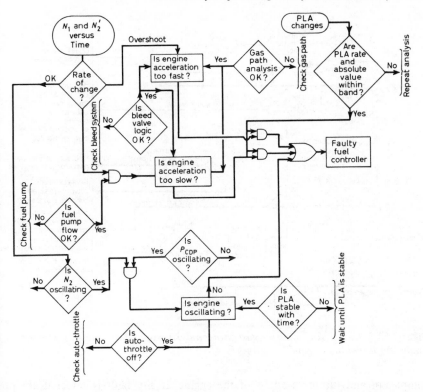

Figure 13.6 *Performance trend logic diagram for fuel controller based on gas thermodynamics modelling*

Although not shown in Figure 13.6 the logic for cases (3) and (4) would complete the fuel controller diagnostics. Deceleration can be sensed through an engine surge which will cause a high vibration level, a decrease in pressure ratio below that requested by the PLA setting, and exhaust gas temperature overlimits. Overtemperature, due to a faulty fuel controller, would be established through measurements of exhaust gas temperature and logical comparison to discern whether the gas path analysis, bleed valve logic, engine pressure ratio, and inlet guide vane scheduling (if applicable) were all within limits.

13.1.8 Turbine gas path – diagnostic logic

An example of logic diagnostics using EECAS techniques with thermodynamic modelling, applied to a single spool turbine engine was given by Moses [9]. Having already established the relative performance shifts in independent parameters, the analysis proceeds by logic comparisons to correlate the implications of these shifts as shown in Figure 13.7. (K represents allowable values of performance limit shifts.)

It then proceeds to analyse and correlate the out-of-tolerance performance shifts to logically isolate a component fault. This particular limited example checks the

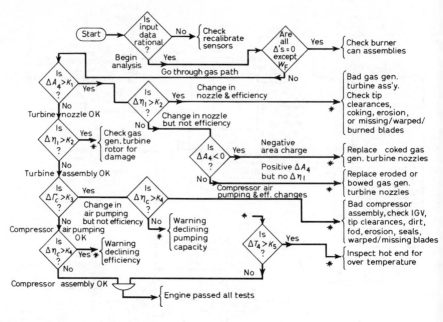

Figure 13.7 *Diagnostic logic – gas path analysis applied to the compressor of a single spool turbine engine*

compressor and turbine sections of the engine. If any fault is detected, the programme also checks turbine inlet overtemperature and will request a hot end inspection in an affirmative case. The matrix formulation and solution is precomputed once for any EECAS installation and then stored in the computer along with the baseline values. Routine diagnostic cycles are then reiterated using this initial information.

13.1.9 Failure/repair decision analysis

Mechanical failures exhibit many failure modes with differing effects on the functional output as measured at the part-component-system level and also on the required maintenance action. Failures due to wear are time-dependent and are assumed to be of a gradual type. Chance failures are sudden and not expected from prior observations.

Failures may be defined in terms of their effect, i.e. whether there is:

(a) complete lack of required function,

(b) partial lack of required function.

These will influence the repair situation according to whether a repair is necessary but may be postponed.

Analysis of the situation may then follow a study of the repairability. In the marine situation, Mathieson [10] quoted that the availability of spare parts and

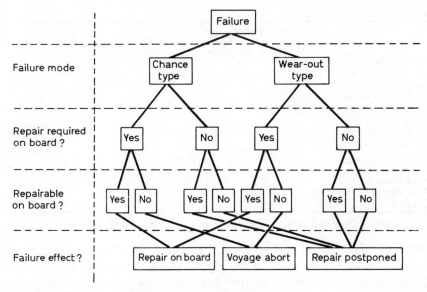

Figure 13.8 *Sequence of failure decision logic [10]*

tools, the qualification of the maintenance personnel, etc., are factors determining the repairability. In this connection, the time elapsing between intervention and the occurrence of the failure may be a decisive factor. The resulting logic model for such a situation is shown in Figure 13.8.

13.1.10 Criticality analysis

Before carrying out a reliability assessment of a system it is useful to list the various sub-systems and consider their critical importance. Critical importance or 'criticality' may be taken as a measurement of the essentiality of an item, the support it gives to the rest of the system, the effect of its unreliability and the influence it has on safety. The technique of criticality rating was described in full by Venton and Harvey [11].

The first step in criticality analysis is to form a dependency chart such as that for ship's machinery, Table 13.1 [12]. This chart, in a rather simple form, lists all the functions or systems being considered as part of the study. The example shown includes the systems for the main engine, the transmission, oil fuel, compressed air, electrical generation and distribution, steering gear, deck machinery, sea water, bilge and ballast, ventilation, exhaust, control and finally steam, for a hypothetical ship — in this case, a trawler. The chart shows at a glance the interdependencies between these systems.

The next step is the determination of the criticality rank for various phases of the mission. Considering the case of the trawler with the passage phases listed as in Table 13.2.

Table 13.1 *Dependency chart*

System	(1)	(2)	Dependency* (3)	(4)	(5)	(6)	(7) (8)	(9)	(10)	(11)	(12)	(13)
(1) Main engine			C	C			C		C	C	C	
(2) Transmission												
(3) Oil fuel					C							
(4) Compressed air					C			C				
(5) Electrical generation			C	C				C	C			
(6) Steering gear					C							
(7) Deck machinery					C							C
(8) Sea water					C							
(9) Bilge and ballast					C							
(10) Ventilation					C							
(11) Exhaust												
(12) Controls					C							
(13) Steam			C						C	C		

* The numbers correspond to the systems as numbered on the left-hand column

The criticality for the total installation is combined in a final table (Table 13.3) which has been prepared on the assumption that this particular vessel has two phases, the passage phase and the trawling phase, each of which are given a 0.5 phase loading.

If reliability analysis resources are scarce, some judgement has to be made as to which systems are going to be studied, and in which order. They can be guessed at random or some sort of systematic, even though subjective, assessment can be made to give the order in which they are to be analysed. The procedure given above is one way of arriving at a solution to this problem before an analysis of individual system reliabilities has been carried out.

13.1.11 Safety assessment

A practical example of the use of failure data to assess the possible hazard (or safety) provided by a design was reported by Stewart and Hensley [13]. The objectives of the assessment were to determine:

(1) All reasonable fault conditions which could possibly lead to hazardous conditions;

(2) The ability of a proposed high integrity protective system to prevent hazardous conditions resulting;

(3) The probable rate at which hazardous conditions might still occur and to compare this this a target specification of 1 in 33 000 years.

Where a protective system is used to minimize hazards there is a relationship between the probability of failure of the system, i.e. its fractional dead time (f.d.t.), the frequency with which it is called upon to operate (the demand rate) and the rate at which the undesirable plant hazard will occur. This relationship (which applies to each fault condition separately) can be expressed as follows:

$$\text{Hazard rate} = \text{Demand rate} \times \text{Protective system f.d.t.}$$

Table 13.2 Phase criticality chart

Essentiality (a)	(b)	Essentiality rating (c)	Support (d)	Support rating (e)	Unreliability rating (f)	Criticality rating (g)	Criticality index (h)	Rank (i)
(1)	✓	10		1	12	121	50.4	3
(2)	✓	10		1	2	22	9.2	10
(3)	✓	10	1, 5, 13	6	10	160	66.6	2
(4)	✓	4	1, 5	4	14	112	46.6	4
(5)	✓	10	3, 4, 6, 7, 8, 9	10	12	240	100.0	1
(6)	✓	10	10, 12	1	1	11	4.6	12
(7)	✗	1		1	4	8	3.3	13
(8)	✗	2	1	2	8	32	13.3	7
(9)	✗	2		1	8	24	10.0	9
(10)	✓	10	1, 4, 5, 13,	8	6	108	45.0	5
(11)	✓	10	1, 5, 13	6	1	16	6.7	11
(12)	✓	10	1	2	4	48	20.0	6
(13)	✗	2	7	2	8	32	17.3	7

(a): system being analysed;

(b): essentiality of each system assessed on a 'yes-essential', 'no', 'standby' condition;

(c): an arbitrary rating of the essentiality of the system based on a subjective 1 (inessential) to 10 (vital) scale;

(d): data from Table 13.1 to show which other sub-systems are interdependent, the numbers being those assigned to identify the subsystems;

(e): each particular system is given a support rating based on the information given in the previous column. This rating again varies from 1 to 10;

(f): system unreliability rating (subjective) varying from 1 (lowest reliability) to 20 (greatest reliability);

(g): criticality rating, a combination of the factors derived from columns c, e and f. Thus for (3) oil fuel system the criticality rating for (g) is, i.e. $(10 + 6) \times 10 = 160$;

(h): the criticality index in relation to the whole system as a sum of the sub-systems. Thus the sub-system with the largest criticality rating in (g) is (5) electrical generation with a rating of 240. Hence the relative criticality index of (3) oil fuel system is given by $(160 \div 240) \times 100 = 66.6\%$;

(i): the order of critical rank according to the highest percentage in (h). Thus the system (5) electrical generation is the highest rank (No. 1) with 100% total criticality rating followed by system (3) oil fuel system with 66.6% criticality rating down to deck machinery (No. 13) with only 3.3% criticality rating.

Table 13.3 *Total criticality chart*

(a)	Passage (0.5 phase loading) (b)	Trawling (0.5 phase loading) (c)	Initial criticality index (d)	Safety rating (e)	Total criticality rating (f)	Total criticality index (g)	Rank (h)
(1)	61	82	143	8	1144	47.7	2
(2)	11	17	28	8	224	9.3	7
(3)	80	80	160	6	960	38.4	3
(4)	56	56	112	1	112	4.7	11
(5)	120	120	240	10	2400	100.0	1
(6)	6	12	18	10	180	7.5	9
(7)	4	24	24	1	28	1.2	13
(8)	16	16	32	8	256	10.7	6
(9)	12	12	24	7	168	7.0	10
(10)	54	54	108	4	432	18.0	5
(11)	8	8	16	3	48	2.0	12
(12)	24	24	48	4	192	8.0	8
(13)	16	32	48	10	480	20.0	4

(b) and (c): criticality ratings for the individual phases loaded by a factor of 0.5;

(d): summation of the phase criticality indices to give the final criticality index;

(e): safety rating allocated arbitrarily from 1 to 10 according to the essentiality of their role in the safety of the vessel;

(f): product of (d) × (e);

(g):(f) expressed in terms of a percentage of the maximum value. Thus in this particular column the maximum value in (f) is 2400 so that all items are obtained as a percentage of this value e.g. (2) Transmission = (224 ÷ 2400) × 100 = 9.3%;

(h): ranking order of the items based on the Total criticality index, (g). This indicates the order in which the system should be analysed if reliability resolves are scarce; it also shows which systems are critical in the operation of the vessel.

The demand rate is the rate at which fault conditions (that could result in the hazard materializing) are expected to occur. The fault conditions themselves need to be identified and their transient behaviour studied, possibly with the use of a logic diagram. The protective system must be designed to protect against faults and the acheivement of this function carefully assessed. This is a measure of the reliability of the system and although it is often taken to mean the likelihood of failure due to component failures under terminal conditions (such as resistors going open or short circuit), real system reliability is dependent on many other factors including accuracy, response, maintenance, operator error, etc. Before the numerical value of system reliability is calculated it should be shown that the system is capable of operating satisfactorily when in the working state by virtue of its accuracy and response.

From the failure data information given and from the logic fault diagram it was shown that the main cause of hazard in the plant being assessed lay in an excess concentration of oxygen gas. Measurement of oxygen concentration accordingly formed a most important trip initiator.

13.1.11.1 Oxygen concentration fault monitor assessment

The equipment and sampling arrangements used involved three tapping points so that each sample, which is tapped off at a high pressure, passes in turn through an isolation valve, a pressure reducing valve followed by a relief valve and on to a sampling panel. The design of the analyser is such that it responds to a relatively small sample flow. To minimize sampling lag the sample passes from the tapping point to the sampling panel at a much higher rate than is required, the bulk of it by-passing the analyser and flowing through a flowmeter fitted with a loss-of flow detector. A small proportion (1%) of the total sample flows through the analyser and then through a sample flowmeter also fitted with a low flow detector, back to the bypass line and into a purged header.

Fail-to-danger faults in this channel could be caused by:

(1) An incorrect flow of sample gas (undetected);

(2) An unrepresentative gas sample being used;

(3) A 'test' gas sample being injected into the analyser (undetected); or

(4) Component failures in the analyser, trip amplifier or associated wiring.

For the sampling arrangement shown, it was estimated that the sample failure rate was approximately 2 faults/year arising from faults in the various valves and pipework. Normally, these failures would be detected and a trip signal given either by one of the low flow detectors unless these also failed. As no information was available on the failure rates of these devices, a prediction of their fail-danger fault rates was made. This was done by considering the effects of failure of each component and then summating the component failure rates for the different fault modes – a technique which is now well established and has been described elsewhere for electronic and mechanical equipments [14, 15]. The component failure rates used in the prediction were as given in Table 13.4.

13.1.11.2 Shut-down system (H.I.S.S.)

An assessment of the safety of the shut-down system was carried out by using failure rates obtained for similar equipment in other plants. From these sources, fail-danger fault rates of 0.11 and 0.01 faults/year for the pneumatic and solenoid valves, respectively, were obtained and by making allowance for faults in the wiring and inter-connecting pipework, a failure rate of 0.2 faults/year per valve unit was derived.

13.1.12 Systems failure analysis

A simple system analysis was used to consider how the variables of frequency, consequences and feasibility can be used in respect of a propulsion system meeting a defined despatch reliability target. In an explanation of the method, Davies [16] used an example which might be regarded as typical for a gas turbine-powered installation:

Table 13.4 *Reliability of the oxygen concentration protective channel*

Component

Type	Failure rate (%/1000 h)
Capacitors – electrolytic,	0.1
aluminium foil	0.1
Lamps, indicator	0.5
Photo-cells	0.5
Potentiometers, carbon track	0.2
Rectifiers, metal	0.5
Relays, general – per coil	0.03
per contact pair	0.02
Resistors, composition, grade 2	0.02
Switches, limit	0.02
Transformers, per winding	0.03
Transistors, alloy silicon	0.05

Equipment

Types of equipment	Failure rates (faults/year)		Observed overall
	Predicted		
	Fail-danger	Overall	
Low flow detector	0.12	0.28	–
Oxygen analyser	0.43	1.51	–
Trip amplifier	0.09	0.39	0.14–0.28

Instrument line

Failure mechanism	Fractional dead time
Loss of sample	1.0×10^{-4}
Incorrect sample	6.0×10^{-4}
Analyser and/or amplifier failure	1.7×10^{-2}
All faults	$\sim 1.8 \times 10^{-2}$

Protective channel

(3 Instrument lines in a 2-out-of-3 trip configuration)
Fractional dead time $\simeq 8.6 \times 10^{-4}$

Assuming a target reliability figure of 99.9% for the complete system which implies a failure of 0.1%, i.e. of one unit in 1000 h with an average operational period of 4 hours, on the basis of the reliability

$$R = e^{-\lambda t}$$

$$= 1 - \lambda t + \frac{(\lambda t)^2}{1.2} - \frac{(\lambda t)^3}{1.2.3} \ldots \text{etc}$$

and using the approximation $R = 1 - \lambda t$

$$= 1 - 4\lambda = 0.999$$

hence
$$\lambda = 0.00025 = 0.25/1000\,\text{h}.$$

Table 13.5 *Some Observed Failure Rates*

Equipment	Sample population	Failures observed (in 1½ years)	Overall failure rate	
			Observed	Predicted
Temperature trip amplifier	48	22	0.31	0.39
Pressure switch	57	3	0.04	0.14
Differential pressure transmitter	27	5	0.17	0.54
Level switch	12	0	0	0.21
Oxygen concentration protective channel				
Low flow detector	30	9	0.2	0.28
Oxygen analyser	15	5	0.22	1.51
Trip amplifier	15	4	0.18	0.39

This value of λ is the complete system failure rate, hence:

Failure rate of complete propulsion system = 0.25/1000 h.

Mean time between failures, MTBF = 4000 h.

13.1.13 Permissible failure rates

A 100% reliability is unachievable and all types of failures have to be conceded as possibilities, even catastrophic failures. The best which can be done is to limit the frequency of failure to an acceptable rate.

In a simple system analysis by Davies [16] the variables of frequency, consequences and feasibility were related in respect of a propulsion system reaching a despatch reliability target of 99.9%; the following procedure was given:

(1) The target reliability is 99.9% for each propulsion system, implying that one delay per power unit can be tolerated in 1000 flights;

(2) The average flight length is known to be four hours;

(3) From $e^{-\lambda t}$, system failure rate is 0.25/1000 h;

(4) Attach importance factors to each of the eight sub-systems, weighting according to the safety hazard, or cost and time penalties of subsequent recovery action (1 is the highest rating);

(5) Include a complexity factor for each sub-system, weighting according to total number of parts and general state of the art; take the product of importance and complexity factors for each sub-system to arrive at the combined factor;

(6) The maximum permissible failure rate per 1000 hours for each sub-system is given by:

$$\frac{\text{System failure rate} \times \text{Combined factor}}{\text{Sum of combined factors}} = \frac{0.25 \times \text{Combined factor}}{150}.$$

Applied to a typical system this gives the values in Table 13.6.

Table 13.6 *Assessment of maximum permissible failure rate*

Sub-system	Import-ance factor	com-plexity factor	Com-bined factor	Maximum permissible	
				Failure rate/1000 h	MTBF hours
Basic engine	1	24	24	0.04	25 000
Oil system	3	2	6	0.01	100 000
Fuel & control	2	24	48	0.08	12 500
Fire system	2	15	30	0.05	20 000
Accessory drives	2	6	12	0.02	50 000
Electrical	3	2	6	0.01	100 000
Variable intake	2	6	12	0.02	50 000
Variable nozzle	2	6	12	0.02	50 000

13.1.14 Failure rate and MTBF

An example of the method of calculating failure rate and mean time between failures (MTBF) given in a very useful introductory text to reliability engineering by Lewis [17] refers to the reliability testing of 100 components over a period of 4000 hours in which the following results occurred.

Table 13.7

No. components	1	1	4	5	3	86
Failure time (h)	250	300	415	800	1200	No failure

From these figures:

$$\text{total survival} = (250 \times 1) + (300 \times) + (415 \times 4) + (800 \times 5)$$
$$+ (1200 \times 3) + (4000 \times 86)$$
$$= 353\,810 \text{ component-hours.}$$

$$\text{total failures} = 1 + 1 + 4 + 5 + 3$$
$$= 14$$

$$\text{mean time between failures} = \frac{353\,810}{14}$$
$$= 25\,270\text{h.}$$

$$\therefore \quad \text{failure rate } (\lambda) = \frac{1}{\text{MTBF}}$$
$$= \frac{1}{25\,270}$$
$$= 3.957 \times 10^{-5}/\text{h}$$
$$= 3.957\%/1000\,\text{h.}$$

The reference quoted includes a number of similar worked examples which should help with further study of this subject.

13.2 Availability

No matter how carefully a system may be designed and manufactured it is impossible to be absolutely certain that it will not fail — unusual loads or a combination of unusual loads and freak operating conditions may combine to frustrate the availability of the system. Thus, in order to calculate the availability of a vehicle to survive a mission of x miles on the understanding that no repair or maintenance action last longer than τ hours two factors must be known:

(1) Readiness or availability of the vehicles — i.e. the percentage of vehicles which can be expected to be in use at x miles assuming repair or maintenance does not exceed τ hours.

(2) Mission availability — the probability that conditions will prevail without freak conditions such that the required survival can be achieved.

The quantitative assessment of these conditions involves the application of statistical theory to deal with random effects and involves the inter-relationship between the probable reliability of the vehicles (or systems) with the lapse of time and the increasing amount of time required to be spent on repair and/or maintenance as a vehicle or system deteriorates following a prolonged period of use.

13.2.1 Availability concepts

The concept of availability combines reliability and maintainability and is one basis for analyzing system effectiveness. The two concepts of availability are:

(1) Point availability (A): The probability that an item will be in an operable condition at any time within a given period.

(2) Average availability (A_{av}): The fraction of a given time period during which an item will be in an operable condition.

$$A_{av} = \frac{1}{t} \quad A(t) \, dt.$$

Both availabilities asymptotically approach a steady-state value.

The extent to which a system is available for use can be expressed in terms of the availability (A) given by

$$A = \frac{\text{MTBF}}{\text{MTBF} + \text{MTTR}}$$

where   MTBF = mean time between failures, (x),

          MTTR = mean time to repair.

Note: at any time (t) the reliability can be calculated in terms of MTBF as $R = e^{-t/x}$.

13.2.2 Marine machinery availability

In the course of a study of the reliability and maintainability of a ship's machinery installation, Wilkinson and Kilbourn [18] of the British Ship Research Association published the following figures on component failures which had been collected from the records on 51 motor tankers over a period of 1 and 4 years, representing a total period effectiveness of 100 ship years.

Table 13.8 *Typical component failure rates (marine)*

Components	Failure rate Failures /1000 h	Mean time Between Failures, h
Turbo-alternator and controls – alternator end	0.007407	135 000
– turbine end	0.05556	18 000
Diesel alternator and controls – alternator end	0.008850	113 000
– diesel end	0.09091	11 000
Turbo-alternator system sea water circulating pump	0.005405	185 000
Diesel alternator sea water circulating pump	0.01	100 000
Condenser	0.01176	85 000
Air ejector	0.00444	225 000
Extraction pump	0.02857	35 000
Boiler feed pump	0.025	40 000
Waste heat economiser-type boiler	0.01471	68 000
Oil-fired boiler	0.70	1 400
Composite boiler circulating pump	0.04167	24 000

The reason for the very low MTBF for the oil-fired boiler was that the estimated use of the boiler was between 30 and 40 days per year. Twenty-eight failures of this component were recorded and although the records refer to 45 ships years experience, the total operating period was only 40 000 hours.

The failure information on diesel alternators was obtained from the records of sets that operated for widely different periods of time during a calendar year. The number of failures experienced per year in auxiliary diesel engines did not alter greatly with equipment usage, resulting in a wide range of MTBF values. Unfortunately, the quantity of data collected was insufficient to produce any definite relationship between MTBF and equipment usage.

13.2.3 Computer evaluation of system reliability and availability

Three different computer programs were used by Mathieson to predict different reliability characteristics of a marine lubricating oil system [10]. Markov analysis methods were used with the programs written in the high-level language NUALG and implemented on the UNIVAC-1108 of the Computer Centre, Trondheim. This computer can handle a maximum of 150 states using only the core memory.

Programs were named RELAN, REAVAN and STAVAN as acronyms respectively for *REL*iability *AN*alysis; *REL*iability/*AV*ailibility *AN*alysis; *ST*eady-state *AV*ailability *AN*alysis.

Program	Computations				
	Reliability		Pointwise availability	Steady-state availability	Mean waiting times in steady-state
	With repair	Without repair			
RELAN		X			
REAVAN	X	X	X	(X)	
STAVAN				X	X

Figure 13.9 *Analysis capabilities of different programs*

The different analyses which these programs can perform are shown in Figure 13.9. The values for system reliability and system availability which were produced are given in Tables 13.9 and 13.10.

Table 13.9 *Reliability*

Program	Time (hours)				Accuracy
	0	4000	8000	16 000	
RELAN	1.00	0.5746948	03295048	0.1076727	1×10^{-16}
REAVAN (without repair)	1.00	0.5746949	0.3295044	0.1076718	1×10^{-8}
REAVAN (with repair)	1.00	0.5754198	0.3311079	0.1096325	1×10^{-8}

Table 13.10 *Availability*

Program	Time (hours)			Accuracy
	0	80	160	
REAVAN	1.00	0.9993267	0.9993256	1×10^{-8}
STAVAN	–	–	–	0.9993250	1×10^{-16}

13.2.4 Availability replacement stock levels

The probability of having r or less failures in time t for a single piece of equipment with constant failure is given [19] by the Poisson distribution

$$P(r) = \sum_{t=0}^{r} \frac{(\lambda t)^i}{i!} \exp(-\lambda t) \tag{13.1}$$

The number of failures which will occur is equal to the number of spares required at the chosen confidence level. Graphs and tables are available [19] for the calculation of the number of failures r from equation 13.1. Stocks can be minimized if spares are always reordered as soon as they are used but the delivery time sets a lower limit to the period which must be covered.

In an example to demonstrate the use of equation 13.1 or equivalent graphs, Buffham, Freshwater and Lees [20] considered a works which uses 100 differential pressure transmitters, the failure rate of which is 0.5 failure/year and proceeded to calculate the amount of stock needed to ensure at the 90% confidence level that there would always be sufficient transmitters for replacements, if the delivery time were six months. Thus

$$t = 0.5 \text{ year} \quad \lambda = 0.5 \text{ failure/year} \quad P(r) = 0.9$$

From equation 13.1 or equivalent graphs it was found that $r = 1.4$ per transmitter so that it would be necessary to stock 140 for 100 transmitters.

13.2.5 Availability – complement of maintenance personnel

According to the availability required of a system it is possible to evaluate the number of maintenance personnel required. One approach is to use the Markov model equations, to obtain system availabilities for schemes involving different numbers of, or different organizations of, repairmen. For example, for a system of n identical components served by a single repairman, it is possible to calculate the probabilities that $0, 1, 2, .., n$ components will be in their failed state. Assuming that all the components must be operating for the system to be operational, the steady-state availability (A) of the system is:

$$A(\infty) = \frac{(\mu/\lambda)^n}{n! \sum_{i=0}^{i=n} \frac{(\mu/\lambda)^i}{i}} \tag{13.2}$$

13.2.6 Availability – preventive maintenance

Chance failures without an element of wearout cannot be reduced but deterioration due to wearout can be reduced by the use of preventive maintenance. True wearout occurs when, if the mean life to wearout is M and the standard deviation is σ, [21], the time interval between overhauls, T_0, is:

$$T_0 = M - 6\sigma \tag{13.3}$$

With such a policy true wearout is virtually eliminated. As a rule of thumb the standard deviation is often taken as:

$$\sigma = 0.1 M \tag{13.4}$$

so that equation (13.3) becomes:

$$T_0 = 0.4M. \tag{13.5}$$

A second type of wearout is that in which the overall constant failure rate λ is the sum of the chance failure rate, λ_c, and the steady-state wearout failure rate, λ_w:

$$\lambda = \lambda_c + \lambda_w. \tag{13.6}$$

The rational design of a preventive maintenance system therefore requires data on wearout failures.

If a particular maintenance policy has been adopted then it is possible to calculate the number of overhauls which will be required and hence the maintenance work load. If it is assumed that the equipment is overhauled whenever it fails and also is subject to overhaul at fixed times, then:

$$R(T_0) = \exp(-\lambda T_0) \tag{13.7}$$

and the average times between all overhauls, T_0^*, is:

$$T_0^* = \int_0^{T_0} R(t)\, dt = \int_0^{T_0} \exp(-\lambda \tau)\, dt = -\frac{1}{\lambda}[\exp(-\lambda t)]_0^{T_0}$$

$$= \frac{Q(T_0)}{\lambda}. \tag{13.8}$$

An example used to illustrate the application of this principle [20] considered a works with 6 chromatographs for which the mean time between failure of the chromatographs was 600 hours and the planned overhaul interval two weeks, it was required to calculate average time between overhauls. The solution accordingly gives $T_0 = 2$ weeks $= 366$ hours and $M = 600$ hours and therefore $Q(T_0) = 0.429$. Accordingly $T_0^* = 257$ hours for one chromatograph and therefore for 6 chromatographs the average time between overhauls should be $256/7 = 43$ hours.

13.2.7 Utilization factor (v)

The utilization factor of a component or a system is the time-based utility, i.e. the rates between the actual time in use or operation and the total time needed to ensure that it is available for use. The total time is the sum of the maintenance time, including scheduled and unscheduled (due to failure) maintenance, idle time, which may occur between completion of maintenance and use, due to administrative difficulties, etc., and actual operating time, hence:

$$v = \frac{t}{t_1 + t_2 + t_3} \tag{13.9}$$

where $t_1 =$ operating time,

$t_2 =$ maintenance time,

$t_3 =$ idle time.

The utilization factor is greatest when the idling time (t_3) is zero and the maintenance time (t_2) is small. The value of the maximum utilization factor is the 'availability' (A), thus with $t_3 = 0$,

$$A = v_{max} = \frac{t_1}{t + t_2'} \qquad (13.10)$$

where t_2' = minimum value of the maintenance time. The dependability of a system (or component) is the availability when the maintenance time is only that period needed for unscheduled overhauls due to failure, i.e. t_2' omits time spent on scheduled overhauls.

13.2.8 Availability function

Whether or not a system is available for use depends essentially on its state. The Markov approach is regarded as the model on which calculations are based [22]. Thus for a single piece of equipment which may be in a good state (0) or a bad state (1) with a repair rate μ given by $1/T$ where T is the replacement time then

$$\dot{P}_0(t) = -\lambda P_0(t) = +\mu P_1(t) \qquad (13.11)$$

$$\dot{P}_0(t) = \lambda P_0(t) - \mu P_1(t) \qquad (13.12)$$

where λ = failure rate.

Assuming that the initial conditions are $P_0(0) = 1$, and $P_1(0) = 0$, the availability function, $A(t)$, is:

$$A(t) = \frac{\mu}{\lambda + \mu} + \frac{\lambda}{\lambda + \mu} \exp\left[-(\lambda + \mu)t\right]. \qquad (13.13)$$

It is usually the long-term or steady-state availability which is of interest, which for this case is:

$$A(\infty) = \frac{\mu}{\lambda + \mu} \qquad (13.14)$$

In many systems the reliability depends not only on the failure rate but on the time required for the replacement of failed components. For such systems it is necessary to ensure that both detection and repair are rapid. Since rapid repair requires immediate availability of components and repairmen, it is costly and some compromise may be necessary.

13.3 Failure prediction/reliability assessment

The reliability of each system, assembly or component can be assessed by the use of reliability logic or mathematical models in order to predict the overall reliability of a design. The failure rates assigned to each system, assembly or component, are usually based on reports of service experience collected over a period of time, or from a standard handbook [23], and so the reliability prediction obtained will be representative of a mature piece of equipment after a period of service.

A basic assumption which appears to be substantiated by experience is the 'Pareto' law that 'the number of different problems making up 50% of the total trouble in service remains reasonably constant for a given product'. The number of problems involved by four families of engines are shown in Table 13.11.

Table 13.11 *Constant number of problems to cure half the trouble*

Engine type	First entered commercial service	No. of problems	% of overall basic unscheduled removal rate
1967 Data			
Avon	1958	7	50.7
Conway	1960	12	56.8
Dart	1953	10	51.4
Spey	1964	14	53.7
1968 Data			
Avon	1958	7	52.8
Conway	1960	5	54.0
Dart	1953	11	53.2
Spey	1964	12	53.6
	Total	78	426.2
	Average	9.75	53.3
	i.e. 9 problems account for 50% of the trouble.		

The problems are of course continually changing as each individual one is solved, but new ones continue to appear. The absolute levels of unreliability normally continue to fall with time. From this we see that the nine worst problems on any engine should cover about half of the total trouble affecting an engine. An examination of the date of initial entry into service shows that there does not appear to be any trend suggesting that the older engines have consistently fewer or more problems than nine to make up 50% of the total trouble. The prediction of initial in-service reliability is now simplified to that of finding the nine most troublesome areas on test to date, possibly modified by any other inputs from stress engineering or design failure analyses that suggest that trouble may strike in an area which has not yet shown up on test.

An examination of the data for one of the engines that Rolls-Royce have developed shows that the most frequent occurrences of trouble during the early phases of development were in the following nine areas on the engine (Table 13.12).

To correlate component degradation after test with unscheduled removals in service, an examination of 'hot end' components produces the most realistic evaluations. Historical experience shows that the worst turbine blade trouble experienced to date produced a basic engine unscheduled removal rate of 0.098 per thousand hours. This would give an exchange rate (taking this very pessimistic view) that the 25.6% defects to date on the test bed are equivalent to an in-service unscheduled removal rate of 0.098 per thousand hours and hence the in-service unscheduled removal rate for 50% of the trouble will be 100/25.6 × 0.098 = 38 per thousand hours. Doubling this number to give 100% of the unscheduled engine removal rate the early in-service unscheduled removal rate is estimated as 0.76 per thousand hours.

Table 13.12 *No. of occurrences* * *of a problem during initial development*

Problem	No. of occurrences * (See Note 1)
1. Combustion liners	27.2%
2. h.p. turbine blades	25.6
3. h.p. nozzle guide vanes	13.4
4. External wheelcase bearings	12.5
5. h.p. stator rings	5.7
6. Stators and guide vanes	5.4
7. l.p. stage 1 rotor blade	5.1
8. Oil leaks in turbine area	2.9
9. l.p. nozzle guide vanes	2.2
Total	100%

* Occurrence can mean failure but usually means some sign of degredation which would eventually lead to failure if no action were taken.
Note 1: The number of occurrences is expressed as a percentage of the total number of occurrences for the 9 problems.

Repeating this exercise for the two worst unscheduled engine removal rates for each of the 'hot end' problems in Table 13.12 [problems (1), (2) and (3)] produces the estimates shown in Table 13.13.

Table 13.13

Problem	No. of occurrences on development	Two-worst in-service unscheduled removal rates for similar problems (rate/1000 h)	Predicted early in-service basic engine unscheduled removals (rate/1000 h)
Combustion liners	27.2	(a) 0.060 (b) 0.034	0.44 0.25
H.P. turbine blades	25.6	(a) 0.098 (b) 0.030	0.76 0.23
H.P. nozzle guide vanes	13.4	(a) 0.046 (b) 0.025	0.69 0.37

For the engine under consideration, the predicted early in-service basic engine unscheduled removal rate is between 0.23/1000 hours and 0.76/1000 hours.

13.3.1 Failure investigations – facts

The scientist investigating industrial failures must be critical of every piece of evidence offered to him, it has been said 'If a material fails there are as many explanations as there are interested parties', background information and eye-witness accounts can be very misleading. Police experience suggests that if five people witness an accident and describe it you often think they were describing five different accidents.

The facts concerning the circumstances of a failure may be misrepresented owing to ignorance or because someone is shielding a colleague. The evidence regarding failures associated with marine engines or agricultural machinery requires very careful scrutiny. A seized engine that proves to have new oil in the sump or clean water in the radiator must be viewed with suspicion. In extreme cases facts can be 'adjusted' to suit insurance or guarantee claims.

In any diagnostic investigation the principal objective should be to identify the source of trouble and to indicate how the problem can be avoided in future. It is not sufficient to report that a bearing failed by fatigue, one must try to establish the cause of the problems which one encountered. A total of 992 failures reported by Wilson [24] and listed in Table 13.14 were found to include 1378 failure modes as listed in Table 13.15.

Table 13.14 *The components that fail [24]*

Components	No.	Category	No.
Plain bearings	130		
Sleeve bearing and bushes	57		
Other bearings including thrust and crosshead slides	18	Bearings	283
Shafts and collars (excluding crankshafts and camshafts)	13		
Gudgeon, crosshead and wrist pins	27		
Ball and roller bearings	38		
Cylinders and cylinder liners	64		
Pistons	74	Cylinders and	
Piston rings	66	piston assemblies	206
Piston ring inserts	2		
Exhaust valves	58		
Cylinder heads, valve seats and pre-combustion chambers	27		
Spark plugs and igniters	35	Combustion zone	130
Inlet valves	7		
Valve guides	3		
Fuel pumps	33		
Carburettors	5		
Fuel tanks and lines	7	Fuel system	66
Injectors	21		
Turbine blades	56		56
Gears	41		41
Boilers, superchargers and boiler tubes	39		39

Table 13.14 (Continued)

Components	No.	Category	No.
Drums and portable containers	3		
Tanks for storage and transport of products	7		
Tank cleaning and ventilating equipment	5		28
Piping and valves (not engine)	13		
Tappets	13		24
Camshafts	11		
Heat exchangers	21		21
Lubricant system components	8		18
Filters, centrifuges and separators	10		
Chassis/foundation	18		33
Bodywork/hull	15		
Pumps	18		18
Compressors	15		15
Cooling system components	12		12
Propellers	10		10
Crankshafts	9		9
Engine compression and turbochargers	9		9
Nuts, bolts, rivets	8		8
Connection and piston rods	6		6
Exhaust system components	6		6
Instruments	5		5
Anchors and mooring equipment	5		5
Seals, gaskets	5		5
Hydraulic system components	3		3
Others	62		62

Table 13.15 *Failure modes [24]*

Mode	Number
* Corrosion(electrolytic, weak organic acids, low temperature generally)	331
* Deposits	222
* Wear (general)	132
High-temperature corrosion (e.g. guttering, burning, metal oxidation)	120
Fatigue	107
Pitting, spalling and surface cracking	106
Scoring/seizing	71
Fracture	66
Erosion (excluding cavitation)	50
Cavitation	33
Abrasion	32
Scuffing	23
Melting	21
Fretting	15
Others	49
Total	1378

* The very large proportion of failures associated with corrosion, deposits and wear should be noted. This high proportion is in accord with a Rolls Royce observation that 70% of their unscheduled engine shut-downs in service were associated with some form of contamination [24].

In very few instances are there more than two failure modes involved in any one investigation, so in nearly 400 cases there were judged to be two major contributory factors to the failure. The failure mode (as distinct from the prime cause of failure, i.e. fatigue as distinct from the cause of the fatigue) was identified in all except 49 cases.

Table 13.16 shows that 729 failures were on engine components, and lists these according to engine type and application. An additional 136 failures were directly associated with transport industries or power-generating equipment. Thus, under the marine heading, problems such as corrosion in sea-water circulating systems, anchor breakages, propeller damage and propeller shaft sealing problems are included. 98 problems relate to specific types of engine still under development that may find application in more than one field. The proportion of failures associated with the power and transport industries suggests that the sample may not be wholly representative, but may not be badly distorted.

The 127 failures not defined in Table 13.16 cover a wide range, including refrigeration plant, air-conditioning equipment, industrial and domestic heating plant, metal-working and machining equipment and hydraulic machinery.

13.4 Hazard rate curve

The 'bath tub' hazard rate curve shown in Figure 13.10 relates failure rate and time in accordance with the three stage (1) early failures (2) random failure (3) wear-out failure. None of the usual mathematical distribution functions (exponential, normal, log-normal, Weibull, extreme-value) have this bath-tub shaped hazard curve but an approximation may be obtained by selecting an appropriate probability density function for each of the three stages [25]. The equation for such a hazard rate curve then becomes

$$f(t) = \sum_{i=1}^{i=3} f_i(t) \cdot p_i$$

where p_i = probability of the particular failure characterizing the time interval i.

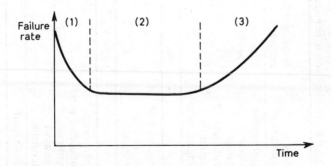

Figure 13.10 *'Bath-tub' hazard rate curve*

Table 13.16 Engineering failures – field of application [24]

Field of application	Spark-ignition engines	Diesel engines	Gas engines	Gas turbines	Steam turbines	Dual-fuel engines	Other engines	Components not on engines	Total
Automotive	96	107		2				25	230
Aviation	85			30				33	148
Marine	5	124	1	11	34			78	252
Railroad		39		2					42
Stationary (engines only)		35	2	44	11	2	1		95
	186	305	3	89	45	2	1	136	767
Engines with no specific application	4	85		5	2		2		98
	190	390	3	94	47	2	3	136	865
Failures not connected with power or transport industries									127
Grand total									992

13.4.1 Hazard rate curve – with repair

The repair of renewable items alters the shape of the hazard rate curve by introducing the mathematical probability arising from renewal. Thus by considering instantaneous renewal by a replacement with an identical unit or maintenance action completely restoring the original properties, the instants of failure (or renewal) as shown in Figure 13.11 are:

$$t_1 = \tau_1$$

$$t_2 = \tau_1 + \tau_2 \quad \text{etc}$$

$$t_n = \tau_1 + \tau_2 + \tau_3, \ldots, \tau_n$$

This constitutes a random flow called a renewal process.

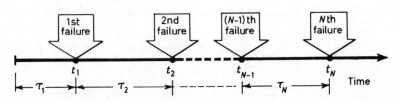

Figure 13.11 *A renewal process*

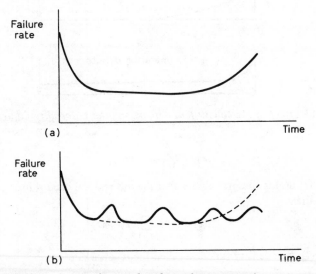

Figure 13.12 *Effect of repair (renewal on hazard rate curves)*
(a) Bath-tub curve
(b) effect of renewal

Applying the theory of distribution of sums of random variables (convolution algebra), the density function of the time of the *n*th failure is given by the multiple convolution integral written as:

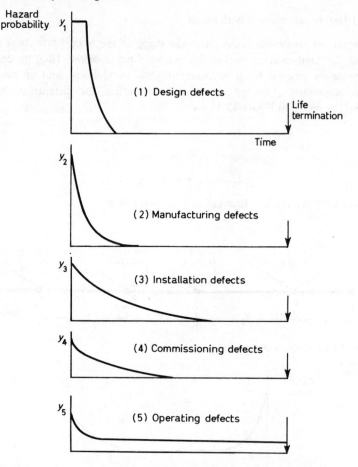

Figure 13.13 *Effect of the constituent defects on infant mortality failure*

$$f(t_n)(t) = h^{(n)}(t)$$

Since $f(t_n)(t)\,d(t)$ is the probability that the nth renewal occurs during the interval $[t, t + dt]$ then

$$m(t)\,dt = \sum_{M=1}^{\infty} f(t_n)(t) \qquad (13.15)$$

where $m(t)$ = a density function called the renewal rate.

On the basis of an asymptotic renewal process the value of $m(t)$ will tend towards a constant value as the cumulative operating time becomes large [26], hence

$$\lim_{t \to \infty} m(t) = \frac{1}{\text{MTTF}}$$

where   MTTF = mean time to failure and is the expectation of time to the first failure.

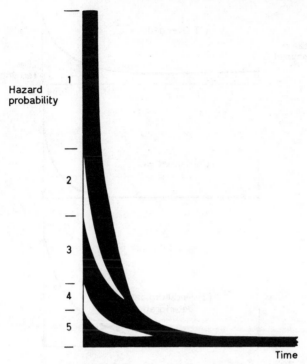

Figure 13.14 *Summation of infant mortality hazard probabilities*

The effect of this renewal process on the bath-tub curve is shown in Figure 13.12 and was particularly discussed by Krohn [27]. Constant renewal does not necessarily imply a Poisson process with exponential distribution. It should be assumed that the bath tub hazard rate curve applies to complex repairable systems during the intermediate life regions where it postulates a constant failure rate.

13.4.2 System life expectancy

An attempt to relate the hazard rate change to the needs of condition monitoring and the criticality decisions preceding catastrophic failure were presented by the author in a study of the transition between infant mortality and terminal failure conditions [28].

The hazard probabilities for the factors contributing to both infant mortality and terminal failure are shown in Figures 13.13 and 13.14 together with the cumulative effects in Figures 13.15, 13.16 and 13.17. By combining these cumulative curves a whole-life hazard probability can be speculated and related by an equation such as

$$y = \frac{K_1}{(t)^M} + \frac{K_4}{(t_1 - t)^P}$$
$$= rK_4\{(x t_1)^{-M} + [t_1(1-x)]^{-P}\}$$

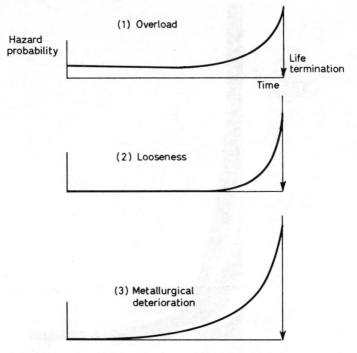

Figure 13.15 *Effect of individual terminal mortality factors*

where $x = t/t_1$ $r = K_1/K_4$

       $t_1 = 1.0$ $K_1 = 1.$

From this relationship the time at which the lowest failure hazard exists can be assessed, henceforth the failure potential will increase up to catastrophic terminal failure.

13.5 Complex system reliability – Monte Carlo simulation

Failure rates or maintenance policies for a complex system may be determined by using Monte Carlo simulation techniques, particularly when the configuration is too complicated for the use of analytical methods.

The use of Monte Carlo simulation involves a series of simulation runs [25, 29]. In a given run the failure of a particular component in the system is determined by its probability of failure and by a random number, so that sometimes the component fails and sometimes it does not, but on average its individual probability of failure is correct. Failure of the system in a given run then depends on the way in which the components fail. Again on average the correct probability of failure of the system is obtained.

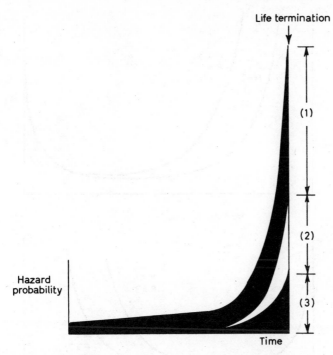

Figure 13.16 *Summation of individual terminal failure probabilities*

13.6 Hazardous chemical plants – high integrity protective systems (HIPS)

Protective systems of great sensitivity and reliability have been designed by the Heavy Organic Chemical Division of ICI Ltd. and successfully proved on a large chemical plant in which pure oxygen is reacted with hydrocarbon to produce oxygenated material [13].

The objective of the system is to reduce the numerically expressed potential hazard of the plant to the average value for non-hazardous plants without excessive interruption to production. The design involved the need to ensure that every possible failure of the plant and the H.I.P.S. had been considered; because of the need for the extremely high degree of reliability; new design techniques were developed. To ensure that all possible failures were covered a very large number of trip-initiating parameters were required; if these were just single trip initiators they would produce a large number of spurious plant shutdowns. Such interruptions to production being unacceptable, a method of achieving this extraordinary safety was devised involving triplicated initiators with majority voting.

Conductors from trip-initiating monitors are taken in pairs from pressure, temperature or other sensors, and 'farmed out' to a number of 'guard lines' 'Majority voting', i.e. plant close-down, results when any two of the three circuits operate switches. This equipment achieves reliability by including its own redundancy and includes triplicated power supplies. Majority voting means that if only one of the

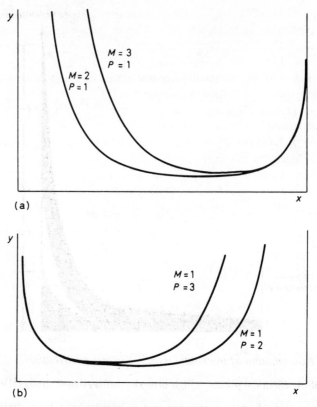

Figure 13.17(a) *Effect of differing indices on whole life expectancy, keeping termi-nal failure index constant at P = 1 and changing infant mortality index M. r = 1;*
(b) *Effect of different indices on whole life expectancy, keeping infant mortality index constant at M = 1 and changing terminal failure index P. r = 1.*

three trip initiators (H.I.T.I.s) sends a trip signal the plant is not tripped but an alarm is given.

The benefits of high integrity voting are:

(a) Spurious trip does not close down the plant

(b) The inability of one trip to operate does not endanger the plant

(c) The initiators can be tested without shutting down or 'disarming'

(d) The ratio of fail-safe probability to fail-danger probability is increased.

References

1   Jordan, W.E. (1972), 'Failure modes, effects and criticality analyses', *Proc. Annual Reliability and Maintainability Symposium,* San Francisco, California.

2   Coutinho, J. de S. (1964), 'Failure effect analysis', *Trans. N.Y. Acad. Sci.,* Series (II) **26**, No. 5.

3 Johnston, M.M. (1973), 'Reliability assurance; genesis by numbers', *Marine Engineers Review,* October, 21–24.

4 Anderson, F.E. (1966), 'Specifying reliability for shipboard electric rotating equipment', *J. Am. Soc. Naval Engineers,* 78, 210.

5 Eisner, R.L. (1971), 'Reliability considerations in design trade studies', *Proc. Ann. Symposium on Reliability,* Washington D.C.

6 Green, J. (1969), 'Systematic design review and fault tree analysis', *Conf. Safety & Failure of Components,* I.Mech.E.

7 Schrøder, R.J. (1970, 'Fault trees for reliability analysis', *Proc. Annual Symposium on Reliability*, Los Angeles 198–205.

8 Eagle, K.H. (1969), 'Fault tree and reliability analysis comparison', *Proc. Annual Symposium on Reliability,* Chicago 12–17.

9 Moses, H.J. (1971), 'Electronic engine condition analysis system', SAE Paper 710448, May.

10 Mathieson, Tor-Chr. (1973), 'Reliability engineering in ship machinery plant design', Report IF/R.12, Division of Internal Combustion Engines, University of Trondheim, Norwegian Institute of Technology.

11 Venton, A.D.F. and Harvey, B.F. (1973), 'Reliability assessment in machinery system design', *Proc. I. Mech. E.*

12 Bridges, D.C. (1974), 'The application of reliability to the design of ship's machinery', *Trans. I. Mar. E,* 86, Part 6.

13 Stewart, R.M. and Hensley, G. (1971), 'High integrity protective systems on hazardous chemical plants', Report SRS/COLL/303/2, UKAEA, Systems Reliability Service, Culcheth, Warrington, Lancs, May.

14 Eames, A.R. (1966), 'Reliability assessment of protective systems', *Nuclear Engineering,* March.

15 Hensley, G. (1970), 'The reliability prediction of mechanical instrumentation equipment for process control', Riso Report, Danish Atomic Energy Commission, February.

16 Davies, A.E. (1973), 'Principles and practice of aircraft powerplant maintenance', *Trans. I. Mar. E.,* 85.

17 Lewis, R. (1970), *An Introduction to Reliability Engineering,* McGraw-Hill, London.

18 Wilkinson, H.C. and Kilbourn, D.F. (1971), 'The design of ship's machinery installations', *Shipping World & Shipbuilder,* August.

19 von Alven, W.H. (1963), *Reliability Engineering,* Prentice-Hall Inc. Englewood Cliffs, N.J.

20 Buffham, B.A., Freshwater, D.C. and Lees, F.P. (1971), 'Reliability engineering – rational technique for minimising loss', *Proc. Symposium on Major Loss Prevention in the Process Industries,* I. Chem. Eng. 34.

21 Bazovsky, I. (1961), *Reliability Theory and Practice,* Prentice-Hall Inc, Englewood Cliffs, N.J.

22 Sandler, G. (1963), *System Reliability Engineering,* Prentice-Hall Inc, Englewood Cliffs, N.J.

23 ———— 'Reliability, stress and failure rate data for electronic equipment', MIL-Handbook-217A.

24 Wilson, R.W. (1972), 'The diagnosis of engineering failures', *South African Mechanical Engineer,* November 22, No. 11.

25 Shooman, M.L. (1968), *Probabilistic Reliability – An Engineering Approach,* McGraw-Hill Inc.

26 Barlow, R.E. and Proschan, F. (1965), *Mathematical Theory of Reliability,* John Wiley & Sons. Inc, New York.

27 Krohn, C.A. (1969), 'Hazard versus renewal rate of electronic items', *I.E.E. Transactions on Reliability,* R-18, No. 2.

28 Collacott, R.A. (1975), 'Component life concepts related to a theory for whole-life expectancy', *Quality Assurance, Institute of Quality Assurance,* December.

29 Hajek, J. and Dupac, V. (1967), *Probability in Science and Engineering,* Academic Press, New York.

I4 Reliability/failure concepts

14.1 Introduction

A precise definition of failure is obscured by the many different mechanisms and modes, particularly when failures result from deterioration processes causing a gradual degration of performance. Thus an engine cylinder liner may be subject to wear. Excessive wear beyond a certain limit (which is difficult to define) may result in an interaction with mating parts causing blowby of gas due to cylinder piston ring collapse, piston seizure and drastic increase in wear. This may be classi-fied as a failure although the engine (system) could continue to function at reduced load with an increased consumption of lubricating oil.

14.1.2 Mechanical reliability

Of the several definitions used to describe reliability the fact emerges that, with equipment, a quality of performance is expected over a specified time. This aspect of reliability can be expressed as a statistical probability which attempts to forecast the failure pattern for a system.

In the preparation of this section, particularly valuable assistance was obtained from the book *Mechanical Reliability* by Professor A.D.S. Carter [1]. Not only was the text matter clear and readable, but the problems and their solution provided a means of developing confidence that an adequate understanding of the subject had been achieved. As a supplement to the subject matter on fault diagnosis this book is warmly recommended.

Reliability logic is based on mathematical models; it is unfortunate that decision logic is not so easily evaluated.

14.1.3 Theoretical failure rate

This is the instantaneous failure rate at a specific point in the life of an equipment. It is quantified as the number of failures which occur during unit operating time.

Practical failure rates can only be measured over a finite time interval. For a large sample a long operating time is needed to provide the greatest confidence in the calculated failure rate. Practical failure rate is normally calculated as No. failures/h or as No. failure/1000 h.

14.1.4 Failure distributions

Probability distributions specify the dependence ratios between random variables and in reliability terms they attempt to mathematically express the time of a failure of a unit and the restoration time. Theoretical distribution functions are defined by one, two or three population parameters which may be estimated from maintenance or failure data.

14.1.5 Basic principles of reliability

Reliability theory is based on the concept that any piece of equipment will fail to run for a long enough period of time. Even the most reliable equipment will fail eventually — this failure does not necessarily occur through wearout. If a number of identical components are operating in an identical environment a definite pattern of failures will become established. These follow standard statistically predictable patterns such as the exponential and the Poisson distribution.

14.1.6 Reliability and probability

There is a close link between the theory of reliability and that of probability based on the probability that the equipment or system will operate for a specific period of time without failure. Availability is an indication of the proportion of the operating time that a system or equipment is available to fulfill its function. It is usual to assume that the repair distribution is of the negative exponential form. It is more likely that the log-normal form of distribution is characteristic of the repair probability. For simplification of the arithmetical analysis many calculations nevertheless assume a negative exponential failure for the repair probability distribution.

14.2 Probability of reliability and failure [1]

Let R = probability of reliable functioning for a specified period of time,

F = probability of failure occurring during the same period of time,

then $R + F = 1$.

14.2.1 Failure rate

The number of failures occurring in unit time is known as the 'failure rate'. An average number is obtained from tests on a sample of components and the resulting proportional failure rate known as the failure rate (λ). This can be calculated by the formula [2]:

$$\lambda = \text{Sample tested} \times \frac{1}{\text{Test time}}$$

thus if 10 failures occur during a 5000 h test to which 1000 components are subjected,

$$\lambda = \frac{10}{1000} \times \frac{1}{5000}$$

$$= 2 \times 10^{-6} \text{ failures/h}$$

$$= 2 \text{ failures}/10^6 \text{ h}$$

14.2.2 Mean time between failures (MTBF)

For an exponential failure rate, mean time between failures = 1/failure rate and this indicates the average interval of time between failure of the equipment.

The mean time between failures is not the time for which equipment will be expected to operate before failure. For example, if equipment is to operate for a period equal to its MTBF, then the probability of the equipment lasting for that period of time without failure is about 38 per cent. Obviously if the equipment is used for a shorter period of time the probability of survival is higher.

The MTBF is therefore the average time between failures and in practical terms it can be obtained by calculating:

$$\text{MTBF} = \frac{\text{total operating time of population}}{\text{No. of failures to occur}}.$$

For example, suppose there are 100 identical pieces of equipments in operation in various systems, each of which has been in operation for about 8000 hours – this means that the total operating for the population is 800 000 hours – and also suppose that during this period of time 80 failures have occurred and have been repaired so that each runs 8000 hours: the MTBF is therefore 800 000/80 = 10 000 hours. The failure rate is the reciprocal of this and is therefore 0.0001 failures/h or 0.01 failures/1000 h.

14.2.3 System reliability components in series [1]

Considering a system with a number of components in series for which:
(1) The failure of any one component causes complete system failure;
(2) The failure of any component is independent of all other failures.
Let $R_1, R_2, R_3, \ldots R_N$ = probability of reliability of individual components.
The reliability probability of the complete system

$$R = R_1 \times R_2 \times R_3, \ldots R_N = (\bar{R})^N$$

where R = mean component reliability (if R_1, R_2, etc. are of about the same magnitude).

A typical concept can be taken from the use of lamps in series – if one fails, all fail.

$\therefore$ system reliability R = Reliability of lamp 1 $\times$ reliability of lamp 2 . . . etc. The reliability can be calculated from failure rate data from the equation $R = e^{-\lambda}$. Consider a number of components such as lamps, say, arranged in series and each individual component, or lamp of reliability $R_1, R_2, R_3, \ldots$, etc.

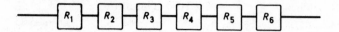

Figure 14.1 *Components in series*

$$\text{System reliability } R = R_1 \times R_2 \times R_3 \times R_4, \ldots R_N$$
$$= e^{-(\lambda_1 + \lambda_2 + \lambda_3 + \ldots \lambda_N)}$$

where λ = constant failure rate of each component.

14.2.4 System reliability – components in parallel [1]

Considering a system comprising a number of components in parallel for which:
(1) The failure of any one component causes failure of the complete system;
(2) The failure of any component is independent of all other failures.
Let F_1, F_2, F_3, . . . F_N = probability of failure of individual components.
The failure probability of complete system, $F = F_1 \times F_2 \times F_3, \ldots F_N$

i.e.
$$(1 - R) = (1 - R_1)(1 - R_2)(1 - R_3), \ldots (1 - R_N)$$
$$= (1 - \bar{R})^N.$$

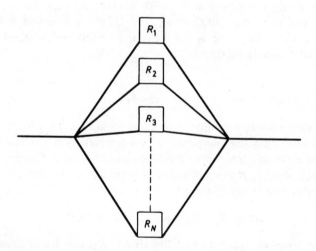

Figure 14.2 *Components in parallel*

14.2.5 Influence of component reliability [1]

As shown by Lusser [3] a very high component reliability is required in a series system involving many parts. Thus in order to achieve a system reliability of 80% with N components in series, each component needs the following reliability:

Systems Reliability, R %	80	80	80	80	80	80	80
No. components, N	1	10	50	100	200	300	400
Component Reliability, $\bar{R}$ %	80	97.79	99.56	99.77	99.88	99.93	99.96

Applying the same reasoning to a system with components in parallel

System Reliability, R %	80	80	80	80	80	80	80
No. components, N	1	10	50	100	200	300	400
Component Reliability, $\bar{R}$ %	80	14.87	3.17	1.60	0.81	0.53	0.39

A scientific study of weapons for which accurate data exists led Chaddock [4] to suggest there was no correlation of a simple probabalistic nature and it was stated by Carter [1] that reliability is achieved when the weakest or least adequate individual component of a system is capable of coping with the most severe loading – on the basis that the 'strength of a chain is that of its weakest link'.

The influence of components in series or parallel as part of a reliability chain was investigated by Howard [5].

14.2.6 System reliability – components in triplicate

Considering a triplicated system of identical components as in Figure 14.3, the same probability of reliability and failure relationship used in connection with parallel systems can be applied, accordingly

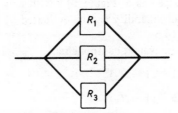

Figure 14.3 *Components in triplicate*

$$R = 1 - [(1 - R_1)(1 - R_2)(1 - R_3)]$$
$$= 1 - (1 - R_1)^3 \dots \text{for a triplicated system } R_1 = R_2 = R_3$$
$$= 1 - (1 - 3R_1 + 3R_1^2 - R_1^3)$$
$$= 3R_1 - 3R_1^2 + R_1^3$$
$$= 3e^{-\lambda_1} - 3e^{-2\lambda_1} + e^{-3\lambda_1}$$

14.2.7 System reliability – components in series/parallel

The reliability of the system can be determined from the failure probability of the arrangement as a consequence of $R + F = 1$, $R = 1 - F$. Considering the arrangement in Figure 14.4,

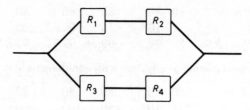

Figure 14.4 *Components in series/parallel*

Failure of R_1, R_2 = 1 – Reliability probability

$$= 1 - R_1 R_2$$

Failure of R_3, R_4 = $1 - R_3 R_4$

∴   System reliability = $1 - [(1 - R_1 R_2)(1 - R_3 R_4)]$

$$= 1 - [1 - R_1 R_2 - R_3 R_4 + R_1 R_2 R_3 R_4]$$

$$= R_1 R_2 + R_3 R_4 - R_1 R_2 R_3 R_4$$

$$= e^{-(\lambda_1 + \lambda_2)} + e^{-(\lambda_3 + \lambda_4)} - e^{-(\lambda_1 + \lambda_2 + \lambda_3 + \lambda_4)}$$

14.2.8 *System reliability – complex system with redundant components*

Considering a system arranged as in Figure 14.5 with some components duplicated and triplicated, the overall reliability can be evaluated by preparing an equivalent series system as in Figure 14.6.

Thus, dealing with the equivalent components:

$$R_A = 1 - (1 - R_2)^2$$

$$R_B = 1 - (1 - R_5)^4$$

$$R_C = 1 - (1 - R_7 R_7)^2 = 1 - [1 - (R_7)^2]^2.$$

Hence the reliability of the whole system is

$$R = R_1 R_A R_3 R_4 R_B R_6 R_C$$

$$= R_1 [1 - (1 - R_2)^2] \quad [R_3 R_4 (1 - (1 - R_5)^4] \quad R_6 [1 - 1 - (R_7)^2]^2.$$

The expression can then be extended in terms of the exponential relationship between the values R_1, R_2 etc. and their associated failure rates λ_1, λ_2 etc.

14.3 Failure pattern – exponential distribution [6]

This represents the constant failure rate case in which the equipment failure rate is independent of age.

Let $f(t)$ = total number of failures at time t.

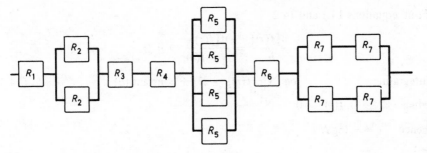

Figure 14.5 *System with redundant components*

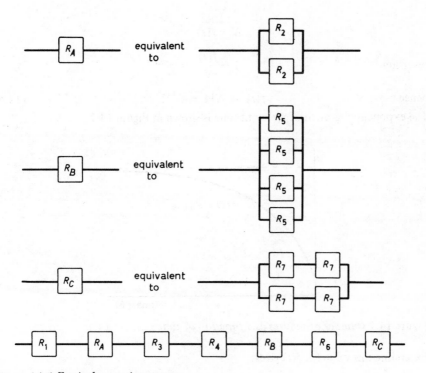

Figure 14.6 *Equivalent series system*

then
$$\frac{\mathrm{d}}{\mathrm{d}t}[f(t)] = n\lambda \tag{14.1}$$

where n = number of components exposed at time t,

λ = constant failure rate of each component in the sample n,

hence
$$n = N - f(t) \tag{14.2}$$

where N = number of components initially at time $t = 0$.

From equations 14.1 and 14.2

$$\frac{d[f(t)]}{n} = \frac{d[f(t)]}{N-f(t)} = n \, dt.$$

Integrating $\qquad$ $\log [N - f(t)] = \lambda t + \text{constant} (c)$ $\qquad$ (14.3)

when $\quad t = 0, f(t) = 0,$

hence $\quad c = -\log N$

Substituting in 14.3 $\qquad$ $\log \dfrac{N}{N - f(t)} = \lambda t,$

$$\frac{N}{N-f(t)} = e^{\lambda t}$$

reversing $\qquad$ $1 - \dfrac{f(t)}{N} = e^{-\lambda t}$

hence $\qquad$ $f(t) = N[1 - e^{-\lambda t}].$ $\qquad$ (14.4)

The exponential growth of $f(t)$ with time is shown in Figure 14.7.

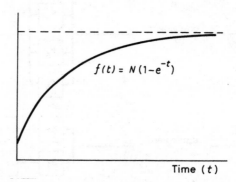

Figure 14.7 *Number of failures as a function of time*

Extending this value for $f(t)$ gives:

$$\text{Unreliability}, F = \frac{f(t)}{N}$$

$$= \frac{N[1 - e^{-\lambda t}]}{N}$$

$$= 1 - e^{-\lambda t}. \qquad (14.5)$$

$$\text{Reliability}, R = 1 - F$$

$$= e^{-\lambda t}. \qquad (14.6)$$

This may be considered as a 'limit' distribution defined by the following expressions for the differential probability functions:

$$h(t) = -\lambda$$

$$f(t) = -\lambda e^{-\lambda t} \quad \lambda > 0, t > 0$$

$$R(t) = -e^{-\lambda t}$$

The single parameter λ has contributed to the popularity of exponential failure distribution and greatly simplified analysis.

Drenick [7] established and proved the following theorem in support of exponential distribution:

'Consider a complex equipment consisting of many components, each having independent times between failures. Suppose that equipment failures result from the failure of any component, then the sequence of equipment failures in time is simply the sequence of all individual component failures'.

The characteristic feature of exponential distribution is the time independence of failure rate, which are sometimes called 'random' failures. Sudden failures of a random nature are well described by an exponential distribution.

Failures which arise as the result of a deterioration process do not obey the exponential law. For such 'wear out' failures distribution with time – dependent failure rates are more realistic.

14.3.1 Binomial law of reliability [6]

When n equal components are in parallel and R and F are the reliability and unreliability of each component, then

$$(R + F)^n = 1$$

which may be expanded into

$$R^n + nR^{n-1}F + \frac{n(n-1)}{2!}R^{n-2}F^2 + \ldots \, ^n(Cr)R^{n-r}F^r + \ldots F^n = 1.$$

If all n components must fail to produce complete system failure

$$R = 1 - (F)^n$$

If only x components must fail to produce complete system failure

$$R = R^n + nR^{n-1}F + \ldots \, ^n(Cx - 1)R^{n-x+1}F^{x-1}$$

and since $R + F = 1$

$$Q = \, ^n(Cx)R^{n-x}Q^x + \ldots Q^n.$$

14.3.2 Multiple failures [1]

A component may have several independent non-interacting failures yet a single failure mode may result in total failure. Nevertheless, as shown by Carter the overall failure rate will be the arithmetic sum of all the individual failure rates.

Thus the overall reliability (R_0) will be related to the reliability with respect to N failure modes $(R_1)(R_2) \ldots (R_N)$ by the product rule

$$R_0 = R_1 \times R_2 \times R_3 \ldots R_N.$$

Taking logarithms

$$\log R_0 = \log R_1 + \log R_2 + \ldots \log R_N = \sum_1^N \log R$$

and differentiating with respect to time

$$\frac{(d/dt)(R)}{R_0} = \sum_1^N \frac{(d/dt)(R)}{R}.$$

These terms represent the overall failure rate and the individual failure rates, hence

$$\lambda_0 = \sum_1^N \lambda,$$

i.e. the overall failure rate is the arithmetic sum of the individual failure rates.

14.3.3 Reliability concepts [8]

The failure distribution function $F(t)$ is based on the probability of failure of a unit which begins to function at time $t = 0$ and failure occurs at the instant time $t = \tau$. If τ is assumed to be a random variable with a cumulative distribution function $F(\tau)$ then the probability of failure within the time interval $[0, \tau]$ can be written as $P(\tau \leqslant t) = F(t)$ in mathematical symbols.

The probability density function or failure density function is given by:

$$f(t) = \frac{dF(t)}{dt}. \tag{14.5}$$

This means that the probability that a unit which has not failed at time t will fail in the interval $[t, t + dt]$ will be

$$f(t) \, dt$$

The hazard rate function or failure rate function $Z(t)$ is the conditional probability density function of time to failure, it is the probability that a unit might fail in the time interval $[t, t + dt]$ given that it has not failed at time t and is given by

$$Z(t) \, dt = \frac{f(t) \, dt}{R(t)},$$

i.e.
$$Z(t) = \frac{f(t)}{R(t)} \qquad (14.6)$$

where
$$R(t) = \text{reliability} = 1 - F(t) \qquad (14.7)$$

From 14.5 and 14.7
$$\frac{dF(t)}{dt} = -\frac{dR(t)}{dt} \qquad (14.8)$$

Substituting 14.8 in 14.6

$$Z(t) = \frac{dF(t)/dt}{R(t)}$$

$$= -\frac{dR(t)/dt}{R(t)}$$

$$= -d[\log R(t)]/dt \qquad (14.9)$$

Integrating equation 14.9

$$R(t) = \left[\exp -\int_0^t Z(t)\, dt\right]. \qquad (14.10)$$

14.3.4 Empirical failure pattern – Weibull distribution [1]

A simple empirical expression suggested by Weibull [9] has found to conveniently represent any failure pattern which is amenable to calculation and is based on the relationship

$$\lambda(t) = \left(\frac{\beta}{z} \frac{t - t_0}{z}\right)^{\beta - 1}$$

where β = a shaping constant which controls the shape of the failure curve,

z = a scaling constant which stretches the time axis.

The effects of β on the failure pattern for $t_0 = 0$, $z = 1.0$ is shown in Figure 14.18

Although the Weibull distribution is an entirely empirical relationship it has been effectively used to relate to practical random and normal patterns. The associated reliability and failure equations are:

$$R(t) = \exp\{-[(t - t_0)/\eta]^\beta\}$$
$$F(t) = 1 - R(t) = 1 - \exp\{-[(t - t_0)/\eta]^\beta\}$$

14.3.5 Failure distribution – time dependent failure rates

Three theoretical distributions are used in reliability predictions in which the failure rate is 'weighted' with a time-dependence, namely:
(1) Normal distribution;

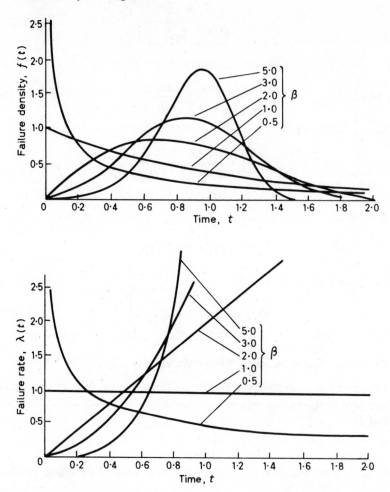

Figure 14.8 *The effect of β on failure density and failure rate for the conditions* $t_0 = 0$, $z = 1.0$ *in the Weibull distribution [1]*

(2) Log-distribution;
(3) Weibull distribution;
(4) Extreme-value distribution.

The distribution functions for these distributions are compared with those for the exponential distribution in Table 14.1.

14.4 Load and strength — statistical distribution [1]

There is a variability or scatter both in the strength of a component and the duty to which it will be subjected.

Table 14.1

Function	$R(t)$	$f(t)$	$h(t)$	$E(t)$
Exponential	$e^{-\lambda t}$ $\lambda > 0, t > 0$	$\lambda e^{-\lambda t}$	λ	$\dfrac{1}{\lambda}$
Normal	$\dfrac{1}{(2\pi)^{1/2}} \displaystyle\int_t^\infty e^{-z^2/z}\, dz$ $z = \dfrac{t-\mu}{\sigma}$ $\mu - 3\sigma > 0$ as $t \geqslant 0$	$e^{-z^2/z}$	$\dfrac{e^{-z^2/z}}{\displaystyle\int_t^\infty e^{-z^2/z}\, dz}$	μ
Log normal	$\dfrac{1}{(2\pi)^{1/2}} \displaystyle\int_t^\infty e^{-z^2/z}\, dz$ $z = \dfrac{\ln t - \mu}{\sigma}$ $t > 0$	$\dfrac{1}{(2\pi)^{1/2}} e^{-z^2/z}$	$\dfrac{e^{-z^2/z}}{\displaystyle\int_t^\infty e^{-z^2/z}\, dz}$	$e^{(\mu + \sigma^2/2)}$
Weibull	$e^{-t^\beta/\alpha}$ $\alpha > 0, \beta > 0$ $t \geqslant 0$	$\dfrac{\beta}{\alpha} \cdot t^{(\beta-1)} e^{t^\beta/\alpha}$	$\dfrac{\beta}{\alpha} \cdot t^{\beta-1}$	$\alpha^{1/\beta}(1 + 1/\beta)$
Extreme	$e^{-e^{-z}}$ $z = \dfrac{t-\mu}{\beta}$ $t \geqslant 0$ $\mu > 0, \beta > 0$	$\dfrac{1}{\beta} e^{-z + e^z}$	$\dfrac{1}{\beta} \dfrac{e^{-z}}{e^{e^z} - 1}$	$\mu + 0.5772\beta$

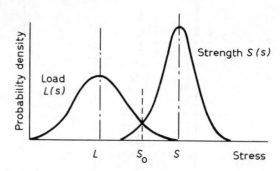

Figure 14.9 *Distribution of load and strength*

A typical distribution for load and strength is shown in Figure 14.9 where

$\bar{S}$ = mean or 'nominal' strength of the component,

$S(s)$ = probability density function, strength,

= probable occurrence of any particular strength in a family of products,

$\bar{L}$ = mean or 'nominal' load applied to a component,

$L(s)$ = probability density function, load,

= probable occurrence of a particular load on a component,

$\bar{S}/\bar{L}$ = safety factor.

The difference between $\bar{S}$ and $\bar{L}$ represents the safety margin and although expressed in terms of the standard deviations σ_S and σ_L to allow for the probable scatter or distribution of $S(s)$ and $L(s)$, can be expressed more precisely as

$$\text{safety margin} = \frac{\bar{S} - \bar{L}}{(\sigma_S^2 + \sigma_L^2)^{1/2}}$$

where σ_S = standard deviation of the strength,

σ_L = standard deviation of the load.

The limit of reliability, onset of failure results from a zero safety margin, i.e. $\bar{S} = \bar{L}$.

Cases which may arise in practice are represented by Figure 14.10 in which

(1) There is a considerable scatter in the strength of the component but the load is limited with little scatter, i.e. $\sigma_L \ll \sigma_S$. This may be described as 'smooth loading' in which $R = (\bar{R})^N$ for a series arrangement.

(2) The strength is consistent with small scatter but the load is variable with a wide distribution, $\sigma_L \gg \sigma_S$. This may be described as 'rough loading' for which $R = \bar{R}$ for a series arrangement.

(3) Both the strength and load are distributed, $\sigma_L = \sigma_S$. This is an intermediate situation between smooth and rough loading.

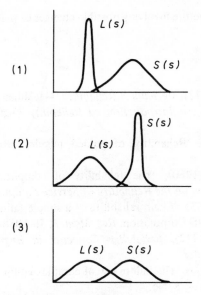

Figure 14.10 *Influence of the standard deviation*

14.5 Reliability assurance – BS 9000 system

The BS 9000 system of specification has been designed to provide a measure of assurance that components released in accordance with it have the requisite degree of reliability irrespective of the location in which they function.

The system, applicable to continuous mass production techniques is based upon three basic distributions and one important theorem [10], namely:

(1) Gaussian distribution, for the assessment of measured and recorded values and for other function concerned with standard errors;

(2) Binomial distribution, for the assessment of attributes where the AQL has a sensible magnitude;

(3) Poisson distribution, for the assessment of failure rate where the probability of failure is extremely small;

(4) Bayes theorem, using the knowledge from previous production to apply that currently running.

From this system of interpretation of the performance of test batches three classes of electronic equipment are specified:

Class 1: Very expensive components in small quantities and manufactured in small batches with limited testing,

Class 2 : Components such as transformers and switches made continuously on a line but with minor variations such as turns ratio or contact combination such that no two consecutive batches are identical although all employ a general design and constructional pattern,

Class 3: Components made in large batches with long periods intervening between batch manufacture involving possible changes of personnel and adjustment of jigs.

References

1 Carter, A.D.S. (1972), *Mechanical Reliability,* MacMillan, London.
2 Lewis, R. (1970), *An Introduction to Reliability Engineering,* McGraw-Hill, London.
3 Lusser, R. (1954), 'Reliability of guided missiles,' Redstone Arsenal, September.
4 Chaddock, D.M. (1960), 'The reliability of complicated machines', *Inter-Services Symposium on the Reliability of Service Equipment,* I. Mech. E.
5 Howard, W.J. (1953), 'Chain reliability — a simple failure model for complex mechanisms', RAND Corporation, *Res. Mem. R.M.* **1058.**
6 Brook, R.H.W. (1972), *Reliability Concepts in Engineering Manufacture,* Butterworths, London.
7 Drenick, R.F. (1960), 'The failure law of complex equipment', *J. Soc. Industril Applied Mathematics,* **8,** No. 4.
8 Mathieson, Tor-Chr. (1973), 'Reliability engineering in ship machinery plant design', Report IF/R, 12 Division of Internal Combustion Engines, University of Trondheim, Norwegian Institute of Technology.
9 Weibull, W. (1951), 'A statistical distribution of wide applicability', *J. Applied Mechanics,* **18,** pp. 293–297.
10 Lovelock, R.T. (1973), 'Minimising the cost of reliability assurance', *Electronic Components,* 29 June 555–557.

15 Reliability data sources

15.1 Introduction

Data used in data collection programmes may obtained in several ways as follows:

(1) Data from laboratory testing where simulated conditions are applied to a number of parts or components;

(2) Field data on similar parts or components obtained during controlled development or operational test programmes;

(3) Field data from similar parts or components obtained during normal operational use;

(4) Field data from generically similar parts or components used in a variety of applications.

For mechanical components only (3) and (4) are feasible. Analyses of field data show that most failures are not part-related or even component-related either in mechanical or electronic applications. Factors of importance when considering applications of field data are:

(1) Similar or identical components are used under different operational care, different environment, different application and different maintenance policy;

(2) Components are used in applications exposing different requirements on component parameters;

(3) A component may exhibit many different failure modes causing different effects upon performance;

(4) Most failures are due to interacting effects at both part, component and system level;

(5) Additional variability is introduced through design, manufacturing and installation;

(6) Mechanical components are low population items, e.g. small sample sizes for evaluation of reliability characteristics;

(7) Mechanical components are subject to continuous efforts for design and manufacturing improvement, which may be initiated from the same failure data collected for prediction purposes.

15.1.1 Failure data imperfections

Data obtained during the course of collection may contain imperfections from a variety of causes, such as:

(1) inadequacies of the data forms in an effort to compromise between administrative expediency and technical comprehensiveness;

(2) Erroneous or incomplete data due to the complexity of causes and mechanisms, time restraints in reporting, inadequacy of personnel training;

(3) Ambiguity of replies, omission of statements of fact committing persons responsible for primary causes of failure arising from improper operation or omitted maintenance;

(4) Incomplete descriptions, particularly of components, failure definition and description;

(5) Inability to provide accurate operating times.

15.1.2 Failure data collection systems

Reference to the following systems used in the field to collect failure data was made by Matheison [1].

U.S.A.

A comprehensive review of sources was made by Harrington Coats and Farley [2]. The Maintenance Data Collection Subsystem (MDCS) forms a major part of the 3-M System used by the U.S. Navy and is implemented on board 800 ships covering all types of machinery ranging from washing machines to main turbines.

West Germany

A data collection system is used at the Staal. Ingenieurschule, Flensburg based on 140 ships [3]. Since 1968 a data collection system has been implemented by MAN, Ausburg on ships equipped with main engines of their make [4].

Japan

Investigations have been undertaken since 1966 by the Ship Research Institute, Japan based primarily on the engineer's log [5]. Statistical analysis of the failure data is limited to computations of failure rate at component and system level except for the main engine and boiler where they are also performed at part level.

U.S.S.R.

Propulsion plant failure investigations from 42 Russian ships were reported by Zhenovak [6]. Information was collected from engine room logs, log-books etc. but only included pumps, starting air compressors, purifiers and heat-exchangers. Data analysis was based on an assumed renewal (repair) process with Gamma-distributed times between failures, namely

$$f(t) = \lambda \frac{(\lambda t)^{k-1}}{(k-1)!} e^{-\lambda t}$$

where the parameter k can describe both exponential ($k = 1$) and the normal ($k > 5$) distribution.

Sweden

Investigations by the Swedish Shipbuilding Research Foundation, Gothenburg were initiated on a small sample of 10 ships [7]. This was too small a sample for effective data analysis.

Norway

A data collection system from a number of different ships has been used by the Ship Technical Research Institute to provide experience in methodologies [8]. Det Norske Veritas has also implement a data collection service on the so-called EO-ships.

15.2 Systems reliability service (SRS)

The Systems Reliability Service is part of the Safety and Reliability Directorate (SRD) of the United Kingdom Atomic Energy Authority (UKAEA) and is a Government owned and sponsored undertaking with no part of interest in the commercial design, manufacture or sale of equipment or systems.

In a 1972 report [9] there were 17 Associate Members making contributions in the form of reliability techniques, computer programs, event and reliability data as follows:

Airworthiness Division of the Civil Aviation Authority, UK. (formerly Air Registration Board).
Central Electricity Generating Board, U.K.,
Imperial Chemical Industries, U.K.,
Gas Council Engineering Research Station, U.K.,
Ministry of Defence, U.K.,
FTL, Military Electronics Laboratory, Sweden,
Development of Reliability Analysis Methods, DRAM Project, Norway,
Technical Research Centre Finland,
African Explosives and Chemical Industries, South Africa,
Oak Ridge National Laboratories, U.S.A.,
Commission des Communautes Europeenes, Brussels,
Atomic Energy Commission, Denmark,
Comitato Nazionale per l'Energia Nucleari, Roma, Italy (Security and Control Division),
Comitato Nazionale per l'Energia Nucleari, Bologna, Italy (Fast Reactor Division),
Junta de Energia Nuclear, Spain,
Atomic Energy Board, South Africa,
Commisseriate a l'Energie Atomique, Saclay, France.

Projects which have been undertaken by SRS include reliability and assessment involving availability, fault modes, sensitive or critical areas, maintenance and system operational viability including applications to:

Nuclear submarine propulsion reactors,
High pressure die casting machines,
Criticality monitoring and alarm systems,
Normal and standby electrical supply and distribution systems,
Chemical plant automatic protective systems,

High pressure relief and protective systems,
Electronic and electro-mechanical logic sequence circuits and systems,
Hazardous gas alarm systems,
Medical engineering equipment,
Life support systems,
Plant measurement and control systems,
Cooling water systems and their associated controls,
Investigations of repair and maintenance characteristics,
Actuator systems,
Fire detection and control systems,
Marine engine control systems,
Chemical plant hazard evaluations,
Plant availability studies,
Boiler feed systems and sequence control systems,
Electronic and control equipment evaluations.

15.2.1 Reliability – systems reliability service (SRS)

Reliability data has been accumulated for a wide range of components and equipment providing the basis of a data bank which is an integral part of SRS. Reliability assessments have been performed on a very wide range of industrial plants including mass production engineering plants, oil refineries, chemical plants, ships and protective systems, any of which can be regarded as an assembly of components liable to periodic breakdown. From information on the failure characteristics of various components, it is possible, by applying the techniques developed by the United Kingdom Atomic Energy Authority, to forecast quantitatively the likely reliability and/or availability of a plant.

The commercial and data bank activities of the Systems Reliability Service are complemented by an Applied Research Unit which undertakes the research and development of system reliability technology in the national interest. The R and D programme is directed to the solution of reliability and availability problems that arise in industry. The Centre is at Culcheth near Warrington under the control of the UKAEA Director of Safety and Reliability.

15.2.2 S.R.S. reliability data service (SYREL)

Reliability data acquisition, analysis, storage and distribution has been an integral part of the work of S.R.S. and the earlier U.K.A.E. Authority Health and Safety Branch (A.H.S.B.), the exact and ideal requirements for such a scheme being defined in report A.H.S.B.(S) R.138 'Data store requirements arising out of reliability analysis'.

The S.R.S. Data Bank, which now exists, incorporates two facilities (1) an 'event' data facility, and (2) a 'reliability' data facility.

The event data bank allows every individual event and its time of occurrence (such as breakdown, maintenance, modification, testing, etc.), on a components, equipment or system to be recorded, stored and analysed. The results of such analyses which lead to particular reliability parameters of interest (such as failure characteristics, failure-rates, repair times, maintenance activities, etc.) are transferred to the Reliability Data Bank where they are stored for subsequent use by system reliability analysts and for subsequent dissemination to the contributors to the Data Bank.

The S.R.S. reliability data bank is continually expanding and currently contains some 4000 entries relating to different components and items of equipment. There is a fairly large content of electronic component information from a number of different applications ranging from land-based computers to control and instrumentation equipment in transportation systems of various types. There is also a wide coverage of information on electrical, electro-mechanical, and mechanical components and items of equipment. A selection of a few of the typical components on which reliability data is stored in the S.R.S. Reliability Data Bank is given in S.R.S. Leaflets SRS/TECH/1/3 and SRS/TECH/1/4 together with a small selection of the type of reliability information which is stored in connection with each component.

15.2.3 'SYREL' reliability data

Participation in the exchange of information with the Systems Reliability Service 'SYREL' Data Bank is provided at 3 levels; full particulars are given in [10].

(1) Basic grade: This requires that for a plant item only its identity, the time of operation, and the number of its failures or breakdowns in that period are known, or are obtainable from the maintenance records. This will enable a 'failure rate' in terms of failures/year or failures /10^6 hours to be calculated. It will usually be necessary to utilize an 'Operating Time Function' for the item in order to obtain operating time as a fraction of 'History Time', unless these are judged to be identical for the plant and item concerned. This simplest form of information can be obtained by manual inspection of records, by machine sorting of punched cards, or by interrogation of computerized storage. Where manual inspection is necessary, economy in specifying the item inventory is desirable in view of the labour likely to be involved. Note that this basic grade takes no account of the qualifying conditions that may affect the behaviour of the item. The minimum information is:

Inventory No.,
IDEP No. (a code based on the Inter-Department Data Exchange Programme),
Description,
Population in use,
Range/size,
Operating time,
Number of failures in the period.

(2) Intermediate grade: For the intermediate grade of reliability data collection, the following data are required:

Identity,
IDEP No.,
Description,
Population in use,
Range/size,
Manufacturer,
Quality (standard specification applicable),
Component material (if relevant),
Environment,
Date of manufacture/entry into service,
Operating time/cycles,
Number of failures in the period.

This grade introduces some qualifying information in order to differentiate between the performances of superficially similar items which may differ in age, environment, or manufactured quality.

(3) High grade: For high grade reliability data collection one is interested not only in breakdown information but also in 'event' information, where an event can be any of the following:

In situ maintenance,
Workshop maintenance,
Breakdown,
Test/calibration,
Modification.

In addition it is necessary to know certain facts associated with these events, e.g.:

Maintenance interval,
Repair time
Outage time
} for each event originating with the item,
Effect on the system of which the item is a part,
Effect on the plant or installation,
Operating conditions of the item at time of failure (as opposed to design conditions),
Replaced parts,
Operating time/cycles,
Number of failures and the mode of failure and cause of failure in each case.

This fuller treatment enables not only average failure rates to be deduced, but also the derivation of distributions for failure rates, repair times and outage times and a more meaningful analysis of modes and causes of failure.

15.2.4 Lloyds Register – defects data store

Defects and damages trends/patterns in a whole fleet class can be identified from an

IBM 370/155 computerized technical information data store which has been established by Lloyds Register of Shipping [11].

The technical records of ships held in the data store are the most comprehensive of their kind in the world and comprise full technical details of hull and machinery constructions together with detailed reports on failures, defects and damages. Regular scanning of the defect store relating to all those areas of a ship known to be troublesome in practice will result in a printout being produced when defects have been reported on either a given number of a pre-determined percentage of the total number of each item at risk. By this means an early warning of a tendency to failure will be obtained before the matter becomes serious. Acting on such information, the appropriate department of Lloyd's Register may suggest amendments to the Rules or changes in survey procedures. Within the limits of confidentiality, (since details of particular ships cannot be published without the owner's consent) the information may also be made available to the shipping industry.

The ability of the system to distinguish between failures which are part of a pattern and those which are a random phenomenon is of assistance to designers when considering a new ship design. It will also enable appropriate fields of research to be embarked upon at the earliest possible moment. The data bank is continuously up-dated. Particulars of new ships are included under their classification as well as alterations or modifications which have occured to existing ships, together with information furnished by surveyors in the field concerning details of damages found, causes (where known) and their recommendations for repair.

15.2.5 EXACT reliability data system

The desirability of setting up an international organization for the exchange of information on the reliability of components and the feasibility of the system was examined by a working group within OECD (Organization for Economic Cooperation and Development, Paris) in 1965. This working group was composed of experts from Belgium, Canada, France, Germany, Japan, the Netherlands, Sweden, Switzerland, United Kingdom and United States. In 1967 a suitable scheme was named EXACT, International Exchange of Authenticated Electronic Component Performance Tests Data. A Central Office was set up in Stockholm and was sponsored by the Swedish Board for Technical Development and Forsvarets Teletekniska Laboratorium, FTL (Military Electronics Laboratory) to the extent of £25 000. The expenses for the National Centres borne by interested organizations in the relevant countries [12].

The information exchanged within EXACT contains the results from: type tests; comparative tests; failure analysis; product analysis; endurance tests; incoming inspection tests; application tests. other tests. Well-known specifications such as IEC, should be used for testing but other specifications may be used if the conditions are clearly explained in the report. Other information of interest for component users may be included, if such information is not readily available on the market. The official language is English. The collected information is distributed monthly and a cumulative EXACT Information Index published twice each year. The organizational chart for EXACT and GIDEP as at March 1974 is shown in Figure 15.1.

15.2.6 EXACT index number codes

To provide a standard identification code for various types of components in order to facilitate immediate access and retrieval of reliability information from the data bank EXACT (Exchange of Authenticated Component Test data, Fack S-10450, Stockholm 80, Sweden) have established a generic coding system developed by GIDEP (Government Interagency Data Exchange Program, H.W. Naval Material Command, Code HAT 03426, Washington DC 20390 U.S.A.).

Each index number contains nine digits by which the component is classified. The first three digits define the major part classification while the remaining three pairs of digits define sub-classifications, thus an index number 742.10.40.96 is made up as follows:

Major classification:	semiconductors, transistors	742
Function:	general purpose (1 watt)	10
Material and polarity:	silicon, PNP	40
Process:	planar	9
Construction:	metal can	6

15.2.7 Failure rate data (FARADA)

The FARADA (FAilure RAte DAta) Programme established at the Naval Ordance Laboratory, Corona, California, U.S.A. by a special group within the U.S. Naval Fleet Missile Systems Analysis and Evaluations Group implements and promotes the exchange of component part failure rate data among prime contractors and major subcontractors engaged in the design, development, and production of equipments for military and space programmes, with a view to improving reliability. A paper describing this facility was prepared by Pollock and Richard [13]. The immediate and continuing FARADA aim is to receive component parts failure rate data from prime contractors, major subcontractors, and those activities of the Navy, Air Force, Army and N.A.S.A., who participate in the programme; then quickly analyse, compile, and distribute the data in usable form to all the participants. At present, there are 125 contractors and 36 government activities participating, and more in each category will be added as the programme progresses.

The ultimate and over-all aim of the FARADA Program is to provide design engineers with the failure rate information that will assist them in making accurate reliability predictions and in developing ever more reliable systems and equipments.

15.2.7.1 Programme publications

One of the first N.O.L.C. projects in implementing the objectives for the

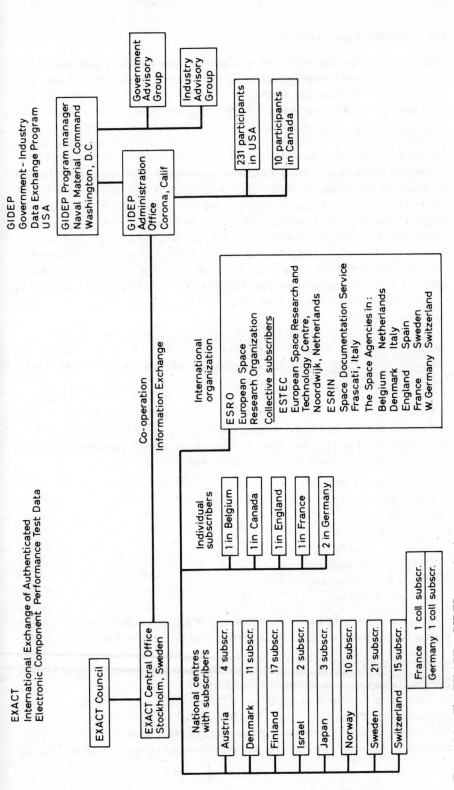

Figure 15.1 *The EXACT – GIDEP organization*

programme was the preparation of a 'Standard Operating Procedure' (SP 63-47). This document, which was distributed to all programme members in March 1962, establishes the administrative procedures that apply to all the participants, and provides basic guidelines for collecting and recording the failure rate data generated by the many military and space projects. As in all active programmes this document by no means presents the ultimate and final procedure. It will be modified as new concepts and techniques are developed.

Another requirement of the programme was achieved by the publication of a handbook. N.O.L.C. because of its role as Information Centre and its experience in the Interservice Data Exchange Program and the Guided Missile Data Exchange Program, was in a unique position for consolidating and publishing the collected failure rate data. Therefore, when the FARADA programme got under way, the Laboratory screened, summarized, and compiled the wide selection of information received from the Army, Navy, Air Force, and N.A.S.A. participants, and consolidated it into the *Failure Rate Data Handbook* (SP 63-470). It was distributed to all programme participants in June 1962, and revisions and additions have been distributed quarterly.

The information in the handbook is presented in three volumes. Volume 1, which presents data in tabular form, has two sections: Table A, Electrical and Electronic Parts (printed on yellow pages) and Table B, Mechanical, Hydraulic, Pneumatic, Pyrotechnic, and Miscellaneous Parts (printed on blue pages). Volume 2 presents data in the form of stress curves. Volume 3 contains the background information on failure rate data provided by the contributing sources, and describes the environmental and usage conditions under which the measurements were observed. The included form, when completed by a participant, gives the required information under the following headings:

Description of equipment,
Conditions of test,
System statistics,
Failure reporting system,
Component part failure modes,
Component part failure times,
Special environment conditions,
Pertinent remarks,

The data in the tables of Volume 1 were originally assembled manually, but the continued expansion of the FARADA Programme necessitated the use of computer techniques for storage, manipulation, retrieval, and reporting. The computer system developed at N.O.L.C. is oriented to a master file, which contains detailed failure rate data in major categories and environments.

There are now more than 300 different major classifications of component parts represented in the tabular collection. Of the total of some 10 000 data entries, 50% are for electrical or electronic component parts (Table A) and 50% are for mechanical, hydraulic, pyrotechnic, and miscellaneous component parts (Table B).

Of the total entries, about 70% are accompanied by sufficient statistical information to constitute the nucleus of the analysis effort included in the plans that
have been initiated for improving the FARADA Programme.

This three-volume handbook is intended to provide design engineers with failure
rate information in a convenient form for use during the design phase. If properly
applied, the data should provide the means of numerically assessing the probability
of survival (i.e., the reliability) of a system prior to, or simultaneously with, the
construction of hardware. After sufficient hardware has been manufactured and a
designated number of systems have been operated for a specified length of time, the
accuracy of the survival estimates made on the basis of failure history can be
checked. As experience is gained in the use of this method, refinements can be
made, and improved design should be the result.

It must be emphasized again that the failure rates compiled by the FARADA
Information Centre at N.O.L.C. are based upon long histories of component part
failures obtained from specific engineering data and from test results. The validity
of the data has been, and will continue to be, examined carefully. As a result, a
degree of confidence in the data is being developed that is sufficiently high to
warrant their practical application as guides to the design engineer.

Because of the technical veracity and the currency of the data, the handbook
provides a reference for use in all military and space programmes in lieu of the
inapplicable, outdated, inaccurate, or inconsistent data that have often been used
as a basis for system reliability assessments.

15.3 Failure data

Typical data by Ablitt in an introduction to the 'Syrel' reliability data bank [10]
indicating the more statistically precise nature of this information included:

Boiler Feed Pumps

Size	7000 HP	6000 HP
Operating pressure	2500 psi	2500 psi
Population	6	10
History time	$3\frac{1}{2}$ years	5.1 years
Failure rate	6 faults/pump/year	4 faults/pump/year

Diode, rectifier, silicon

History time	11 300 h
Mean operating time	1 980 h
No. items	1400
No. faults	1
Items × operating time	$2.8.10^6$ item-h
Failure rate (operational)	0.357 faults/10^6 h (mean)

Electrical motors (industrial)

Description	Three phase motor for feed pump frequenty 60/cycles sec 26 HP 415 volts
No. of items	12
Mean operating time	2.28 years
Sample size (operating time)	27.4 item years
Mean history time	10 years
No. of faults	2
Mean failure rate (operating time)	8.33 faults/10^6 h
Upper failure rate (operating time)	30.1 faults/10^6 h
Lower failure rate (operating time)	1.01 faults/10^6 h
Confidence band	95%
Distribution	Poisson

Gas turbine (industrial)

Application:	Average industrial
Range/size/power	6400 HP
Data year	1970
Failure rate of fuel control unit of the gas turbine	
No. of items	3
Items × operating time	20 item years
No. of faults	2
Failure rate (mean)	
(Based on operating time)	11.4 faults/10^6 h
(lower limit)	1.38 faults/10^6 h
(upper limit)	41.2 faults/10^6 h
Confidence band	95%

Nuclear reactor period meter (with trip facility)

History time	26 000 h	
Mean operating time	11 400 h	
No. of items	18	
No. of faults	22	
Items × operating time	205 200 item h	
Failure rate (operational)	mean value	107 faults/10^6 h
	lower confidence value	67 faults/10^6 h
	upper confidence value	162 faults/10^6 h
Confidence band	95%	
Assumed distribution	exponential	

Transistor, silicon, PNP, planar

History time	12 600 h	
Mean operating time	2 160 h	
No. of items	1088	
No. of faults	9	
Items × operating time	2.35×10^6 item h	
Failure rate (operational)	mean value	3.82 faults/10^6 h
	lower confidence limit	1.75 faults/10^6 h
	upper confidence limit	7.28 faults/10^6 h
Confidence band	95%	
Assumed distribution	exponential	

Sample figures relating to failure rate (λ) as a quantity per 1000 h of operation or per 1000 cycles are given below

Mechanical

Device	λ Low	High	MTTF (10^6 h) High	Low
Accelerometer	0.0003	0.021	3.3	0.0477
Clutch (N.slips)	0.00007	0.0009	*	—
Counter (mechanical)	0.0002	0.0002	5.0	5.0
Joint-hand soldered	0.00002	0.00008	*	—
Thermostat	0.0002	0.0005	*	—
Hydraulic				
Piston	0.00008	0.00035	*	—
Pump	0.001	0.005	1.0	0.2
Tubing	0.0002	0.0002	*	—
Valve (float)	0.0056	0.011	*	—

* MTTF for these are not calculated as the failure rate refers to the number of cycles.

15.3.1 Failure rate – Mechanical components

Component		Failure rate/10^5 h
Bellows metal		0.5
Bourdon tubes	Leakage of	0.005
	Creep of	0.02
Diagphragms	Metal	0.5
	Rubber compound	0.8
Gaskets	Leakage of	0.05
Grub screws	Loosening of	0.05
Hair springs	Breakage of	0.01

Component		Failure rate/10^5 h
Joints	Pipe	0.05
	Mechanical	0.02
	Ball	0.1
Nuts, bolts etc.	Breakage or loosening of	0.002
O-ring seals	Leakage of	0.02
Pipes	Fatigue or leakage of	0.02
Pressure Vessels	Small, general	0.3
	Small, high standard	0.03
	Large, high standard	0.01
Springs	Breakage of	0.2
Valves, relief	Leakage of	0.2
	Blockage of	0.5

15.3.2 Failure rates – electrical components

Component		Failure rate/10^5 h
Boards	Power station distribution	0.02
Brushes	Rotating machines	0.1
Bus bars		
Cables	Less than 1 kV/mile	0.1
	1-33 kV/mile	0.7
Circuit breakers	General less than 33 kV	0.2
	132 kV	0.4
	275 kV	0.7
Contactors	Low voltage	0.2
	3.3 kV	0.4
Fuses	Low voltage	0.02
Rotary equipment	Alternators	7.0
	Generators	5.0
	Motors	2.0
	Tachometers	0.5
Overhead lines	10-33 kV/mile	2.0
	132-275 kV/mile	0.5
Transformers	Less than 15 kV	0.06

15.3.3 Failure rates – electronic components

Component		Failure rate/10^5 h
Capacitors	Paper	0.1
	Synthetic film	0.05
	Mica	0.03

Component		Failure rate/10^5 h
	Polystyrene	0.01
	Electrolytic aluminium foil	0.2
	Electrolytic tantalum pellet	0.05
Diodes	Germanium point contact	0.05
	Silicon high power	0.05
	Silicon low power	0.002
Potentiometers	Wirewound general purpose	0.3
	Wirewound precision	0.6
Resistors	High stability carbon	0.05
	Composition grade 2	0.01
	Metal film	0.01
	Oxide film	0.005
Transistors	Germanium high power	0.1
	Silicon low power	0.005
Valves	Diode	1.2
	Triode	1.9
	Pentode	2.3

15.3.4 Failure data – aircraft propulsion

Data regarding the reliability of aeronautical propulsion systems published by Davies [15] included:

Subsystem	Failure sets/100 h	Mean time between failures (h)
Accessory drives	0.02	50 000
Basic engine	0.04	25 000
Control system	0.08	12 500
Electrical system	0.01	100 000
Fire systems	0.05	20 000
Fuel system	0.08	12 500
Lubricating oil system	0.01	100 000
Variable intake	0.02	50 000
Variable nozzle	0.02	50 000

15.3.5 Failure data – turbine components

References to failure modes and times were made during a discussion [16–18] as follows:

Turbine blades: failure due to stress corrosion cracking detected at 60 000 h,

Turbine discs: radial cracks 2.5 in long due to stress corrosion cracking after 14 to 18 years service.

15.3.6 Failure data – marine lubricating oil system components

In the course of analysing the reliability of the lubricating oil system for a 9-cylinder crosshead diesel engine Mathieson used the following component failure data [1].

Component	Mean time between failures (h)
Bearing, engine, crankshaft	200 000
Bearing, engine, crosshead	330 000
Bearing, engine, main	200 000
Cooler, lubricating oil	100 000
Pump, lubricating oil	150 000
Valves, piping etc.	75 000

15.3.7 Failure data – marine power components

From a record of 51 motor tanker vessels over periods ranging from 1 to 4 years the following failure data was presented by Wilkinson and Kilbourn [18]:

Component	Mean time between failures (h)
Air ejector	225 000
Boiler-circulating pump	24 000
Boiler-feed pump	40 000
Boiler (complete) – oil fired type	1 400
Boiler (complete) – waste heat, economiser type	68 000
Condenser	85 000
Diesel alternator – alternator	113 000
Diesel alternator – sea water circulating pump	100 000
Diesel alternator – diesel engine	11 000
Extraction pump	35 000
Turbo alternator – alternator	135 000
Turbo alternator – sea water circulating pump	185 000
Turbo alternator – turbine	18 000

15.3.8 Gas turbine deterioration

The gas turbine is basically a simple machine with one or two large moving parts. According to Wert [20] it can be considered extremely reliable if maintained to schedule. He quotes the following figures:

Major inspections	30 000 operating hours
Turbine buckets (nominal life)	100 000 operating hours

15.3.9 Failure data – piston rings and pistons

According to Brandt [21] the failures occur as follows:
    (1) 10% of all marine diesel piston rings fail in under 3500 h;
    (2) 20% of all marine diesel piston rings fail in under 5000 h;
    (3) If the top rings are changed at 2500 h the probability of failure at 5000 h is reduced to 11.6%.
    In the case of direct drive engines it was recorded [22]:
    (4) Pistons need regrooving every 20 000–40 000 h;
    (5) Top rings require replacement every 5000–8000 h;
    (6) Burning of the piston crown limits total piston life;
    (7) Top rings peripheral wear rate = 10 × linear wear rate.

15.4 Environmental influences on instrument failure rates

Studies by Anyakora, Engel and Lees [23] show that the failure rate of chemical process instruments and control systems depends considerably on environment, the best environment being a location in a control room and the worst location being in a plant in direct contact with aggressive process materials.

The following table relates published data with values obtained from three works:

Works A: Producers of heavy organic chemicals, some fluids are relatively clean (air, steam, water) while others are contaminated or corrosive or liable to promote blockages;

Works B: Producers of heavy inorganic chemicals involving an acid plant, sintering plant, furnace, water treatment. This is a more severe environment than works 'A' as it includes acids, solids handling and high temperatures;

Works C: Glass works with similar conditions to works 'B' but with a large amount of the instrumentation consisting of equipment for the flow control of furnace fuel oil and air, it is housed in clean rooms adjacent to the control room.

15.4.1 Failure rate – instruments

Instrument	Published data	Works A	Works B	Works C
Control valve	0.25	0.57	2.27	0.127
Controller	0.38	0.26	1.80	0.32
Differential pressure transmitter	0.76	–	–	–
Pressure transducers (pressure)	–	0.97	2.20	–
Thermocouple	0.088	0.40	1.34	1.00
Flowmeters (fluids)	0.68	1.90	1.68	1.22
Analyser – O_2	2.5	–	1.45	7.0

Instrument	Published data	Works A	Works B	Works C
H_2O	–	–	8.00	–
CO_2	–	–	–	10.5
pH	–	17.1	–	4.27
Flame failure detector	–	10.0	–	1.37
Pressure gauge	0.088	–	–	–
Pressure switch	0.14	0.30	1.00	–
Power cylinder	–	0.64	1.45	–
Air supply	–	0.046	0.11	0.046
Pneumatic connections	–	0.014	0.014	–
Electrical connections	–	0.024	0.062	–

The aggressiveness of the process materials is of primary importance, a severe environment appears to increase the failure rate by a factor of up to four.

15.5 Failure data – confidence level

The confidence which can be assigned to any failure data is dependent upon the nature and source of the data and the number of samples (population) from which it has been obtained.

The confidence level can be computed from the equation

$$R^N = 1.00 - C,$$

where R = percentage of the number of units tested which do not fail,

N = number of units tested,

C = confidence level.

Thus if a reliability of 90% is required with a confidence of 95% the number of units which must be tested and found to operate satisfactorily would be given by:

$$(0.9)^N = 1.00 - 0.95 = 0.05,$$

$$N = \frac{\log (0.05)}{\log (0.9)} = 38.8$$

References

1 Matheison, Tor-Chr. (1973), 'Reliability engineering in ship machinery plant design', Report IF/R.12., University of Trondheim.
2 Harrington, R.L., Coats, J.W. and Farley, F.E. (1970), 'Assessment of marine systems reliability and maintainability', *Annals of Reliability & Maintainability* **9**.

3   Harrington, R.L., Coats, J.W. and Farley, F.E. (1969), 'Technischer Bericht der Forschungs-u-Erprobungsstelle für Schiffsbetriebstechnik, Zuverlässigkeits-technik in der Sieschiffart' Staatl. Ingenieurschule, Flensburg.

4   ——— (1971), *Diesel Courier,* 26, MAN, Augsburg, March.

5   Tamaki, H. (1968), 'Failure investigation of ship propulsion plants', *Japan Shipbuilding & Marine Engineering,* 3.

6   Zhenovak, A.G. (1969), 'The degree of reliability of ship's auxiliary machinery in service', *Trans. Res. inst. Merch. Marine U.S.S.R.* 112, (B.S.R.A. Translation 3340).

7   Albert, L. and Strandberg, I.L. (1966), 'Drifsstörningar, reparationer och under-häll i Fartyg,' Swedish Shipbuilding Research Foundation, Report 48, Gothen-burg.

8   ——— (1967), 'Data Fra Skip', Ship Technical Institute of Norway Report M85, Trondheim.

9   ——— (1972), 'Information of the background, experience and current capa-bilities of the systems reliability service', Report SRS/Tender/1031, UKAEA Culcheth, Warrington, Lancs.

10 Ablitt, J.F. 'An introduction to the SYREL reliability data bank', Report SRS/GR/14, UKAEA, Culcheth, Warrington, Lancs.

11 ——— (1974), 'Warning system detects machinery failures', *Marine Engin-eering Review,* February.

12 Gussing, T. (1972), 'EXACT – The international system for the exchange of information on electronic components', *Electronic Components,* December.

13 Pollock, S.I. and Richards, E.T. (1964), 'Failure rate data (FARADA) program conducted by the U.S. Ordnance Laboratory, Corona, Calif.,' *Proc. 3rd. Re-liability & Maintenance Conference,* A.S.M.E.

14 Green, A.E. (1969), 'Reliability Prediction', Paper 3, *I. Mech. E. Conference,* September.

15 Davies, A.E. (1972), 'Principles and practice of aircraft powerplant mainten-ance', *Trans. I. Mar. E.* 84, 441–447.

16 Kalderon, D. (1972), 'Steam turbine failure at Hinckley Point', *Proc. I. Mech. E.* 86, No. 31.

17 Gray, J.L. (1972)', 'Investigation into the consequences of the failure of a turbine generator at Hinckley Point 'A' Power Station, *Proc. I. Mech. E,* 186, No. 32.

18 ——— Discussion paper D.117, *Proc. I. Mech. E.* 186, Nos. 31, 32.

19 Wilkinson, H.C. and Kilbourn, D.F. (1971), 'The design of ships machinery installations', *Shipping World and Shipbuilder,* August.

20 Wert, W.G. (1971), 'Maintenance', *Shipping World and Shipbuilder,* July.

21 Brandt, H. (1969), 'Reliability of marine engines – method for determining the time-dependent failure rate of engine components', M.T.Z. November.

22 Burtenshaw, R.P. and Lilly, L.R.C. (1972), 'Towards wear reduction in engines using residual fuel', *Trans, I. Mar. E.* 84.

23   Anyakora, S.N., Lees, F.P. and Engel, G.F.M. (1971), 'Some data on the reliability of instruments in the chemical plant environment', *The Chemical Engineer*, November.

Index